Milestones in Microbiology

translated and edited by

THOMAS D. BROCK
E. B. Fred Professor of Natural Science
Department of Bacteriology
University of Wisconsin

AMERICAN SOCIETY FOR MICROBIOLOGY
Washington, D.C.

American Society for Microbiology
1913 I St., N.W.
Washington, D.C. 20006

Originally published in 1961 by
Prentice-Hall, Inc., Englewood Cliffs, N.J.

Reprinted in 1975 by the American
Society for Microbiology

Library of Congress Catalog Card Number: 75-29589
ISBN 0-914826-06-9

Preface

This is not a history of microbiology. This collection of articles supplements the work done by undergraduate students in their first year of microbiology. It is further intended to be an interesting and useful reference for advanced students and practicing microbiologists.

There was a time when most microbiologists were acquainted with at least some of the papers presented here. But present generations are less likely to read any of the articles which form the foundations of the science. Although the history of microbiology is always treated in introductory texts, and names like Leeuwenhoek, Pasteur, and Koch are cited, there is little opportunity for the student to become aware of just what these men did and what their experimental work was actually like. To a great extent microbiology is taught as a body of pre-existing facts, and the student is not aware of the way in which current theories have developed out of the past.

I feel that the undergraduate student does not learn enough about the experimental aspects of science. He is not encouraged to consider the ways in which experiments are designed until he has become an advanced student. However, by reading papers reporting actual experiments, he will be forced to think of experimental design.

Further, through reading original papers, the student may begin to feel some of the excitement and interest of microbiology. When reading an original paper, he may feel that he is in the laboratory of the investigator, following the investigator's mind and hands. He may realize that microbiology is a vital, moving, developing thing.

It is difficult for a student to read current research papers. They are generally only fragments of a completed work, written in a highly special language. To understand them properly requires a considerable background knowledge of basic concepts and terms.

However, the historically important papers offer ideal material for the student. They were written in earlier times, before microbiology had become fragmented into sub-sciences. Their language is generally simpler—at least, the number of technical words is smaller. And it is possible to select articles which we know were milestones in the development of the science.

This is reason enough for the present volume. But in addition to its use by students, I hope that microbiologists also may find it interesting. Few of them have had the occasion or the opportunity to read the articles presented here, although they are surely aware of some of them. I think that anyone working in either pure or applied microbiology will find many things here which will be of interest to him. In addition, by thinking about the ways in which various workers have attacked key problems in the past, the professional microbiologist may be given ideas for approaches to his own work which he would not get otherwise.

The above has been a defense of history as a subject which has something useful to offer. In a real sense, however, history does not need such a defense. It can stand alone, I believe, as a subject worthy of study in and for itself. Those curious to learn how microbiology developed will read in this volume some historical facts. I hope readers will find this as entertaining as I have. It should be emphasized, however, that these papers were not translated, edited, and annotated with the professional historian in mind. My main concern has been to present the material so that it will be as understandable to the undergraduate taking his first course in microbiology as it is to those more advanced in the field.

Microbiology developed into a science almost exclusively during the nineteenth century. The research papers are therefore published predominantly in German and French, with a few in English. This language barrier would be enough to deter most people from reading Koch and Pasteur. I have saved them this difficulty by translating all of the German and French papers into English. I am fully aware that without this translation the present book would find few readers, even among those who have passed language examinations for the doctorate. Unfortunately, it is almost impossible to reproduce in translation the nineteenth-century flavor of the originals. I would like to encourage those who can do so to read some of the articles in their native tongues, just to experience the flavor of the writing.

Most of the articles were translated by me. Therefore, I am solely responsible for any errors in translation. Fortunately, most of the articles were available to me locally. Many workers would not have available to them the originals of these articles, as I have had in the excellent facilities of the Cleveland Medical Library Association. Without this library, the present project would have been difficult to execute.

Almost all of the articles have been abridged in some way. This has been necessary to keep the size of the book within reasonable limits. It can also be justified on other grounds. Most nineteenth-century workers were verbose, and long, historical surveys and discussions were common. In addition, portions of a paper may present data which seemed interesting or important at the time but which are of little historical significance. I believe that few people will miss the material omitted and will know it is gone only because of the ellipses (. . .) indicating deleted material. In addition, references cited by the authors have been omitted. Anyone sufficiently interested in following up these citations would want to read the originals anyway and could then obtain the correct citations.

I must further credit Bulloch's *The History of Bacteriology*, which provided the necessary initial historical orientation, as well as many of the literature citations. The articles here could be quite effectively read as a companion to Bulloch.

The articles selected for this volume have been chosen from among the large number of articles available. In many cases, the selection of a specific article was obvious, since it was clearly a major advance in microbiology. In other cases, the selection may have been less obvious. It is at times difficult to point to a given article of an author and be assured that this is his most important work. In such cases, my selection has been guided by a consideration of the educational value of the article, as well as of its ability to stand alone. At times it was difficult to select one article which would read well when taken out of the body of literature of its time. If I have left out any really important articles, I hope that they will be called to my attention.

There are no papers on the fungi, algae, or protozoa in this collection. Except for yeast, the fungi have been studied primarily by botanists, using different points of view. The same can be said for the algae. The protozoa are usually studied by zoologists, and here again the emphasis is different. Since these organisms are studied only briefly or not at all in microbiology, their "Milestones" are not presented here. This collection is restricted to bacteria, viruses, and yeasts.

I am aware that there is a definite medical orientation to this collection. Every attempt has been made to present as many papers in general microbiology as possible. But the nineteenth-century emphasis was primarily medical, and this inclination has affected the present collection. In another 100 years, the historical perspective of current work will be evident, and a quite different group of "Milestones" could be collected. But for the present, the medical bias must remain.

All of the work on this book was done while I was at Western Reserve University, Cleveland, Ohio. The book could not have been completed without the capable assistance of my wife, Louise Brock, who served as both typist and editor.

THOMAS D. BROCK

CONTENTS

SPONTANEOUS GENERATION AND FERMENTATION (*Continued*)

PART II

THE GERM THEORY OF DISEASE

PART III

IMMUNOLOGY

PART IV

VIROLOGY

VIROLOGY (*Continued*)

PART V

CHEMOTHERAPY

PART VI

GENERAL MICROBIOLOGY

GENERAL MICROBIOLOGY (*Continued*)

Historical Introduction

THE PRACTICAL AND THE SCIENTIFIC aspects of microbiology have been closely woven. Perhaps it is for this reason more than any other that microbiology as a *field* of study did not develop until the twentieth century. Nineteenth-century workers were chemists or physicians, a few were botanists, but none called themselves bacteriologists or microbiologists. Indeed, even to this day microbiology is often not taught as a body of knowledge, in the way botany and zoology are, but as an applied science.

One reason for its practical orientation is that microbiology developed to solve practical problems. The two main areas of development were in the fields of fermentation and medicine; the pure science of microbiology did not exist.

Perhaps more than most sciences, the development of microbiology depended on the improvement of a tool —the microscope. Although there are no papers on the development of the microscope in the present collection (that is really a branch of physics), the quality of all of the work reported here reflects the quality of the microscopes used. Since bacteria cannot be seen individually with the unaided eye, their existence as individuals can only be known from microscopic observations. And since most bacteria are very small, ranging in size down to the limits of resolution of the light microscope, their existence and characteristics are quite difficult to decipher. Indeed, it would be interesting to speculate on how bacteriology might have developed if the limits of resolution of the microscope were poorer, so that all bacteria would have been invisible. Bacteriology could undoubtedly have evolved under these conditions, but surely at a slower rate, and surely in quite a different manner.

From the time Leeuwenhoek first saw bacteria, until the time Koch proved a bacterium to be the cause of a disease, there was a tremendous development in the light microscope. From Koch's time, when the Abbé condenser was introduced (latter nineteenth century), to the present there have been only minor developments, and few of them can be said to have had a crucial impact on bacteriology. But as he reads the articles, the reader should stop to consider in each one: how do the observations made and conclusions drawn reflect the quality of the microscope? Many controversies may have really been due to the use by the investigators of different microscopes. This is a good lesson for today: When workers disagree, isn't it usually due to variations in the methods used?

Although medicine and fermentation presented the practical problems which stimulated the development of microbiology, the first studies that put the subject on a scientific basis arose from a problem in pure science. This was the controversy over spontaneous generation, a dispute that aroused men of a philosophical nature, men who studied science for science's sake, and consequently led to the first scientific approaches to microbiology.

The early crude ideas of spontaneous generation (mice from old rags, maggots from meat) were dispelled by Redi in the seventeenth century. But the more subtle forms of spontaneous generation, such as that protozoa and bacteria (then generally and loosely known as *infusoria*) can arise from vegetable and animal infusions, were still believed in the nineteenth century. Indeed, the controversy also involved fermentations, since it was considered that the yeast fermentation (especially in uninoculated grape juice) was of spontaneous origin. But this also involved the problem of whether or not yeast caused alcoholic fermentation. The interrelationships here were quite complex.

As anyone can verify, when a meat or vegetable infusion is allowed to stand at room temperature, certain changes occur. At the level of the senses, the liquid becomes cloudy and odors develop. At the chemical level, organic materials of the infusion decompose. At the microscopic level, infusoria (protozoa and bacteria) appear. These changes are known as putrefaction. If, instead of an infusion, a fruit juice is used, characteristic changes of another sort occur. The sugar in the juice decomposes into alcohol and carbon dioxide, and yeast appears. These changes are called *fermentation*, or *alcoholic fermentation*. In the case of fermentation, so much yeast is formed that it settles out and can be collected and transferred to a pure sugar solution, where a further alcoholic fermentation is set up. Fermentation in pure sugar solutions does not occur spontaneously.

These observations on the ready decomposability of organic materials have probably been made as long as man has attempted to keep food for future use. They obviously have many practical aspects, and affect commercial life in a number of ways. There would naturally be considerable speculation about their nature, and their causes. Also there would be speculation on the origin of these microscopic beings, bacteria, yeasts, and protozoa. The question of cause and effect was a burning one. Which came first, the putrefaction or the infusoria? Did putrefaction occur because of the action of the infusoria on the organic material, or did the infusoria arise because of the decomposition of organic material via putrefaction?

Although many workers became involved in the study of fermentation and spontaneous generation, Pasteur stands out as a giant. Coming into biology from the field of chemistry, he seems to have been able to remove all of the philosophical hurdles that blocked the thinking of others. Within a period of four years after he began his studies, he had clarified the problems so well that the controversy eventually died a natural death.

Pasteur states that he became interested in fermentations because certain products formed were optically active compounds. He started with the hypothesis that optically active organic compounds were always formed as the result of the activity of living organisms. With this preconception, he studied the formation of amyl alcohol and lactic acid during fermentation. He was able to show to his satisfaction that the lactic fermentation was due to the action of a living organism. He established the same fact for the production of ethyl alcohol by yeast. It was only a natural step for him to tackle the problem of spontaneous generation, since, as we have mentioned, the two problems were highly interrelated. The experiments he devised to solve the problem of spontaneous generation are marvels of simplicity. His work was so clear and concise that it could readily be repeated by anyone. Soon the controversy over spontaneous generation subsided.

Pasteur was able to go easily from fermentation into the field of medical microbiology, which occupied him in the latter part of his life. His contributions in this area were numerous, and although he was overshadowed in some respects by Koch, his work has been the basis of many modern medical practices. Pasteur's development from chemist to medicine man has been told in many biographies.

It should be emphasized that the development of sterilization methods, so necessary to the solution of the spontaneous generation controversy, was essential to put the science of microbiology on a firm foundation. Workers did not set out to develop sterilization methods. They set out to solve the controversy of spontaneous generation. That sterilization methods were evolved was a bonus that was received for solving the other question.

The second main area of development was in medicine. The microbiological aspects of medicine arose out of considerations of the nature of contagious diseases. Probably contagious diseases were recognized as such far back into antiquity. Smallpox has been known to be contagious for a long time. But the nature of this contagion was little understood. It was known that a disease could be transmitted from one person to another, but it was not known how this could occur. Naturally, supernatural forces were proposed. But it was probably the introduction of syphilis into Europe which served to crystallize thinking. Here was a disease which was only transmitted by contact. It was obviously contagious. What was it that was transmitted?

Without data, one could only speculate. It is interesting that Fracastoro, the man who gave syphilis its name in the sixteenth century, actually came as close as he did to devising a germ theory of disease. Although his ideas were conceived in mediaeval terminology and reasoning, we can see the glimpse of an idea. This idea attracted a number of workers all the way down to the nineteenth century.

As microscopes were developed and improved, it was natural that people would speculate on the relationships of the microorganisms they were seeing to disease. In the eighteenth century, the cause of a certain itch was shown to be a microscopic, although multicellular, animal. Here was a disease caused by a living agent which had previously been too small to see. Did this mean that other diseases were caused by even smaller organisms? This speculation occurred often, but was often denied. Only in the early nineteenth century were there sufficient facts available to begin to see how the problem might be approached.

By the late 1830's Schwann and Cagniard-Latour had shown that alcoholic fermentation and putrefaction were due to living, organized beings. Although these results were often denied, there were many who believed them. If one accepted the idea that decompositions of organic materials were due to living agents, it was only a step further to reason that disease, which in many ways appears to be a decomposition of body tissues, was due to living agents. Even Fracastoro in the sixteenth century had noticed the similarity between putrefaction and decay on the one hand and disease on the other. Henle, in 1840, further commented on this similarity, and with the new-found knowledge of the nature of fermentation, he proceeded to draw rather clear conclusions. But Henle's genius was to point out what experimental proofs would be necessary to clinch this hypothesis. Bacteriologists who have been raised on Koch's postulates for proving the cause of disease may be startled to find that Henle enunciated these ideas many years earlier and with even more clarity than did Koch in his first papers.

Henle's statement of the requirements for proof that microorganisms can cause diseases included the important point that the organism should be isolated from the host. This operation of isolation of the organism is at the basis of all research in experimental infections. But at Henle's time, microorganisms were too little understood, and techniques for isolation did not exist. Henle recognized that the methods for this proof were not available. It would be necessary to isolate pure cultures.

It is interesting that pure cultures were first developed for the fungi. These microorganisms are much larger than bacteria and are more easily studied. The isolation of single fungal spores was accomplished by Brefeld, and he watched a spore develop into a new fungal mycelium. He stated that to prove the purity of a culture, it must be shown that it arose from a single spore or single cell. Brefeld's work undoubtedly influenced Koch in his studies on animal diseases.

When Koch first worked on anthrax, and studied the life cycle of this organism from rod-shaped bacterium to spore, and from spore to bacterium, he did not obtain pure cultures in the manner of Brefeld. He isolated his bacteria from drops of blood from anthrax-infected mice. He was indeed lucky that these mice did not contain other bacteria, since he made no precautions to insure that his cultures arose from single spores or single cells. Other people repeating his first experiments were not so lucky, and actually had at least two different species of bacteria in their cultures from the blood of mice. They merely inoculated drops of bacteria-containing blood into a liquid nutrient fluid, and examined them later for growth. If more than one type is present, each type may grow. But it may be that one of the two types is far in excess of the other. This may mean that even a close microscopic examination will reveal only one type, while there are in reality two types. Lister was the first to point out, in 1878, two years after Koch's historic first paper, that the size of sample seen under the microscope was too small to be certain that the whole volume from which the sample was taken actually consisted of a pure culture. Microscopic examination was not enough. But Koch had determined the purity of his cultures in 1876 only by microscopic examination. It was not until he began to work on other diseases, those much harder to study than anthrax, that Koch was forced to develop really rigid methods for isolating pure cultures.

Lister in 1878 showed the way to this work by his study of lactic fermentation of milk. By diluting his bacteria-containing samples sufficiently he obtained dilutions in which a single drop had the probability of containing a single cell only one-half the time. The other half of the time, a drop from this dilution was sterile, and did not bring about a lactic acid fermentation. This experiment of Lister's was of great theoretical interest, since it showed without question the particulate nature of the causal agent of milk souring. In addition, his technique was the first method for the isolation of a pure culture. However, Lister's method was only useful in isolating a pure culture of the predominant organism in a suspension, since the less frequent organisms would be diluted out earlier. Also, it required the use of liquid media, which have a number of disadvantages, as Koch pointed out.

The development of solidified culture media by Koch was without a doubt the most important single development in the history of microbiology after the perfection of sterilization techniques. The technique was developed out of Koch's observations that bacteria grow in discrete colonies on potato slices. When he had convinced himself that each colony could and probably did arise from a single cell which had fallen on the potato slice, it was obvious that here was a universal method for the isolation of pure cultures. Also, by carrying bacteria on solid media, it was possible to tell easily when a culture was contaminated, since a contaminant usually developed as an isolated colony which could be differentiated from the desired culture with the naked eye. It only remained for Koch to find a way to solidify the more common liquid media, so as to avoid using potato slices. This he did with gelatin, and developed his technique to such perfection that it could be readily duplicated by anyone. The later substitution of agar for gelatin was a minor, although important, modification.

Although the pure culture technique using solid media was the most important single advance of Koch, other developments of his were also important. Anthrax bacilli are rather large and easily seen under the microscope. However, other bacteria are much smaller and less easily visualized. Koch developed the technique of staining with aniline dyes to visualize bacteria in diseased tissues. He made use of the Abbé condenser to improve his microscope. He developed the simple smear method of preparing bacteria for staining. With these techniques, he was now able to make his most famous discovery: the discovery of the bacillus that causes tuberculosis.

It was in this work that he was forced to develop his techniques to their ultimate precision. The tubercle bacillus is one of the most difficult to work with, since it grows slowly and has many odd properties. Here we see Koch actually advance *for the first time,* his famous postulates (Koch, 1884, p. 116). In all cases of human and animal tuberculosis, the same characteristic bacterium could be found in the tubercles. To prove that these bacilli are the cause of the disease, they must first be isolated in pure culture and allowed to reproduce through a number of transfers until it is certain that no products of the animal body remain. Then these cultured bacilli must be reinjected into healthy animals and induce the same disease in these as one obtains when experimental animals are injected with fresh tubercle material. And finally, these bacteria must again appear in the artificial disease, and be reisolated. It was only in this rigid way that the bacterial nature of tuberculosis could be proved. Such rigor was not needed in Koch's earlier work on anthrax. However, in his 1884 paper on tuberculosis, where he sums up all of this work, he discusses these postulates first with relation to anthrax, and shows how they can be used to prove the causal organism of anthrax. This is only after-the-fact reasoning, since Koch had "proved" that a bacterium caused anthrax eight years earlier with much cruder methods.

Now that Koch opened the doors, a flood of research developed in which the causal organisms for most of the common infectious diseases were isolated. By the beginning of the twentieth century (20 years after Koch) the medical aspects of infectious diseases were on a firm bacteriological footing.

At some future time it might be interesting to examine the development of bacteriology in the Kaiser's Germany. Surely this era, from 1871, when Imperial Germany began, to the end of World War I, when this same empire died, was one of the most productive eras in the history of bacteriology. In this time Germany had Koch, Ehrlich, von Behring, Loeffler, Gaffky, Buchner, Brefeld, Cohn, von Nageli, and many others. And many of these were employed or supported by the government itself, rather than by the universities. In this same era, the German dye industry developed, and there was a very close relation between this industry and bacteriology, resulting in the development of familiar staining methods as well as in the development of chemotherapy. Chemotherapy developed in a straight line from Ehrlich to Domagk, and culminated in the discovery of the sulfa drugs.

The general term for a living infectious agent which had been used for years was "virus." Any infectious agent was described with this word. Although many infectious diseases were easily shown to be caused by bacteria, it was not always possible to do so. Sometimes it was possible to transmit a disease from host to host even after the infectious fluid was filtered through pores so small that all bacteria were removed. When it could be shown that this agent could pass through a filter, then reproduce in a new host, and after being filtered again, infect another host, it was obvious that here was a living agent. Beijerinck called this a "liquid (or fluid, that is, non-particulate) living agent of disease." Others called these "filterable viruses." Gradually the word "virus" became restricted to filterable viruses, so that now the adjective filterable has been dropped. We know now that these viruses are particulate, and since Stanley succeeded in crystallizing one, we know that they are quite unlike other living organisms, so that many refuse to consider them living. It is interesting to read both Beijerinck (plant virus) and Loeffler (animal virus) to see how preconceived notions must later be discarded under the weight of incontrovertible facts. Because the viruses are obligate parasites, virology has developed much slower than bacteriology. At the present time, virology is expanding at a rapid rate. Much of this expansion is due to the development of concepts which had their origin in the discovery of bacterial viruses. These bacterial viruses, or "bacteriophages" were discovered in the early part of the twentieth century. Although they have had little practical application, their theoretical implications have influenced a wide area of molecular biology and genetics. The paper by d'Herelle presented here describes the first discovery of a bacterial virus.

One of the most important applied developments in microbiology was in the understanding of the nature of specific acquired immunity to disease. That such immunity is possible had been known for a long time, and this knowledge was crystallized and developed into an amazing prophylactic treatment for smallpox by Jenner long before the germ theory of disease had been established. Later workers developed additional methods of increasing the immunity of an individual to a disease, but the most dramatic triumph in this phase of microbiology was the discovery of the diphtheria and tetanus antitoxins. With this discovery by von Behring and Kitisato, later developed into a practical tool by Ehrlich, it was now possible to specifically cure a person suffering from these diseases by injecting into him some antitoxic serum prepared by earlier immunization of a horse or other large animal. This led to the development for the first time of rational cures for infectious diseases, and was more than anything responsible for Ehrlich's later conception of chemotherapy.

The antibiotics era has only been documented here by Fleming's first paper, which should be sufficient to show the promise that was in store. It would be difficult indeed to select one or two additional papers in the field of antibiotics as further milestones along this road. At some future time, a whole book will be needed to document the historical development of the antibiotic age.

Although the bulk of the papers in this collection deal with applied material, there are a number of studies which can only be classified as fundamental in nature. They were not begun with any practical goal in mind, but only because the investigator was interested in the subject. These have been included here under the heading of General Microbiology. Also in this section are several papers dealing with methods, which have been placed here for want of a better place.

The further development of general microbiology is shown here through the papers of Winogradsky and Beijerinck who were primarily interested in microorganisms as objects of study, with little regard for practical problems. Kluyver was Beijerinck's successor and pioneered the field of microbial physiology and biochemistry.

The first clear-cut discovery of a growth substance which can be considered to be a vitamin was made using a microorganism (yeast) as a test organism. Although later work with mammals was necessary to establish the vitamins as important substances in nutrition, the work with microorganisms has always been a great stimulus to vitamin research. Wildier's early paper on the Bios substance has been included here.

Another whole area of microbiology has been completely omitted in the present collection. This is the area of microbial genetics. Although some of the most important developments in microbiology in recent times have occurred in this field, it has not yet reached the stage where it can be presented intelligibly to undergraduates through isolated articles. In addition, there are two collections which include most of the significant work in this area: Lederberg's "Papers in Microbial Genetics" and Peters' "Classic Papers in Genetics."

The articles in the present collection were selected primarily to show the beginnings of the science of microbiology, which arose through practical problems. Today it is becoming increasingly evident that microbiology is at last becoming a true biological science, able to exist apart from its applications. It is also becoming increasingly clear that bacteria are probably the most useful living organisms for studying many processes in the life of the cell. This is indeed encouraging. No science can ever thrive if it is built only on purely practical grounds. It is to be hoped that if a new collection of milestones in microbiology is made 100 years from now, it will amply document the fundamental aspects of microbiology, and show that this science has truly come of age.

PART I
Spontaneous Generation and Fermentation

Observations . . . concerning little animals observed in rain-, well-, sea- and snow-water

1677 • Antony van Leeuwenhoek

Observations, communicated to the Publisher by Mr. Antony van Leewenhoek, in a Dutch Letter of the 9th of Octob. 1676, here English'd: Concerning little Animals by him observed in Rain- Well- Sea- and Snow-water; as also in water wherein Pepper had lain infused. *Philosophical Transactions of the Royal Society of London,* Vol. 11, 1677, no. 133, pages 821–831.

. . . 3. May the 26th, I took about ⅓ of an ounce of whole pepper and having pounded it small, I put it into a Thea-cup with 2½ ounces of Rain-water upon it, stirring it about, the better to mingle the pepper with it, and then suffering the pepper to fall to the bottom. After it had stood an hour or two, I took some of the water, before spoken of, wherein the whole pepper lay, and wherein were so many several sorts of little animals; and mingled it with this water, wherein the pounded pepper had lain an hour or two, and observed, that, when there was much of the water of the pounded pepper, with that other, the said animals soon died, but when little they remained alive.

June 2, in the morning, after I had made divers observations since the 26th of May, I could not discover any living thing, but saw some creatures, which tho they had the figures of little animals, yet could I perceive no life in them how attentively I beheld them.

The same day at night, about 11 a clock, I discovered some few living creatures: But the 3d of June I observed many more which were very small, but 2 or 3 times as broad as long. This water rose in bubbles, like fermenting beer.

The 4th of June in the morning I saw great abundance of living creatures; and looking again in the afternoon of the same day, I found great plenty of them in one drop of water, which were no less than 8 or 10000, and they looked to my eye, through the Microscope, as common sand doth to the naked eye. On the 5th I perceived, besides the many very small creatures, some few (not above 8 or 10 in one drop) of an oval figure, whereof some appear'd to be 7 or 8 times bigger than the rest. . . .

Microscopical observations about animals in the scurf of the teeth

1684 • Antony van Leeuwenhoek

An abstract of a Letter from Mr. Anthony Leevvenhoeck at Delft, dated Sep. 17, 1683. Containing some Microscopical Observations, about Animals in the scrurf of the Teeth. . . . *Philosophical Transactions of the Royal Society of London,* Vol. 14, May 20, 1684, no. 159, pages 568–574, 1 pl.

. . . THO MY TEETH ARE KEPT USUALLY very clean, nevertheless when I view them in a Magnifying Glass, I find growing between them a little white matter as thick as wetted flower: in this substance tho I do not perceive any motion, I judged there might probably be living Creatures.

I therefore took some of this flower and mixt it either with pure rain water wherein were no Animals; or else with some of my Spittle (having no Air bubbles to cause a motion in it) and then to my great surprize perceived that the aforesaid matter contained very many small living Animals, which moved themselves very extravagantly. the biggest sort had the shape of A. their motion was strong & nimble, and they darted themselves thro the water or spittle, as a Jack or Pike does thro the water. These were generally not many in number. The 2d. sort had the shape of B. these spun about like a Top, and took a course sometimes on

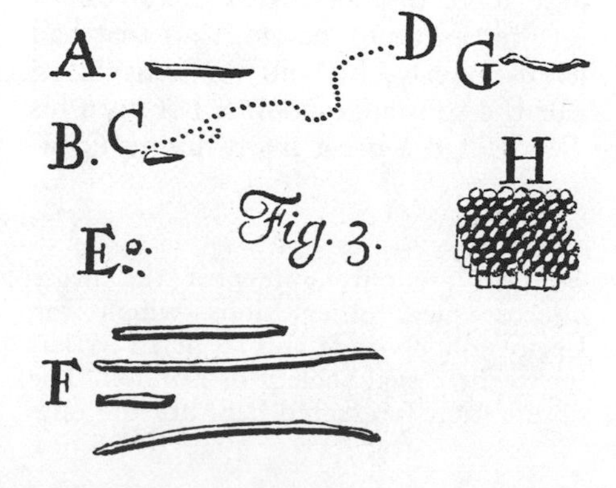

one side, as is shown at G. and D. they were more in number than the first. In the 3d. sort I could not well distinguish the Figure, for sometimes it seemed to be an Oval, and other times a Circle. These were so small that they seem'd no bigger than E. and therewithal so swift, that I can compare them to nothing better than a swarm of Flies or Gnats, flying and turning among one another in a small space [Brownian movement?]. Of this sort I believe there might be many thousands in a quantity of water no bigger than a sand, tho the flower were but the 9th part of the water or spittle containing it. Besides these Animals there were a great quantity of streaks or threds of different lengths, but like thickness, lying confusedly together, some bent, and others streight, as at F. These had no motion or life in them, for I well observed them, having formerly seen live Animals in water of the same figure.

I observed the Spittle of two several women of whose Teeth were kept clean, and there were no Animals in the spittle, but the meal between the teeth, being mixt with water (as before) I found the Animals above described, as also the long particles.

The Spittle of a Child of 8 years old had no living Creatures in it, but the meal between the Teeth, had a great many of the Animals above described, together with the streaks.

The Spittle of an old Man that had lived soberly, had no Animals in it; But the substance upon & between his Teeth, had a great many living Creatures, swimming nimbler then I had hitherto seen. . . . The Spittle of another old man and a good fellow was like the former, but the Animals in the scurf of the teeth, were not all killed by the parties continual drinking Brandy, Wine, and Tobacco, for I found a few living Animals of the 3d. sort, and in the scurf between the Teeth I found many more small Animals of the 2 smallest sorts.

I took in my mouth some very strong wine-Vinegar, and closing my Teeth, I gargled and rinsed them very well with the Vinegar, afterwards I washt them very well with fair water, but there were an innumerable quant'ty of Animals yet remaining in the scurf upon the Teeth, yet most in that between the Teeth, and very few Animals of the first sort A.

I took a very little wine-Vinegar and mixt it with the water in which the scurf was dissolved, whereupon the Animals dyed presently. From hence I conclude, that the Vinegar with which I washt my Teeth, kill'd only those Animals which were on the outside of the scurf, but did not pass thro the whole substance of it. . . .

The number of these Animals in the scurf of a mans Teeth, are so many that I believe they exceed the number of Men in a kingdom. For upon the examination of a small parcel of it, no thicker then a Horse-hair, I found too many living Animals therein, that I guess there might have been 1000 in a quantity of matter no bigger then the 1/100 part of a sand.

Comment

These are only a few of the many microscopical observations which van Leeuwenhoek made and reported by letter to the Royal Society of London. The observations presented here are the ones which seem most likely to represent descriptions of bacteria.

van Leeuwenhoek was a minor city official who built microscopes as a hobby. He became probably the best microscope builder in Europe, and people traveled long distances to look through his instruments, although he kept his construction methods secret.

It is amazing that van Leeuwenhoek was able to see bacteria, since he built microscopes with a single lens, rather than the compound type used today. It was only because of his great skill as a microscope builder that he was able to achieve the high resolving power needed to see bacteria. He made a large number of observations, painstakingly recorded. His work became widely known through its publication in the Philosophical Transactions of the Royal Society, and was very influential on later workers. Eighteenth and early nineteenth century investigators cited his work frequently.

van Leeuwenhoek himself did not speculate on the origin of microorganisms or on their relationship to disease, although a number of workers felt that these organisms might be implicated in infectious diseases. But it was not until the late nineteenth century, through the work of Koch, that this idea was finally shown to be correct.

Observations on the generation, composition, and decomposition of animal and vegetable substances

1748 • Turbevill Needham

A Summary of some late Observations upon the Generation, Composition, and Decomposition of Animal and Vegetable Substances; Communicated in a Letter to Martin Folkes Esq.; President of the Royal Society, by Mr. Turbevill Needham, Fellow of the same Society. Written in Paris, Nov. 23, 1748, read Dec. 15 and 22. 1748. In *Philosophical Transactions of the Royal Society of London,* Vol. 45, 1748, no. 490, pages 615–666.

. . . FOR MY PURPOSE, THEREFORE, I took a Quantity of Mutton-Gravy hot from the Fire, and shut it up in a phial, clos'd up with a Cork so well masticated, that my Precautions amounted to as much as if I had sealed my Phial hermetically. I thus effectually excluded the exterior Air, that it might not be said my moving Bodies drew their Origin from Insects, or Eggs floating in the Atmosphere. . . . I neglected no Precaution, even as far as to heat violently in hot Ashes the Body of the Phial, that if any thing existed, even in that little portion of Air which filled up the Neck, it might be destroyed, and lose its productive Faculty. Nothing therefore could answer my Purpose of excluding every Objection, better than hot roast-Meat Gravy secured in this manner, and exposed some Days to the Summer-Heat: and as I determined not to open it, till I might reasonably conclude, whether, by its own Principles, it was productive of any thing, I allow'd sufficient Time for that Purpose to this pure unmix'd Quintessence, if I may so call it, of an animal body. . . . My Phial swarm'd with Life, and microscopical Animals of most Dimensions, from some of the largest I had ever seen, to some of the least. The very first Drop I used, upon opening it, yielded me Multitudes perfectly form'd, animated, and spontaneous in all their Motions: . . .

I shall not at this present time trouble you with a Detail of Observations upon three or four Scores of different Infusions of animal and vegetable Substances, posterior to these upon Mutton-Gravy; all which constantly gave me the same Phaenomena with little Variation, and were uniform in their general Result: These may better appear at length upon some other Ocasion; let it suffice for the present to take notice, that the Phials, clos'd or not clos'd, the Water previously boil'd or not boil'd, the Infusions permitted to teem, and then plac'd upon hot Ashes to destroy their Productions, or proceeding in their Vegetation without Intermission, appear'd to be so nearly the same, that, after a little time, I neglected every Precaution of this kind, as plainly unnecessary. . . .

. . . It seems plain therefore, that there is a vegetative Force in every microscopical Point of Matter, and every visible Filament of which the whole animal or vegetable Texture consists: And probably this Force extends much farther; . . .

Hence it is probable, that every animal or vegetable substance advances as fast as it can in its Resolution to return by a slow Descent to one common Principle, the Source of all, a kind of universal *Semen;* whence its Atoms may return again, and ascent to a new Life.

Comment

This selection of Needham's is typical of the thinking of the proponents of spontaneous generation. The crucial experimental points are whether or not the heating of the vial in hot ashes was really sufficient to destroy all living things within, and whether the cork was able to keep any living things out of the vial after the heating. Spallanzani deals with both of these problems in the next article, and in so far as it was possible with 18th century techniques, solves them.

Tracts on the nature of animals and vegetables; observations and experiments upon the animalcula of infusions

1799 · Lazaro Spallanzani

Tracts on the nature of animals and vegetables. By Lazaro Spallanzani, R. P. U. P., Edinburgh, 1799. Translation from the Italian. Translator? *Observations and experiments upon the animalcula of infusions,* pages 1–69.

I HERMETICALLY SEALED VESSELS WITH the eleven kinds of seeds mentioned before [kidney beans, vetches, buckwheat, barley, maize, mallow, beet, peas, lentils, beans, hemp]. To prevent the rarefaction of the internal air, I diminished the thickness of the necks of the vessels, till they terminated in tubes almost capillary, and, putting the smallest part to the blowpipe, sealed it instantaneously, so that the internal air underwent no alteration. It was necessary to know whether the seeds might suffer by this inclusion, which might be an obstacle to the production of animalcula. Other experiments had shewn me, 1. vessels hermetically sealed have no animalcula, unless they are very capacious: 2. animalcula are not always produced: 3. when they are produced, the number is never so great as in open vessels. Although I used pretty large vessels, two substances, peas and beans, had not a single animalcule. The other nine afforded a sufficient number; and to these I limited my experiments. I took nine vessels with seeds, hermetically sealed. I immersed them in boiling water for half a minute. I immersed another nine for a whole minute, nine more for a minute and a half, and nine for two minutes. Thus, I had thirty-six infusions. That I might know the proper time to examine them, I made similar infusions in open vessels, and, when these swarmed with animalcula, I opened those hermetically sealed. Upon breaking the seal of the first, I found the elasticity of the air encreased. Seeds contain much air: a great quantity should escape in their dissolution, by heat or maceration, which must, of necessity, render the portion of included air denser and more elastic. However, the elasticity may originate partly from the elastic fluid discovered in vegetables, the nature of which is apparently different from the atmospheric fluid. I examined the infusions, and was surprised to find some of them an absolute desert; others reduced to such a solitude, that but a few animalcula, like points, were seen, and their existence could be discovered only with the greatest difficulty. The action of heat for one minute, was as injurious to the production of the animalcula, as of two. The seeds producing the inconceivably small animalcula, were, beans, vetches, buckwheat, mallows, maize, and lentils. I could never discover the least animation in the other three infusions. I thence concluded, that the heat of boiling water for half a minute, was fatal to all animalcula of the largest kind; even to the middle-sized, and the smallest, of those which I shall term animalcula of the higher class, to use the energetic expression of M. Bonnet; while the heat of two minutes did not affect those I shall place in the lower class.

Having hermetically sealed six vessels, containing six kinds of seeds producing animalcula of the lower class, I immersed them in boiling water for two minutes and a half, three, three and a half, and four minutes. The seals of twenty-four vessels being broken at a suitable time, there were no animalcula of the higher class seen, but more or fewer of the lower. The air was almost always condensed, both in this and in the other experiments.

In vessels immersed seven minutes, I found animalcula of the lower class. They appeared in vessels immersed twelve minutes.

The minuteness of animalcula of the lower class, does not prevent our distinguishing the difference of their figure and proportions.

Boiling half an hour was no obstacle to the production of animalcula of the lower class; but boiling for three quarters, or even less, deprived all the six infusions of animalcula.

We know, that the heat of boiling water is about 212°. These infusions were of this heat at least, as appeared by the marks they exhibited of ebullition, the whole time the surrounding water boiled. Philosophers know, that water, boiled in a close vessel, acquires a greater degree of heat, than when boiled in an open. To know how much less than half a minute the boiling might be abridged, an animalcula of the higher class yet exist, I made use of a second-pendulum, and immersed the vessels in boiling water for a given number of seconds, beginning with 29. In a word, boiling for a single second prevented their existence; and I could only employ a degree of heat less than that of boiling water. . . . Not a single animalcule was seen of the higher class, in vessels hermetically sealed, and exposed to the moderate heat of 113°. This was during the middle of April: the thermomenter in the shade stood at 88°. I took eighteen vessels; nine had been exposed to 99° of heat, and nine to 88°. No animalcula of the higher class were produced in the former; but I found them in the latter. In each vessel, the quantity and kind of animalcula, as in vessels not subjected to heat. The degree of heat fatal to them, was 92°.

Animalcula of the lower class, exist in sealed vessels exposed to the heat of 212°; while those of the higher class hardly appear at 92°: But, when produced, the same intensity of heat that is fatal to the one, also deprives the other of life; and animalcula of the higher, as well as of the lower class, perish at 106°, or at most at 108°.

Two important consequences thence arise. The first evinces the extreme efficacy of heat to deprive infusions in close vessels of a multitude of animated beings; for, in open vessels, are always seen a vast concourse of animalcula. The second consequence, concerns the constancy of animalcula of the lower class appearing in infusions boiled in close vessels; and the heat of 212°, protracted an hour, has been no obstacle to their existence. . . .

We are therefore induced to believe, that those animalcula originate from germs there included, which, for a certain time, withstand the effects of heat, but at length yield under it; and, since animalcula of the higher classes only exist when the heat is less intense, we must imagine they are much sooner affected by it, than those of the lower classes. Whence we should conclude, that this multitude of the superior animalcula, seen in the infusions of open vessels, exposed not only to the heat of boiling water, but to the flame of a blowpipe, appears there, not because their germs have withstood so great a degree of heat, but because new germs come to the infusions, after cessation of the heat. . . .

In my observations, I have particularly inquired, whether animalcula specifically varied as the infusions of vegetable seeds were different, so that each might have peculiarly its own; but I have found nothing constant. It is true, I have often found certain species of animals only, in particular kinds of vegetables, but I have frequently seen the reverse. The animalcula of the same infusion were different, at different times and different places; and it even is not uncommon to see this variety in two infusions made of the seed of the same plant. All this well agrees with the vast variety of animalcular eggs, scattered in the air, and falling every where, without any law. . . . The idea, that animalcula come from the air, appears to me to be confirmed by undoubted facts. I took sixteen large and equal glass vases: four I sealed hermetically; four were stopped with a wooden stopper, well fitted; four with cotton; and the four last I left open. In each of the four classes of vases, were hempseed, rice, lentils, and peas. The infusions were boiled a full hour, before being put into the vases. I begun the experiments 11. May, and visited the vases 5. June. In each there were two kinds of animalcula, large and small; but in the four open ones, they were so numberous and confused that the infusions, if I may use the expression, rather seemed to teem with life. In those stoppered with cotton, they were about a third more rare; still fewer in those with wooden stoppers; and much more so in those hermetically sealed. . . .

The number of animalcula developed, is proportioned to the communication with the external air. The air either conveys the germs to the infusions, or assists the expansion of those already there. . . .

Comment

Spallanzani's genius was in devising experiments that would be able to answer the questions that were the most often raised by proponents of spontaneous generation. His first experiment was to determine whether heating really was able to kill all of the organisms within an infusion. To do this, what simpler way than to set up a large series of flasks and heat different ones for different lengths of time? When this was done, he discovered that one group of microorganisms, which he calls superior animalcula, or animalcula of the higher class, were destroyed by very slight heating. These microorganisms were undoubtedly protozoa. The other group, which he calls animalcula of the lower class, were much more resistant to heating. From their description as being very minute, there is no doubt that these were bacteria. Thus Spallanzani has shown that certain types of microorganisms are much more sensitive to heat than others, and that for certain types to be killed, the liquids must be heated to boiling for close to an hour. Such an experiment is important in establishing the conditions necessary for studying spontaneous generation. Obviously, if the flasks are not heated long enough, microorganisms might appear, and if one were not aware of Spallanzani's experiments, it might be assumed that spontaneous generation had occurred.

Since he has shown that the superior animalcula are very easily killed by heat, he can only infer that any superior animalcula appearing in flasks which have been boiled and then allowed to stand exposed to the air have come into the flasks from the air. This is a very important inference which he then proceeds to test experimentally. By sealing vessels with closures of varying degrees of porosity, he was able to show that the number of organisms present was a function of the porosity. Spallanzani was careful not to conclude that his experiment proved that microorganisms were present in the air, since he also states that the air might be helping those already present to grow. But the experiment was probably sufficient to get people thinking about the possibility that microorganisms were present in the air. This eventually stimulated Pasteur to devise a more critical experiment which answered the question and became one of the turning points in the controversy over spontaneous generation (see page 43).

Preliminary report on experiments concerning alcoholic fermentation and putrefaction

1837 • Theodore Schwann

Schwann, Theodore. 1837. Vorläufige Mittheilung, betreffend Versuche über die Weingährung und Fäulnis. *Annalen der Physik und Chemie*, Vol. 41, pages 184–193.

AT THE LAST MEETING OF THE Versammlung der Naturforscher in Jena, I reported experiments concerning spontaneous generation. I showed that if an enclosed glass sphere which contains air and a small amount of an infusion of meat, is heated in boiling water so that the liquid and air of the sphere are warmed to 80°R.*, then the liquid shows no putrefaction or production of infusoria, even after many months. This is true even when the quantity of meat extract in the sphere is so small that there is no chance that it would have absorbed all of the oxygen out of the air. Even so, it was desirable to modify the experiment so that it would be possible to allow for the entrance of air under conditions where the new air entering the sphere could be heated first. I have accomplished this in the following manner.

A small flask which contained a small piece of meat and was filled one third full of water, was closed with a stopper which contained two thin glass tubes through it. These glass tubes passed for a distance of three inches through molten metal alloy which was continuously heated almost to the boiling point of mercury.† One of the

* [80°R. = 100°C.]

† [357°C.]

glass tubes was connected then to a manometer. Then the liquid in the flask was boiled strongly, so that all of the air which was in the flask and tubes was either driven out or heated to the boiling point of water. After this cooled, a continuous stream of air was passed out of the manometer through the flask, and then out of the flask through the second tube. This air had the chance to be heated considerably on its passage through the tubes. This experiment was performed many times, and in all cases there was no putrefaction or production of infusoria or mold, even after many weeks, and the liquid remained as clear as it had been right after preparation.

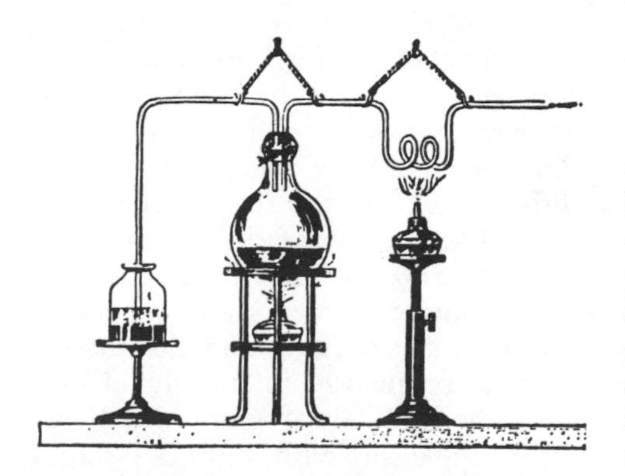

I will argue on another occasion whether this experiment, which has been repeated many times, will end the controversy concerning spontaneous generation. I will only remark here, that this experiment, when viewed from the standpoint of an opponent of spontaneous generation, can be explained as follows: the germs of molds and infusoria, which according to this theory are present in the air, are killed by the thorough heating of the air. In the same way, putrefaction must be explained as follows: the germs which develop nourish themselves from the organic substances in the meat extract, and the phenomenon of putrefaction is due to the destruction of these substances. This opinion is also supported by the observation that those substances which act as poisons for infusoria and molds, like arsenic or mercuric chloride, are also the most effective in preventing putrefaction, while those substances which are only poisonous for infusoria, like *Extractum Nucis vomicae spirituosum*, and do not affect molds, prevent all of the manifestations of putrefaction which are associated with the production of infusoria, like hydrogen sulfide production, and only permit those manifestations of putrefaction which are associated with the production of molds. . . .

I performed experiments on alcoholic fermentation in the following manner. A solution of cane sugar was mixed with beer yeast and four flasks were filled with it and stoppered. The flasks were then placed in boiling water for about 10 minutes, so that all of the liquid in them had reached this temperature. Then they were removed and inverted under mercury, and after cooling, air was allowed to enter in all four flasks to about ⅓ or ¼ of the volume of liquid. This occurred in two of the flasks through a thin glass tube which was brought to red heat. In the other two the air which entered was not heated. An analysis had shown that air which had passed through a red-hot tube still contained 19.4 per cent oxygen. . . . The flasks were then stoppered and then incubated upright at a temperature of 10° to 14°R.* After 4 to 6 weeks a fermentation began in the two flasks which had received the unheated air . . . while the two other flasks remain even now, after twice the time, completely unaffected.

* [10° — 14°R. = 13° — 18°C.]

Thus, in alcoholic fermentation as in putrefaction, it is not the oxygen of the air which causes this to occur, but a substance in the air which is destroyed by heat.

One is forced to think that perhaps alcoholic fermentation is also a destruction of sugar which is caused by the development of infusoria or some other type of plant. Since *Extr. Nucis vom. spir.* is a poison for infusoria and not for molds, while arsenic is a poison for both, I employed these substances first, in order to see if my attention should be directed to infusoria or plants. I found that the *Extr. Nucis vom.* had no effect on alcoholic fermentation, while several drops of a solution of potassium arsenate abolished it completely. Thus I should probably look for a plant as the responsible agent.

Microscopic examination of the beer yeast showed the familiar little grains * which the ferment forms, but the majority of these were connected together in chains. They were partly round, but mostly oval grains of a light yellow color, which occasionally occurred singly, but most often in chains of two to eight or more. From such chains came ordinarily one or more other chains at oblique angles. Frequently I could see between two units of a chain, a small grain attached at one side, as the beginning of a new chain, and generally at the last grain of the chain was a small grain which from time to time became elongated. In brief, the whole appearance was quite similar to many other articulated fungi, and it is without a doubt a plant.

Herr Prof. Meyen, who observed these things at my request, was of the same opinion as myself, and found it difficult to say whether this was more like an alga or a filamentous fungus, but favored the latter because of the lack of green pigment.

The beer yeast consists almost exclusively of these fungi. In freshly pressed grape juice, nothing of this kind can be seen. But if this juice is placed at a temperature of about 20°R.,† one finds already after 36 hours such plants in it, which first consist of only a few units. The growth of these can be watched under the microscope, so that one can see already after ½ to 1 hour the increase in volume of a very small unit which is connected to a larger one. It is only several hours later that one can see the development of gas bubbles, since the first carbon dioxide formed remains dissolved in the water. The formation of these plants increases during the course of the fermentation, and at the end they can be seen in large amount as a light yellow powder at the bottom. They show only slight differences from the fungi in the beer yeast. Only a few of them are identical with those of the beer yeast. In the alcoholic fermentation of grapes the units are rounder and do not remain so frequently in straight chains. Also the number of units which remain single or with only a second small grain is much larger than is the case in beer yeast. The observation of their growth leaves no doubt of their plantlike nature.

From these experiments the following main points can be established:

(1) A boiled organic substance or a boiled fermentable liquid does not putrefy or ferment, respectively, even when air is admitted, so long as the air has been heated.

(2) For putrefaction or fermentation or other processes in which new animals or plants appear, either unboiled organic substance or unheated air must be present.

* [Schwann's word in German was "Körnchen."]

† [20°R. = 25°C.]

(3) In grape juice the development of gas is a sign of fermentation, and shortly thereafter appears a characteristic filamentous fungus, which can be called a sugar fungus.* Throughout the duration of the fermentation, these plants grow and increase in number.

(4) If ferments which already contain plants are placed in a sugar solution, the fermentation begins very quickly, much quicker than when these plants must first develop.

(5) Poisons which only affect infusoria and do not affect lower plants *(Extr. Nucis vom. spir.)* prevent the manifestations of putrefaction which are characteristic of infusoria, but do not affect alcoholic fermentation or putrefaction with molds. Poisons which affect both animals and plants (arsenic) prevent putrefaction as well as alcoholic fermentation.

The connection between the alcoholic fermentation and the development of the sugar fungus should not be misunderstood. It is highly probable that the development of the fungus causes the fermentation. Because a nitrogen containing substance is also necessary for the fermentation, it appears that nitrogen is necessary for the life of this plant, as it is probable that every fungus contains nitrogen. The alcoholic fermentation must be considered to be that decomposition which occurs when the sugar fungus utilizes sugar and nitrogen containing substances for its growth, in the process of which the elements of these substances which do not go into the plant are preferentially converted into alcohol. Most of the observations on the alcoholic fermentation fit quite nicely with this explanation.

* [Note the derivation here of the Latin name *Saccharomyces*, which is the genus name for common yeast.]

Comment

From the simplicity and clarity of Schwann's observations, it would seem obvious that his conclusions should quickly become universally accepted. However, this was not to be, as we shall see. In his first experiment Schwann confirms Spallanzani's experiments, at the same time answering any question that might arise about the availability of oxygen for the putrefaction process. He devises an ingenious system to prevent any unheated air from reaching his vessels, which surely required considerable patience and endurance. Many later workers could not confirm Schwann's experiment, and this may have been due merely to their lack of care in running the experiment. Schwann then proceeds to draw the conclusion that putrefaction is a by-product of the process of the growth of organisms which use the organic materials as food. He makes clever use of general and specific metabolic poisons in deciding what organisms were responsible for what phenomena.

His experiments on alcoholic fermentation are also outstanding. He was the first to observe yeast in the process of growing. He accurately describes this process, and draws the obvious conclusion that this growth which he sees is of a living organism. He shows that alcoholic fermentation and the appearance of yeast cells are always associated events, and draws the conclusion, later denied by Liebig, that the yeast cells make the alcohol.

Schwann was many years ahead of his time, and it was not until Pasteur's work of 1860 that the scientific world generally accepted these early conclusions. Two years after the preceding paper was published, Schwann published his classic work, "Microscopical researches on the similarity in structure and growth of animals and plants," in which he described for the first time the cellular nature of the higher animals.

Memoir on alcoholic fermentation

1838 • Charles Cagniard-Latour

Cagniard-Latour, C. 1838. Mémoire sur la fermentation vineuse. *Annales de chimie et de physique,* Vol. 68, pages 206–222.

IT IS WELL KNOWN THAT WHEN FRESH beer yeast is mixed with a solution of sugar and added to a closed flask . . . and exposed to a temperature of 25° centigrade, very quickly, within a few minutes, the solution undergoes a fermentation which proceeds very rapidly if the amount of yeast added is considerable. However, when the solution does not contain the yeast and the sugar in the solution is pure, the same circumstances do not ensue, even after a very long time. It seemed important to examine under the microscope this material which had the property of causing the fermentation of sugar. This examination, which I have had the honor of reporting in a letter to the Academy, has led me to consider that the particles of which the yeast is composed are globular in form, and I have concluded that it is very probable that these particles are organized. (Footnote: Twenty-five years ago I first examined fresh yeast under the microscope. However, my instrument was very poor, and I concluded that the yeast was like a very fine sand composed of crystalloid particles. These observations were in error. The majority of the microscopic observations indicated in this memoir were performed on a microscope constructed by M. Georges Oberhauser. It enabled me to obtain enlargements of 300–400 times. For measuring the size of the globules, I used an ocular micrometer constructed by M. Charles Chevalier. . . .)

The globules which I have observed are generally simple, transparent, spherical or slightly elongated, almost colorless. I have never seen them exhibit any movement which could be considered an indication of their independent action. On the other hand, the globules of yeast can appear in a liquid before the alcoholic fermentation has begun. But when the globular bodies, that is to say, the non-crystalline bodies, are beginning to produce themselves in a mucous liquid which, before the experiment has begun, does not reveal these globules, and when these bodies do not appear to have any power of locomotion, the microscopist will consider these bodies as simple plants. This is the way M. Turpin has viewed the protospheres which developed in a gelatinous medium, as I have reported to the Institute in a previous letter.

It is highly probable that the globules of yeast are organized, and they appear to belong to the vegetable kingdom. . .

But these plants, if one can give this name to such simple vesicles, are very small, and highly variable in size. The diameter of the largest globules does not exceed a hundredth of a millimeter.* For the most part they are of such a size, that a cake of yeast of one cubic millimeter would contain at least a million globules.

On the assumption that the yeast globules had the faculty to reproduce themselves, I performed a number of experiments to find out more about them. . . .

My earlier observations (see the journal of the Institute, no. 185) had yielded the following principal results: (1) In order to liberate gas, the globules of yeast rise to the surface of the wort (of beer) and many of these globules are within the foam or scum which is produced during the fermentation, and can be readily observed with the microscope. (2) During their action on the wort, the globules diminish in size, and very probably produce seeds or reproductive bodies through this contraction, since one discovers very soon in the wort new globules, smaller in size. These globules, which are not observed at first, seem to have the ability to reproduce by budding or elongation, and because of this can form globules in chains, which confirms my hypothesis that the yeast globules are organized and belong in the plant kingdom.

I examined samples removed hourly from the fermentation vessel. One hour after the addition of the yeast, I already saw that the wort contained single globules each of which contained a small secondary globule attached to it. This small globule seemed to grow, since a little while later most of the pairs seemed to contain globules of the same size. The fourth sample revealed a very few paired globules. In order to assure myself that these pairs of globules were actually attached and not merely touching each other, I applied a force to the cover glass with a small needle while watching the globules under the microscope, and although this force produced a violent agitation of the globules, it did not affect the point of attachment of the pairs. It would appear that the individual globules become separated naturally as they grow to a certain age, since the commercial yeast generally consists of simple globules. . . . In the course of many fermentations which I have carried out with beer yeast, it has been possible to distinguish in certain globules many granules, and sometimes a round or oval body, sometimes central and sometimes lateral, which one can presume to be a scar or umbilical mark which develops during the separation of the two globules. . . .

The globules of ferment are able to develop very rapidly, because when one collects the total amount of yeast that is produced during the fermentation, one finds that this quantity weighs about seven times the initial quantity of yeast added, which agrees with the microscopic examination.

Considering the rapidity with which this increase in yeast is obtained, there is every reason to believe that this is the result of the reproduction of the globules of yeast which have been added. It appears that the globules reproduce in a liquid which contains the nutrients favorable for their growth. All brewers are aware that the beer wort ordinarily produces a larger amount of yeast than that added, but it has been supposed that this increase is due principally to a precipitation of the plant proteins which are present in the wort. . . .

But while the wort is a medium in

* [0.01 mm = 10 microns.]

which reproduction of the ferment proceeds very readily, it is not the same as a simple solution of sugar, since an active yeast does not increase in weight in the latter, and actually loses its activity. . . .

M. Gay-Lussac has demonstrated in various experiments that oxygen has a great influence on the beginning of fermentation in certain liquids, especially in grape juice. But although oxygen is necessary for fermentation to begin, it is not necessary for fermentation to continue. Because of this fact and the knowledge that beer yeast is able to produce a fermentation in sugary materials without the presence of oxygen, M. Gay-Lussac has advanced the opinion that the power of fermentation rests in a large number of substances, but in a different state than in beer yeast. The experiments to be presented below seem to indicate that this opinion is correct.

I expressed the juice from a bunch of grapes while they were enclosed in a bell jar under an atmosphere of hydrogen and allowed this juice to remain under mercury for more than 15 hours. During this time I examined under the microscope a small amount of the deposit which separated from the juice and found it to be practically amorphous. But after I allowed a small amount of oxygen into the jar, which started the juice to ferment, I found a large deposit of yeast globules. One can surmise the following from this: (1) The bodies of these small plants form a part of the material which deposited. (2) They had not germinated because they were enclosed within the berries of the grapes. (3) This germination begins when they are exposed to oxygen and because of this beginning of development, they are able to act like the beer yeast.

It has been reported to me that it is possible to cause the production of globules when egg white is diluted in water and heated, and that the globules of yeast arise from a nitrogenous material of animal or vegetable origin, which is present in the wort, and coagulates to form the globules. If this is so the yeast globules are no more living than those globules obtained during the coagulation of egg white by heat.

In order to clarify this point, I placed in a capsule in a bath heated to 90° centigrade, a mixture of 50 grams of water and one gram of egg white. When a portion of the albumin had been coagulated by the heat, I removed the capsule and after cooling, examined a small amount of the material which had formed at the surface of the liquid under the microscope. I found that this material contained a type of globule with a diameter averaging about a hundredth of a millimeter. But these globules resembled in general something crystalloid, and in none of them could be seen granules or an umbilical mark. It seems to me therefore, that the objection that the globules of ferment are analogous to those of coagulated egg white is not valid.

Furthermore, I have conducted a spontaneous fermentation of wort in a closed container, that is to say, a fermentation without the addition of yeast. This wort, even though it had been filtered, produced during its alcoholic fermentation a deposit of yeast, which upon examination under the microscope, proved to be composed of globules analogous to those in ordinary yeast. This fermentation took place much slower than those carried out in breweries, and one might expect, if the hypothesis that the globules are formed by a type of proteinaceous coagulation, that they would resemble those from coagulated egg white. But this is not the case. The globules in the deposit did not prove to be the same size as those in ordinary yeast, but this does not argue against their living nature, since the fermentation in this case had taken place over a very long time, and the globules may be quite different in age from those in ordinary yeast.

I carried out the same experiment in a flask which had been filled with carbon dioxide. Although the fermentation developed a little slower, the deposit obtained showed the same appearance under the microscope.

Since in these experiments the deposit formed showed the same appearance under the microscope as beer yeast, I have carried out a number of fermentations in closed vessels with various liquids, such as currant juice, grape juice, and prune juice, as well as in a solution of sugar with egg white. In all these cases the liquids were filtered before they were placed in their vessels. Examination of the deposits obtained has shown that in all cases they are composed to a large extent of globules which are analogous to those of beer yeast. . . .

All who carry out fermentations on a large scale, such as brewers and distillers, know that in spite of the greatest care in their operations, their results are extremely variable. These irregularities are favorable to the hypothesis that the alcoholic fermentation is caused by bodies endowed with life, because who knows in how many different ways such similar bodies can be affected. . . .

SUMMARY

I am acquainted with the principal literature concerning alcoholic fermentation, but I have seen no work in which the microscope was used to study the phenomena on which it depends. (Footnote: Leeuwenhoek, in 1680, examined beer yeast under the microscope and stated that it is composed of globules, but he thought they were a part of the grain which was used to make the wort. But this observation did not suggest to this author the most important point, that these globules are capable of germinating and growing in the wort during the fermentation.)

The principal results of the present work are: (1) The beer yeast is a mass of small globules which are able to reproduce, and consequently are organized, and are not a simple organic or chemical substance, as has been supposed. (2) These bodies appear to belong to the plant kingdom. (3) They seem to cause a decomposition of sugar only when they are alive, and one can conclude that it is very probable that the production of carbon dioxide and the decomposition of sugar and its conversion to alcohol are effects of their growth.

I would also like to remark that the yeast considered as an organized body merits the attention of physiologists for the following reasons: (1) It can grow and develop under certain conditions very rapidly, even under an atmosphere of carbon dioxide. (2) Its manner of growth presents details of a type which has not been observed in other microscopic objects composed of isolated globules. (3) It does not die when frozen or in the absence of water.

In conclusion I would like to add that the old question of the cause of alcoholic fermentation, which was proposed by the Institute, appears to be solved from the present results and those I have reported in 1835 and 1836. These results indicate the conclusion that in general ferments, or at least those which cause alcoholic fermentation in the manner of yeast, are composed of organized microscopic bodies which are very simple, and the substances which they induce to ferment are purely chemical substances, since they are sugar and compounds related to sugar.

Comment

This work was performed independently of Schwann's, and confirms his for the most part. The student should note that the real reason these two workers came up with the same observations at the same time was probably due to the production of better microscopes, as Cagniard-Latour mentions. Since the yeast cell is quite small in comparison to plant and animal cells, it would take at least 400 magnifications to be able to discern cellular details. It was this greater magnification which enabled Cagniard-Latour to differentiate yeast cells from globules of coagulated egg albumin.

The author's observations on some of the physiological properties of the yeast are interesting. The yeast cell's ability to grow anaerobically and to withstand freezing and desiccation are well confirmed today. Many other microorganisms, including some pathogenic ones, also possess these properties.

This paper and the preceding one by Schwann gave Henle support for his early germ theory of disease. See page 76.

Concerning the phenomena of fermentation, putrefaction, and decay, and their causes

1839 • Justis Liebig

Liebig, Justis. 1839. Ueber die Erscheinungen der Gährung, Fäulniss und Verwesung, und ihre Ursachen. *Annalen der Physik und Chemie*, Vol. 48, pages 106–150.

. . . I WISH NOW TO DRAW THE attention of natural scientists to a previously undiscovered cause for the changes and decompositions which in general are known as *decay*, *putrefaction*, *fermentation* and *rotting*.

This cause is the ability which a chemical substance possesses, when in the process of decomposition or combination, to cause or enable another substance which is touching it, to undergo this same change which it itself is undergoing. . . .

The generality of this causation can be demonstrated with countless experiences. It will suffice when I indicate only several of these.

Platinum, for example, does not possess the ability to decompose in nitric acid and hence to dissolve in it. Platinum alloyed with silver dissolves very easily in nitric acid. The ability which silver possesses is carried over to the platinum. . . . [Then follow a number of other examples of the same sort drawn from the chemical literature.]

The indicated examples suffice to prove the existence of this characteristic principle. . . .

Before I go into more detailed considerations, it is necessary to establish the essence of the above mentioned experiences for the cases at hand.

By decay, one understands in general the changes which organic materials undergo at ordinary temperatures. These changes take place only under moist conditions, they cease at the freezing point of water, they do not occur when oxygen is withheld.

If air is withheld from a substance in the process of decay, putrefaction sets in.

Decay is an oxidation at low temperature in which the elements of the substance take unequal parts, depending on their congeniality with oxygen.

Putrefaction is a decay in which the oxygen of the atmosphere takes no part. It is an oxidation of one or more elements of the substance, using the oxygen of the material itself, or the oxygen of water, or both.

By withholding oxygen and in a deficiency of water, putrefaction and decay begin simultaneously. This decay process is called rotting.

Fermentation is a putrefaction of vegetable substances which results in little or no unpleasant odor. . . .

The so-called *ferment* arises in the course of a metamorphosis which begins after the entrance of air into a sugar-containing plant juice, and which can continue to a certain point without interruption in the absence of air. In the ferment is found all of the nitrogen of the nitrogen-containing substances of the plant juice. . . .

. . . it appears that the ferment is not a characteristic cause of putrefaction or fermentation. It is rather that putrefaction and fermentation elicit the changes which the ferment undergoes.

The ferment is a substance which is in the process of putrefying or decaying. It transforms oxygen from the surrounding air into carbon dioxide and also produces carbon dioxide out of its own substance . . . its ability to cause fermentation disappears when putrefaction is completed.

To preserve the properties of the ferment . . . the presence of water is necessary. Through simple expression of water by squeezing, its ability to cause fermentation is reduced, and through drying this ability is destroyed. It is destroyed by boiling water temperature, alcohol, sodium chloride, acetic acid, an excess of sugar, mercuric oxide, mercuric chloride, sulfuric acid, silver nitrate, and esters; that is, by all substances which prevent putrefaction.

The insoluble substances, which are called ferment, do not cause fermentation. If one washes beer or wine yeast with boiled cooled distilled water, always being careful that the material is covered with water, the residue does not cause fermentation when placed in sugar water. This ability is retained, however, by the wash water, which will lose the ability in several hours if exposed to air.

The ability to cause fermentation which the soluble part of the yeast possesses, does not depend on an effect brought about through contact with the yeast. The yeast loses its fermentation powers immediately upon contact with alcohol, without having taken up any of it. A hot, clear, watery extract of ferment does not cause any fermentation when mixed with sugar water in a closed vessel. *The fermentation is effected by the soluble material, only after it itself has undergone a decomposition.* If this hot extract is allowed to cool in the air and allowed to stand for several hours, it is able to cause a vigorous fermentation in sugar water. Without first contact with air, no fermentation occurs. During contact with the air, absorption of oxygen begins, and the extract contains after several hours a significant amount of carbon dioxide.

During the fermentation of sugar with ferment, two decomposition processes occur. If one takes a graduated bell jar filled with mercury and adds one cubic centimeter of a beer yeast suspension and 10 grams of a sugar solution containing one gram of pure sugar, after 24 hours at 20° to 25°, one finds about 50–51 per cent of the weight as carbon dioxide.

Thenard found that from 100 parts of cane sugar, he obtained 51.27 parts of carbon dioxide, and 52.62 parts of alcohol. Thus one finds that the carbon of the sugar is distributed almost exactly between the carbon dioxide and alcohol.

The analysis of cane sugar has shown without a doubt that it contains the elements of 4 atoms of carbon dioxide, 2 atoms of ether, and 2 atoms of water.

The products of its fermentation show that ⅔ of the alcohol and ⅓ of the carbon dioxide carbon comes from the sugar, but these products contain 2 atoms of hydrogen and 1 atom of oxygen more than the sugar. It is clear that these additional atoms have arisen from one molecule of water. . . .

A certain amount of ferment is necessary in order to bring a certain amount of sugar into fermentation, but its effect is not one of mass action. Its influence is limited to its necessity to be present until the time when the last atom of sugar has been decomposed. The ferment is not the cause of fermentation. The insoluble part does not possess this property. The extracted part, which arises through the decomposition of the ferment, possesses this property. Both materials are able to cause fermentation from the moment when they suffer changes through the agency of air and water which results in their own destruction. It is however, no characteristic body, substance or material which causes this destruction. This characteristic body or substance is only the carrier of an activity which serves to increase the amount of substance undergoing decomposition. . . .

During the putrefaction of animal substances, one finds the elements of these substances in continual change, in a state of unstable equilibrium, during which even the weakest forces can alter them or modify them. Such a condition seems to be a fruitful base for the development of imperfect and lower animals, the microscopic animals, whose eggs are well known to be widely distributed. They develop in this putrefying material and multiply in myriads, making use of the products developing during the putrefaction as their nutrients.

Many natural scientists view the chemical process of putrefaction as only a result of the productions of these animals. This would be like saying that the cause of the decay of wood and its rotting arise from the plants which use these decaying products as nutrients.

None of these animals develop in putrefying materials, if one prevents these materials from contact with atmospheric air, which is a necessary condition for their presence, in the same way that one can prevent maggots from developing in putrefying cheese, when one protects this cheese from flies. . . .

Organic chemistry recognizes two separate types of phenomena for the behavior of its compounds.

(1) Substances develop from newly developed properties, in which the elements of many atoms of simple compounds become transformed into molecules of a higher order.

(2) Molecules of a higher order can decompose into molecules of a

lower order, as a result of a neutralization of the equilibrium in the attraction of their elements.

The following disturbances will bring about decomposition:

(a) Heat

(b) Contact with different substances

(c) The influence of a substance which is in the process of changing.

Comment

I have translated only a small portion of this long, rambling account. Its inclusion in this collection is only warranted because Liebig was a powerful figure who was able, by sheer weight of prestige, to influence a number of lesser scientists. Because of this the controversy over spontaneous generation was extended for another generation.

It will be noted that there is little experimental work in Liebig's paper. He starts from a position of pure prejudice, and consequently cannot but conclude as he does. This paper should be a continuous object lesson to all scientists, young and old, that one good experiment is worth one hundred bad theories.

This should not be taken to mean that Liebig did not make significant contributions to science. He was a great chemist, and was convinced that all vital activity could be explained in terms of chemistry and physics. Much of the history of biochemistry had its beginnings in the work of Liebig and his school. It is unfortunate that he took the view he did on the putrefactive process, since this probably delayed for a number of years the rise of microbiology as a science.

Report on the lactic acid fermentation (author's abstract)

The original French article appears in the Appendix, pages 265–267.

1857 • Louis Pasteur

Pasteur, L. Mémoire sur la fermentation appelée lactique. (Extrait par l'auteur). *Comptes rendus de l'Académie des sciences,* Vol. 45, pages 913–916, 1857.

My work on the properties of the amyl alcohols and on the very remarkable crystallographic properties of their derivatives has led me to a study of the process of fermentation. Later I will have the honor to present to the Academy observations which show an unexpected relationship between fermentation and the optically active properties of organic molecules found in nature.

The necessary materials for the preparation and production of lactic acid are well known to chemists. It is known that it is only necessary to take a solution of sugar and add chalk, which keeps the medium neutral, a nitrogenous material, such as casein, gluten, animal membranes, etc., in order to have the sugar transformed into lactic acid. But the explanation of this phenomenon is quite obscure, since the way in which the decomposable nitrogenous material acts is completely ignored. Its weight does not change significantly. It does not putrefy. Although it becomes modified and is continually in a marked state of change, it would be difficult to speak of what its composition is.

Careful studies up to the present have not revealed the development of organized beings during the fermentation process. Those observers who have recognized such beings have always established at the same time that they were accidental and harmful to the fermentation process.

The facts appear therefore to be very favorable for the ideas of M. Liebig. In his eyes, the ferment is a substance which is highly alterable which decomposes and in so doing induces the fermentation because of the alteration which it experiences itself, communicating this agitation to the molecular group of the fermentable material and in this way bringing about its decomposition.* According to M. Liebig, this is the principal cause of all fermentations and the origin of the majority of contagious diseases. Each day his opinion receives more favor. . . .

I propose to establish in the first part of this work that, in the same way that there exists an alcoholic ferment, the yeast of beer, which is always found wherever sugar is decomposed into alcohol and carbon dioxide, there also exists a particular ferment, the lactic yeast, which is always present when sugar becomes converted into lactic acid. Furthermore, the decomposable nitrogenous material which is able to bring about the conversion of sugar into this acid is used as a convenient nutrient for the development of this ferment.

It is possible to observe in ordinary lactic acid fermentations, on top of the sediment of chalk and nitrogenous material, a gray substance which occurs at the surface of this deposit. Under microscopic examination it can be barely distinguished from the casein, disintegrating gluten, etc. so that nothing indicates that it may be a special material, nor that it has arisen during the fermentation. Nevertheless, it is this substance which plays the principal role in the fermentation. I will shortly reveal the method for its isolation and preparation in a state of purity.

I have extracted from beer yeast its soluble material by boiling it several times with 15 to 20 times its weight of water. This extract is filtered carefully.† In this are dissolved 50 grams of sugar per liter, and some chalk is added. Then it is seeded with a trace of the gray material which I mentioned above, and a good lactic acid fermentation is obtained of the usual sort. On the next day the fermentation is vigorous and regular. The liquid which had been perfectly clear at the beginning becomes turbid, the chalk gradually disappears, and at the same time a deposit is produced which increases continually and progressively at the same rate that the chalk disappears. In addition, all of the characteristics and symptoms of the well-known lactic acid fermentation are observed. In this experiment it is possible to replace the yeast water with a decoction of any decomposable nitrogenous material, whether fresh or decomposed. Let us now see the nature

* [See pages 24–27 for a first-hand expression of Liebig's views.]

† [This is probably the first description of the preparation of yeast extract, a nutrient that is widely used in culture media today.]

of this substance which is correlated with all of the phenomena that are included under the words *lactic acid fermentation*. Its appearance is similar to that of the beer yeast when it is studied *en masse* and squeezed or pressed. Under the microscope it is seen to form tiny globules or small objects which are very short, isolated or in groups of irregular masses.* These globules are much smaller than those of beer yeast and move actively by Brownian movement. If washed with a large amount of water by decantation, then diluted into a solution of pure sugar, they immediately begin to make acid, but quite slowly, since acid inhibits significantly their action on sugar. If chalk is added so that the medium is maintained at neutrality, the conversion of sugar is considerably accelerated, and even though only a small amount of material is acting, in less than an hour the production of gas is observed and the liquid contains large amounts of calcium lactate and butyrate. Only a small amount of this yeast is needed to convert a large amount of sugar. This fermentation is preferably carried out in the absence of air, since it is inhibited by plants or by parasitic infusoria.

Therefore the lactic fermentation, like the ordinary alcoholic fermentation, is always correlated with the production of a nitrogenous material which has all the properties of an organized body of the mycodermal type, and is probably closely related to the yeast of beer. But the difficulties of the subject are only half solved. There are many complications. Lactic acid is indeed the principal product of the fermentation which has been given its name, but it is far from the only product. Butyric acid, alcohol, mannitol and a viscous material are always found accompanying the lactic acid. The proportions of these materials are highly variable. The circumstances with mannitol are especially mysterious. The proportions which are formed of this substance are subject to large variations. In addition, M. Berthelot has shown that if sugar is replaced by mannitol in the lactic fermentation, mannitol is fermented to alcohol, lactic acid, and butyric acid. How is one to explain the formation of mannitol in the fermentation from sugar, when it should be destroyed at the same rate that it is produced? Let us examine carefully the chemical properties of this new yeast. I have stated that if it is washed well and placed in a pure sugar solution, it acidifies the liquid over a period of time. The conversion of sugar becomes gradually slower under these conditions, since the liquid becomes quite acidic. The liquid can be analyzed successfully only after the acid is neutralized with chalk and the excess sugar is destroyed by beer yeast. One then finds, if the liquid is evaporated, variable proportions of mannitol and a viscous material. However, washed lactic yeast brings about the transformation of sugar into various products including always mannitol, only if the liquid is allowed to become quickly acid. If the same experiment is repeated with the precaution that an amount of chalk is added which is sufficient to continually neutralize the liquid, then neither the gum nor the mannitol are formed, or, perhaps more exactly, are quickly transformed further.

I reported above that M. Berthelot had shown that when mannitol is substituted for sugar in the lactic fermentation, this material is fermented. But it is simple to show that in this case, the fermentation of mannitol occurs when the lactic yeast develops

* [Probably *Streptococcus lactis*.]

and brings about its production. If a solution of pure mannitol is mixed with powdered chalk and fresh washed lactic yeast, within an hour gas appears and the chemical transformation of mannitol begins. Carbon dioxide and hydrogen are released, and the liquid contains alcohol, lactic acid and butyric acid, all productions of the fermentation of mannitol.

Concerning butyric acid, experiments have shown that the lactic yeast acts directly upon calcium lactate and converts it into calcium carbonate and calcium butyrate. But sugar is acted upon first, and if it is present in the liquid, the yeast prefers to ferment it to lactic acid.

In a later communication I will have the honor of presenting to the Academy some general ideas and new methods of experimentation from work on other fermentations.

Comment

This paper announces the entry of Pasteur into the field of fermentation, and represents the beginning of the science of microbiology. Previously Pasteur had worked in crystallography, and had shown that optical isomers of an organic compound differed in crystal structure. He became interested in optical isomers and made many studies. Through these he formulated the idea that optical isomerism existed through the agency of living organisms. His studies on the optical isomers of amyl alcohol led him to the study of fermentation. His first reported studies were on the lactic acid fermentation, as seen above. As he mentions, the doctrine of Liebig was well established by this time, the earlier works of Schwann and Cagniard-Latour (see pages 16 and 20) having been forgotten. Pasteur entered into a battle with Liebig which lasted for a number of years, but which was essentially ended when Pasteur published his paper on alcoholic fermentation (see page 31).

However in this, his earlier work, he discusses lactic acid fermentation, announcing that it is brought about by a living organism. He describes crude microscopical observations of this organism, and manages to separate it from the fermentation broth by washing. He is able to show that this organism can grow in a medium containing yeast extract, sugar, and chalk, and that in this medium it produces lactic acid. These results, only described briefly here, were sufficient to convince him that the organism caused the fermentation.

As he himself comments, there are a number of complications in the natural lactic fermentation. He was not dealing with a pure culture. In fact, the very concept of a pure culture was foreign to him, and it was only later, through more detailed studies on fermentation, that it became obvious that many of the complex changes occurring were not due to the action of a single organism. This is obvious in the present work, and there would be no point in attempting to explain the variable appearance of mannitol, butyric acid, and gum. Pasteur himself later solved this problem well enough (see page 39).

In his publications, Pasteur followed the French habit, still prevalent today, of writing many short papers, with very few pieces of actual data in each paper. In this way the essential features of a process were revealed to the readers only gradually over a period of months or years, leading to many misconceptions and misinterpretations. The number of papers Pasteur published in total runs to over six volumes, but many of these are repetitions of earlier papers, with a few additions. The papers of Pasteur that are included in the present volume will give the reader a brief idea of some of the areas that he worked in, although whole areas of his work are not included because they did not seem to lend themselves well to the purpose of the present volume.

Memoir on the alcoholic fermentation

1860 • Louis Pasteur

Pasteur, L. 1860. Mémoire sur la fermentation alcoölique. *Annales de Chimie et de Physique*, Vol. 58, 3rd Series, pages 323–426.

BY ALCOHOLIC FERMENTATION, I REFER to that fermentation which sugar undergoes under the influence of the ferment which is called beer yeast. It is the fermentation which is the source of alcohol in wine and all alcoholic beverages. . . .

Any hesitation to apply the words *alcoholic fermentation* and to realize their true meaning seems to me impossible, since they have been applied by Lavoisier, Gay-Lussac and Thenard to the fermentation of sugar by beer yeast. . . .

In the first part of my memoir I will study the changes which occur in sugar during the alcoholic fermentation, and in the second part I will consider especially the ferment, its nature and the transformations which it undergoes.

[The first part is concerned with the products formed in the fermentation, such as alcohol, carbon dioxide, glycerol, and succinic acid, and the stoichiometric relationships to the sugar.]

SECOND PART. CONCERNING THE TRANSFORMATIONS OF THE BEER YEAST DURING ALCOHOLIC FERMENTATION

I. Historical summary of the state of the science of beer yeast and its modifications during alcoholic fermentation.

In 1680 Leeuwenhoeck studied beer yeast under the microscope and found very small spherical or oval globules, but the chemical nature of this substance was unknown to him. Fabroni identified the yeast with gluten. This was some progress. It gave an indication that yeast might be an organic product. M. Thenard published a memoir in which he said: All natural sugary juices, in the process of spontaneous fermentation, deposit a substance which resembles beer yeast and which has the power of fermenting pure sugar. This yeast is animal in nature, since it is nitrogenous and yields ammonia upon distillation. . . .

In his observations published in 1835 and 1837, M. Cagniard de Latour introduced a new idea. Before his time, yeast had been regarded as a vegetable product, produced *in situ*, which precipitated out in the presence of a fermentable sugar. M. Cagniard de Latour recognized "that the yeast was a mass of globules which reproduced by budding, and were not merely a simple chemical or organic substance." He concluded that "it is very probable that the production of carbon dioxide and the decomposition of sugar and its conversion into alcohol are effects of the growth of the yeast."

This opinion immediately found a powerful opponent in M. Liebig.

In his eyes, the ferment is an extremely unstable substance which decomposes itself and which causes fermentation as a result of the decomposition which it itself undergoes, during which it communicates this perturbation and disassimilation to the fermentable material. He expresses himself thus: "The experiments which we have revealed demonstrate the existence of a new cause which brings about decomposition and synthesis. This cause is nothing else than the movement which a body in the process of decomposition communicates to other substances in which the elements are held together very weakly. . . . Beer yeast, and in general all animal and vegetable materials undergoing putrefaction, communicate to other substances the state of decomposition in which they find themselves. . . ."

M. Liebig has developed his opinions throughout the majority of his works with such a persistence and conviction that they have gradually triumphed. Today they are accepted generally in Germany and France. His ideas have been applied to other fermentations, such as the lactic, by MM. Fremy and Bautron.

As far as I can see, the reason that the ideas of M. Liebig have become gradually established amongst chemists is the following. There have been discovered during the last twenty years a large number of phenomena which have been placed in the same group with alcoholic fermentation and in which it has been impossible to recognize the existence of particular lower plants, but in which there was a substance which was undergoing decomposition. For example, if one places a sugar solution containing chalk with a nitrogenous material from some animal source, like casein, gluten, fibrin, gelatin, rennet, an animal membrane . . . one sees that the sugar gradually becomes converted into lactic acid. But although these animal materials are very diverse, the effect on the sugar is always the same. There is only one thing which appears to be similar in these nitrogenous materials; this is their gradual decomposition. A correlation is thus demonstrated between the transformation of sugar into lactic acid and an instability of the animal substance, a tendency to decompose.

The work of M. Colin on alcoholic fermentation had already shown in 1825 that the analogous facts existed for it. He showed that these animal materials of diverse origin are able to induce the decomposition of sugar into alcohol and carbon dioxide.

Meanwhile a remarkable circumstance ought to have aroused the attention of, and cautioned, those who were concerned with alcoholic fermentation. Indeed, after the publication of the observations of M. Cagniard de Latour, M. Turpin, who had been in charge of the proceedings of the Academy, studied, at the request of M. Thenard, the deposits which form in the alcoholic fermentation of sugar in the presence of egg white, and found that they consisted exclusively of globules of the beer yeast.

Since one of the materials used by M. Colin, albumin, did not cause an alcoholic fermentation while allowing for the appearance of yeast, it may be assumed that all the other nitrogenous substances behaved similarly, and because of their diversity would prove nothing more concerning M. Liebig's theory.

But I hasten to add that nothing of the like would exist in the case of the very diverse and very numerous lactic fermentations. All of the workers agree that here there is only a chemical alteration of the animal substance. The facts concerning this fermentation and many other phenomena of the same order had therefore a decisive influence on the theory.

The idea of M. Cagniard de Latour, which at first had a certain amount of acceptance, was gradually abandoned. Many people did not contest the idea that beer yeast was organized, but it was believed to be partly destroyed by the fermentation, as had been stated by M. Thenard, and in common with all of the other nitrogenous materials acting as ferments, it was in this way that it acted on sugar. Such is the thought of M. Liebig.

Berzelius did not agree with the ideas of M. Liebig, while rejecting those of Cagniard de Latour and Schwann. For him fermentation was an action through contact. He did not believe at the same time in the existence of a living organism in the yeast. "It is only a chemical product which precipitates in the fermentation and which takes the ordinary form of a non-crystalline precipitate, even inorganic, of small balls which group themselves one after the other and form chains." Elsewhere he explains himself thusly: "It is clear that when the organized bodies decompose in the water, and when the dissolved materials precipitate, the latter should assume a form, and as they do not assume regular geometrical forms, the result must be other forms, depending on the nature of the bodies which influence them. . . . It is thus quite natural that they imitate the forms of the simplest organisms of plant life. Nevertheless form alone does not constitute life."

The chemical composition of yeast as published by M. Payen shows the following figures: Nitrogenous material, 62.73%, cellulose integument, 29.37%, fat, 2.10%, minerals, 5.80%. Elemental analysis by Schlossberger for ale yeast shows: Carbon, 50.05%, Hydrogen, 6.52%, Nitrogen, 31.59%, Oxygen, 11.84%. An analysis of the yeast ash by Mitscherlisch gives the following figures: phosphoric acid, 41.8%, potassium, 39.8%, soda ash, none, magnesium phosphate, 16.8%, calcium phosphate, 2.3%, proportion of ash in total, 7.65%.

II. The nitrogen of the yeast is never transformed into ammonia during the alcoholic fermentation. Instead of ammonia formation, a slight amount of it disappears.

On 18 January, 1858, I placed 100 grams of sugar in a liter of water which contained in it the soluble substances from the beer yeast. To this I added a trace of the globules of fresh yeast. An analysis on a portion of this revealed that it contained 0.038 grams of ammonia per liter. On 5 February, the fermentation was ended. An analysis for ammonia revealed 0.020 grams per liter, or less ammonia than at the beginning.

On the 30th of April I repeated this experiment with 100 grams of sugar, but this time I used a very small amount of ordinary yeast, so that the fermentation could last for a longer time. I added only 1.037 grams of yeast (weight of material dried at 100°). On the 30th of August the fermentation was still proceeding. A tube which lead out of the flask was always immersed in the liquid. The liquid was analyzed on the 27th of November. All of the liquid contained only 0.0008 grams of ammonia, and it is possible that it contained no ammonia and this minimum value was an error in the assay. . . .

The consistency of the results and

the data from many other experiments seem to leave no doubt of the principal fact. Not even the minimum quantity of nitrogen is formed during the alcoholic fermentation at the expense of the yeast. But these results show that there is a disappearance of ammonia from the original liquid. In order to study this phenomenon, I added ammonia from the original liquid. In this experiment I added 100 grams of sugar, 10 grams of washed yeast cake, and 0.200 grams of ammonium tartrate levorotatory, containing 0.0185 grams of ammonia. The fermentation lasted a long time. When the sugar was all gone, the liquid was assayed. It contained 0.0015 grams of ammonia.

I recovered unchanged all of the levorotatory tartaric acid free in the liquid. Therefore almost all of the ammonia added as tartrate had disappeared, as well as that which existed in the 10 grams of yeast. . . .

In summary, we see that instead of the formation of ammonia during the alcoholic fermentation, that which is added disappears, especially in the case where there is insufficient albuminous * material present because only a small amount of beer yeast was added. The studies in the following paragraphs will show us that the ammonia which disappears enters into the constitution of the yeast in the state of albuminous material.

III. Production of yeast in a medium composed of sugar, ammonium salt, and phospates.

The experiments which follow will show all the power of organization of yeast and put to an end all discussions on its nature:

In a solution of pure candy sugar † containing 10 grams, I put the ash of 1 gram of yeast, 0.100 grams of ammonium tartrate (dextrorotatory) and the amount of fresh beer yeast which would fit on the head of a pin, which had been washed and which contained 80% water. Very remarkably, the globules which were added under these conditions developed, multiplied and the sugar fermented, at the same time that the minerals slowly dissolved and the ammonia disappeared. In other words, the ammonia was transformed into albuminous materials of a complex nature which entered into the composition of the yeast, at the same time that the phosphates were taken into the new globules produced. The carbon of the yeast is evidently furnished by the sugar.

In a similar mixture, the vessel was filled up to the neck and was well stoppered, and a gas tube was immersed in the liquid. After 24 to 36 hours, the liquid began to show signs that a fermentation had begun by the production of microscopic bubbles which indicated that the liquid was already saturated with carbon dioxide. . . .

The following day the liquid became progressively agitated, and because of the production of gas, foam filled the neck of the flask. A deposit gradually covered the bottom of the vessel. A drop of the deposit was examined under the microscope and showed an extensive development of yeast. The yeast looked very young, with swollen globules which were transparent, without granules, and amongst them could be distinguished very easily each globule of the small quantity of yeast which had been originally used as seed. These globules have a thick envelope and stand out in a black circle. They are yellowish and filled with granules. But the way in which they are many times sur-

* [Proteinaceous.]
† [Sucrose?]

rounded by young globules, indicates quite clearly that they have produced these globules which form the head of the chain.

It is in the first days after inoculation that it is possible to make these interesting observations. At night, using gaslight for illumination, it is possible to distinguish the old globules from the many more young ones, in the same way that one can distinguish a black ball amongst many white balls.

Gradually the differences disappear and the new globules that form lose all appearance that they are in chains. One no longer sees buds. The globules are now very granular in the manner of adult or spent beer yeast.

Nevertheless, the fermentation using ammonium as a source of nitrogen never becomes as active as when an albuminous material is used as a nitrogen source, such as that from grapes, or beet juice, or the soluble part of ordinary beer yeast. If one seeds into sugar water containing a little albuminous material some fresh yeast, the process proceeds in general exactly like that described above, but the fermentation is perceptibly more active. For example, instead of the first appearance of bubbles of carbon dioxide after 36 to 48 hours, they have already appeared after 12 to 24 hours. In addition, the amount of yeast formed and deposited in the same time is greater. But I repeat, in all respects is the process the same, except that it is more vigorous, and the products formed are exactly the same. . . .

It can be stated with certainty that the ammonium salt is indispensable for the fermentation. When yeast is seeded into a sugar solution containing yeast ash but no ammonium salt, there is hardly any sign of fermentation. Occasionally there is a fraction of a cubic centimeter of gas, but this may be due to the ammonium content of the distilled water or the small amount of albuminous material carried over with the inoculum.

The necessity of sugar as a source of carbon for the yeast globules has been sufficiently proven that it requires no further experiments. Therefore, all that is necessary to bring about the phenomenon of fermentation are these things: sugar, nitrogenous substance, minerals. . . .

IV. Study on the relationship between the yeast and the sugar.

We now arrive at a very delicate point in these researches. I would like to speak of the relationship which exists between the sugar and the yeast.

It will hardly be a question of the relationship between the atoms, but rather the more intimate relationships, the physiological connections.

I would like to indicate first several details concerning the structure of the beer yeast globules.

There can be no doubt that the globules form small vesicles with elastic walls, full of a liquid with which is associated a soft material which is more or less granular and vacuolar which is situated directly within the wall. This vesicle gradually reaches the center as the globule ages.

The wall of the cell is elastic. Indeed, when a sample of water containing young globules is allowed to dry on a microscope slide resting on the microscope stage, the contraction of the sample which occurs as the result of the introduction of air presses the globules together and one can see them become deformed and become more or less polyhedric.

The contents of the globules, especially the central contents, are liquid. This is proven by the presence in most of the adult globules of one or more granules which exhibit movement similar to Brownian movement. It would be very difficult to say if it is true Brownian movement. The cause of Brownian movement, which is probably purely physical, is so little understood, that it cannot be said whether it can exert an effect on the free granules in the center of the globule right through the wall of the yeast.

The budding of the globules constitutes an important discovery of M. Cagniard de Latour. It can be represented, after M. Mitscherlich, as the change from Figure 3 to Figure 4, in which it can be seen that the new globule begins as a simple bulge. I have confirmed these observations of M. Mitscherlich. I have seen this process quite clearly many times. Soon the small bud, while remaining attached, appears to have its own wall, and is itself a true globule. The movements of the liquid only detach it when it has reached the size of the mother globule. Therefore its attachment is close and firm.

Is the bud borne, as many people have felt, by an effect of contact, of a pressure between the internal wall of one of the granules of the globule? I have seen nothing which might confirm this opinion, and I believe it to be inexact.

On one hand, the transparent globules, those without apparent granules, are always those globules which are budding, while on the other hand, the development of granules appears to occur only in older globules. It appears that the older the globule, the less active it is, and the less able to bud.

FIG. 3.

FIG. 4.

FIG. 5.

FIG. 6 [1].

I do not believe the statement of M. Mitscherlich, already advanced by Cagniard de Latour and Turpin, that the globules of yeast are able to burst frequently and empty their granules which then spread through the liquid like seeds which then enlarge and become ordinary globules of yeast.

I can state that I have never observed this phenomenon in the course of three years of careful and frequent study on beer yeast, studies which were carried out under a wide variety of conditions of development. The fact that the volume of the globules of yeast during their action on sugar is quite uniform argues against this theory. Those globules which are smaller in size are not free, but attached to the larger globules in the form of buds. It is clear that if the yeast did reproduce by granules which were liberated into the medium by large globules, then one would find all sizes of globules amongst those which are free. . . .

Sugar never undergoes alcoholic fermentation without the presence of living globules of yeast. Reciprocally, globules of yeast are never formed without the presence of sugar or a carbohydrate material or without the fermentation of this material. Any statements which are contrary to this principle have been derived from incomplete or inexact experiments.

All of the chemical work on alcoholic fermentation indicates that it can be accomplished through two separate circumstances, depending on whether the yeast is added to a solution of pure sugar, or whether the sugar solution is mixed with albuminous material. In the first case, it is said, the ferment acts, but it does not reproduce. In the second case, it acts, but it does reproduce. This second case is what occurs during the manufacture of beer.

M. Liebig has said: "If the fermentation is a consequence of the development and reproduction of the globules, this could not cause the fermentation in pure sugar solution, since this solution lacks the conditions necessary for the maintenance of vital activity. This solution does not contain the nitrogenous material necessary for the production of the nitrogen substances of the globules. In this case the globules cause the fermentation, not because they continue to develop, but because of the metamorphosis of their internal nitrogen which decomposes into ammonia and other products. That is to say, because of a chemical decomposition which is completely the opposite of an organic action."

The facts which I have reported are obviously in opposition to these views, and I am certain that whether the yeast is mixed in a pure sugar solution or in a sugar solution containing albuminous material, the phenomena are in many ways similar. In both cases the yeast is organized and multiplies. Only in the first case, when the fermentation is ended, all of the globules, young and old, are deprived of soluble nitrogenous material. The nitrogenous nutrients are fixed in an insoluble state in the new globules that have been formed. The aggregate of these globules does not therefore have an action on the pure sugar water. There are only enough nitrogenous nutrients for the globules which may still be young enough to act and to multiply. On the contrary, in the case of fermentation in the presence of an albuminous material, there are plenty of globules which are exhausted, but the majority of new globules are filled with nitrogenous material and minerals and with the aid of these nutrients are quite able to act when introduced into a new sugar solution. . . .

V. In all alcoholic fermentations a portion of the sugar is fixed in the yeast as cellulose.

I allowed to ferment 100 grams of sugar in 750 cubic centimeters of water with 2.626 grams (dry weight) of yeast. After 20 days, I recovered 2.965 grams (dry weight) of yeast. I boiled this with sulfuric acid (diluted 20 times) for 6 to 8 hours, and also a sample of yeast before fermentation. The weight of the fermented yeast was 1.707 grams (dry weight) and the weight of the unfermented yeast was 1.730 grams. The residue insoluble in sulfuric acid was collected on a tared filter and dried at 100°. The filtrate was saturated with barium carbonate to neutralize and then the amount of sugar was estimated both by Fehling's solution and by the amount of carbon dioxide released during fermentation. The values were calculated for the original weights of yeast (2.626 and 2.965 grams). It was found that 2.626 grams of yeast gave an insoluble nitrogen residue of 0.391 grams (14.8%) and a fermentable sugar value of 0.532 grams. The 2.965 gram sample gave an insoluble nitrogen residue of 0.634 grams (21.4%) and 0.918 grams of fermentable sugar. These results show that (1) in the fermentation of 100 grams of sugar by 2.626 grams of yeast, it fixes into itself 0.4 grams of carbohydrate material transformable into fermentable sugar by sulfuric acid. (2) There is an increase in the amount of nitrogenous material insoluble in dilute sulfuric acid. This last result is a new proof that during fermentation, there is a fixation into an insoluble state of the albuminous materials of a soluble nature which are present within the active yeast globules.

It is still necessary to determine if the boiling with dilute sulfuric acid has dissolved all of the cellulose. I determined the amount of cellulose in the yeast by the method of M. Schlossberger. . . .

These results indicate that boiling with dilute sulfuric acid has removed all of the cellulose. The amount of cellulose in the 2.626 grams of yeast corresponds to 0.532 grams or 20.2%. . . . The yeast collected after fermentation, 2.965 grams, had an amount of cellulose which corresponds to 0.918 grams of sugar, or 31.9%, so that there was an increase of 11% in the amount of cellulose present in the yeast after fermentation.

This considerable increase in the weight of cellulose during the fermentation of sugar is another proof to add to all that I have presented, concerning the living state of the yeast during the alcoholic fermentation.

Comment

Pasteur's main contribution to the problem of alcoholic fermentation is the application of quantitative methods for determining what has happened to the various substances. By doing this, he was able to show that the yeast actually increased in weight, nitrogen, and carbon content during the fermentation process. This added considerable support to the argument that the yeast was really a living organism. But perhaps the most important contribution in this paper is that the yeast can actually increase extensively in weight and produce alcohol even in a liquid which lacks proteinaceous materials of a natural source. He obtained an active alcoholic fermentation in what we would today call a synthetic (or defined) medium, consisting merely of trace elements, ammonium salt and sugar. The problem became considerably clarified by this observation, since it could be easily shown in such a defined medium that the fermentation always proceeded with the development of the yeast, and the increase in protein in the yeast was accompanied by a decrease in nitrogen of the medium. Pasteur's long and fairly precise paper, of which only a small portion is excerpted here, can be said to have ended the controversy regarding the nature of the alcoholic fermentation.

Animal infusoria living in the absence of free oxygen, and the fermentations they bring about

The original French article appears in the Appendix, pages 267–268.

1861 · Louis Pasteur

Pasteur, L. 1861. Animalcules infusoires vivant sans gaz oxygène libre et déterminant des fermentations. *Comptes rendus de l'Académie des sciences,* Vol. 52, pages 344–347.

THE WIDE VARIETY OF PRODUCTS THAT are formed during the lactic acid fermentation are well known. Lactic acid, a gum, mannitol, butyric acid, alcohol, carbon dioxide and hydrogen all appear simultaneously or successively in highly variable proportions and in a quite capricious manner. I have gradually been led to realize that the plant ferment which converts sugar into lactic acid is different from the one(s) which bring about the production of the gummy material, and that the latter in turn do not produce lactic acid. On the other hand it is equally true that none of the various plant ferments can give rise to butyric acid, if they are separated from the other forms.

Therefore there has to be a distinct butyric acid ferment. I have been occupied with this idea for a considerable length of time. I would like to address myself to the Academy today on the origin of butyric acid in the so-called lactic acid fermentation.

I will not go into all of the details of this research. I would like to indicate first one of the conclusions from my work. This is that *the ferment which produces butyric acid is an infusorium.*

For a long time I was prevented from discovering this fact because I had devoted my efforts to eliminating these small animals which I feared would feed on the plant ferment which I supposed to be the cause of the butyric acid fermentation and which I was looking for in the liquids which I was studying. But after unsuccessfully looking for the cause of the butyric acid fermentation, I was finally struck with the correlation . . . between this acid and the infusoria, and inversely between the infusoria and the production of this acid, circumstances which I had attributed to the necessity of butyric acid for the life of these small animals.

Since then a large number of experiments have convinced me that the conversion of sugar, mannitol, and lactic acid into butyric acid is due exclusively to these infusoria, so that it is necessary to consider them to be the true butyric acid ferment.

Here is their description: They are small cylindrical rods, rounded at the ends, generally straight, singly or in chains of two, three or four, and sometimes more. They average about 0.002 mm in diameter, varying in length from 0.002 to 0.015 or 0.02 mm. They move with a gliding motion. During this movement, their bodies remain rigid or make slight undulations. They spin, balancing or quivering actively the two ends of their bodies. The undulatory nature of their movements becomes very obvious when they are longer than 0.015 mm in length. Frequently they are bent at one end, sometimes at both ends. This latter is seldom seen when they are young.

They reproduce by binary fission. It is apparently because of their mode of reproduction that they occur often in chains. One of the units attached to others may move quickly several times in order to detach itself. . . .*

These infusoria can be inoculated in the same way as beer yeast. They multiply if the medium is suitable for their nutrition. But it should be stated that they can be inoculated into a medium containing only the crystallizable and mineral substances sugar, ammonium, and phosphates, and they can reproduce simultaneously with the rapid appearance of the butyric acid fermentation. The weight of cells that are formed is significant, although always small compared to the total quantity of butyric acid formed, but this is true for all ferments.

The existence of infusoria which are able to bring about fermentations is already a notable fact. But in addition another unusual aspect should be mentioned. This is that these infusorial animals are able to live and multiply indefinitely in the complete absence of air or free oxygen. . . .

These infusoria can not only live in the absence of air, but air actually kills them. If a stream of carbon dioxide is passed through a medium in which they are multiplying, their viability and their reproduction are not affected in the least. On the contrary, under the same conditions, if one substitutes a stream of air for the carbon dioxide, in one or two hours they all die, and the butyric acid fermentation which requires their viability is stopped immediately.

We have arrived therefore at the following double proposition:

1. The ferment which produces butyric acid is an infusorium.
2. This infusorium lives in the complete absence of free oxygen.

I believe this is the first example known of an animal ferment, and also the first example of an animal living in the absence of free oxygen.

We will have to consider how the relationship of the mode of life and the properties of these small animals, together with the same aspects of the plant ferments which can also live without free oxygen, are related to the processes of fermentation. In the meanwhile I would like to make no further comment on the ideas which these new facts suggest until further research has been done.

* [These organisms are not animals, as Pasteur supposed, but small motile bacteria.]

Comment

This is the first report that any organism can live and reproduce in the complete absence of free oxygen. This discovery is quite important for general biology, since it shows that oxygen gas is not a requisite for life. This discovery opened up for Pasteur a new field of study, relating fermentative and biological processes to the presence or absence of oxygen. It led to his discovery that yeast can live either aerobically or anaerobically, and that the yeast differs in function under these two conditions (see page 41).

There are a large number of organisms now known that will grow under anaerobic conditions. Some of these are pathogenic, while other seem to be saprophytic. So far as is known, oxygen is not toxic to these organisms in itself, but its presence brings about certain oxidation processes which allow the accumulation of hydrogen peroxide, which is toxic. Organisms that can grow either in the presence or absence of free oxygen are called facultative organisms. Those which will grow only in the absence of oxygen are called obligate anaerobes. Their culture requires special procedures that eliminate oxygen gas, and consequently they are not studied as often as are aerobic and facultative organisms.

Influence of oxygen on the development of yeast and on the alcoholic fermentation (abstract)

1861 · Louis Pasteur

Pasteur, L. 1861. Influence de l'oxygène sur le développement de la levure et la fermentation alcoolique. *Bulletin de la Société chimique de Paris*, June 28, 1861, pages 79–80 (Résumé).

M. Pasteur revealed the results of his researches on the fermentation of sugar and the development of the yeast cell, depending on whether that fermentation had occurred in the absence or in the presence of free oxygen gas. Moreover, his results differ from those of Gay-Lussac on grape juice in the absence or presence of oxygen.

Yeast forms small buds and develops repeatedly in a liquid medium containing sugar and protein in the complete absence of oxygen or air. In this case only a small amount of yeast is formed, while a large amount of sugar disappears, amounting to 60–80 parts of sugar disappearing for one part of yeast formed. The fermentation is very slow under these conditions.

If the experiment is carried out in the presence of air and with a large surface of the liquid exposed, the fermentation is rapid. For the same quantity of sugar disappearing, a much larger amount of yeast is formed. The air gives up some of its oxygen which is absorbed by the yeast. The yeast reproduces vigorously, but its fermentative character tends to disappear under these conditions. Indeed, it is found that for one part of yeast formed, only 4–10 parts of sugar are removed. Nevertheless this yeast retains its fermentative characteristics and these are again strongly revealed if it is allowed to act on sugar in the absence of free oxygen gas.

It seems reasonable to assume therefore that when the yeast is a ferment, acting in the absence of air, it takes oxygen from the sugar, and this is the origin of its fermentative characteristics.

M. Pasteur described the vigorous

activity that occurs at the beginning of the fermentation under the influence of the oxygen which is dissolved in the liquid when the process begins. In addition, the author has observed that beer yeast, when seeded into an albuminous liquid, such as yeast water,* continues to multiply even when there is no trace of sugar in the liquid, but only if oxygen from the air is present in large amounts. The same experiment can be repeated with an albuminous liquid mixed with a solution of non-fermentable sugar such as crystalline lactose. The results are quite similar.

The yeast formed in this way in the absence of sugar has not changed its nature. It is able to ferment sugar if it is allowed to act on it in the absence of air. It should be remarked always that the development of the yeast is quite poor in the absence of a fermentable material.

In summary, beer yeast behaves exactly like an ordinary plant,† and the analogy may be completed if ordinary plants would have an affinity for oxygen which would permit them to respire with the aid of this element by removing it from compounds that are not too stable, in which case, following M. Pasteur, they would be considered to be ferments for these materials.

M. Pasteur stated that he hoped to achieve this result, by finding conditions in which certain lower plants could live in the absence of air but in the presence of sugar, in this way bringing about the fermentation of this substance in the same way as beer yeast.

* [A water extract of yeast cells.]

† [This presumably means that beer yeast is able to respire under aerobic conditions in the same way that green plants do.]

Comment

This is apparently the first description of a phenomenon known in biochemistry today as the Pasteur effect. It describes the different behavior of yeast when grown aerobically or anaerobically. When yeast grows anaerobically, a large amount of sugar is converted into alcohol, with some of the sugar being converted into yeast protoplasm. The efficiency of the process of anaerobic growth is poor. Under aerobic conditions, although less sugar is used, a greater proportion of the sugar is converted into yeast protoplasm, with little or no alcohol produced. A point, not stressed here by Pasteur, but described in later work, is that under aerobic conditions, the utilization of sugar actually seems to be inhibited. Therefore, although a much greater efficiency of conversion of sugar into yeast protoplasm exists aerobically, so that greater amounts of yeast cells are formed, the actual utilization of sugar is depressed. The reasons for this are quite complex, and even today are not completely explained. This latter phenomenon is the Pasteur effect and has been found in other organisms and in animal tissues, and currently has been studied considerably in relation to cancer tissue metabolism. The implications of a discovery can never be realized fully at the time the discovery is made.

Another observation made by Pasteur in this paper is that yeast seems to be able to grow anaerobically when sugar is present, but not when some other carbon and energy source replaces sugar. In the presence of other carbon sources, the yeast can grow, but only aerobically. The explanation for this observation has only been made in the last 20 years. We know that when sugar is present, yeast cells can obtain energy from this source anaerobically by oxidizing part of the sugar into carbon dioxide, while another part of the sugar is reduced to alcohol. Thus some sugar molecules are electron acceptors, while other molecules are electron donors. During this process, called glycolysis, energy is released which is stored as high energy phosphate bonds. No free oxygen is necessary for this process. However, when other energy sources are used which are not sugars, such as amino acids or organic acids, these compounds cannot be handled in the above manner, and must be oxidized using free oxygen as the electron acceptor. Therefore yeast can only grow when these materials are the energy sources when there is oxygen present.

The discovery that yeast could grow both aerobically and anaerobically is quite important for biochemistry and biology. Pasteur could not see his discovery in the light of modern work, and therefore could only view it in terms of the fermentative process he was seeking to explain. His hope to bring about fermentation of sugar with green plants remains unfulfilled.

On the organized bodies which exist in the atmosphere; examination of the doctrine of spontaneous generation

1861 · Louis Pasteur

Pasteur, L. 1861. Mémoire sur les corpuscles organisés qui existent dans l'atmosphère. Examen de la doctrine des générations spontanées. *Annales des sciences naturelles*, 4th series, Vol. 16, pages 5–98.

Chapter I. Historical. . . .

CHEMISTS HAVE DISCOVERED DURING THE last twenty years a variety of really extraordinary phenomena which have been given the generic name of *fermentations*. All of these require the cooperation of two substances: one which is fermentable, such as sugar, and the other nitrogenous, which is always an albuminous substance. But here is the universally accepted theory for this phenomenon: the albuminous material undergoes, when it comes in contact with air, an alteration, a particular oxidation of an unknown nature, which gives it the characteristics of a *ferment*. That is, it acquires the property of being able to cause fermentation upon contact with fermentable substances.

The oldest and the most remarkable ferment which has been known to be an organized being is the yeast of beer. But in all of the fermentations discovered since the beer yeast was shown to be organized, it has not been possible to demonstrate the existence of organized beings, even after careful study. Therefore, physiologists have gradually abandoned regretfully the hypothesis of M. Cagniard de Latour concerning a probable relation between the organized nature of this ferment and its ability to cause the fermentation. Instead, to beer yeast has been applied the following general theory: "It is not the fact that it is organized that makes the beer yeast active, rather it is because it has been in contact with air. It is the dead material of the yeast, that which has lived and is in the process of change, which acts on the sugar."

My studies have lead me to entirely different conclusions. I have found that all true fermentations–viscous, lactic, butyric, or those of tartaric acid, malic acid, or urine–only occur with the presence and multiplication of an organized being. Therefore, the organized nature of the beer yeast is not a disadvantage for the theory of fermentation. Rather, this shows that it is no different than other ferments and fits the common rule. In my opinion, the albuminous materials were never the ferments, but the nutrients of the ferment. The true ferments were organized beings.

This granted, it was known that the ferments originate through the contact of albuminous materials with oxygen gas. If this is so, there are two possibilities to explain this. Since the ferments are organized, it is possible that oxygen, acting as itself, is able to induce the production of the ferments through its contact with the nitrogenous materials, and therefore the ferments have arisen spontaneously. But, if the ferments are not spontaneously generated beings, it is not the oxygen gas itself which is necessary for their formation, but the stimulation by oxygen of a germ which is either carried with it by the air, or which already exists preformed in the nitrogenous or fermentable materials. At this point in my studies on fermentations, I wanted to arrive at an opinion on the question of spontaneous generation. I would perhaps be able to uncover a powerful argument in favor of my ideas on the fermentation themselves.

The researches which I will report here are only a digression made necessary by my studies on fermentations. . . .

Chapter II. Examination under the microscope of the solid particles which are disseminated in the atmosphere.

My first problem was to develop a method which would permit me to collect in all seasons the solid particles that float in the air and examine them under the microscope. It was at first necessary to eliminate if possible the objections which the proponents of spontaneous generation have raised to the age-old hypothesis of the aerial dissemination of germs.

When the organic materials of infusions have been heated, they become populated with infusoria or molds. These organized bodies are in general neither so numerous nor so diverse as those that develop in infusions that have not been previously boiled, but they form nevertheless. But the germs of these infusoria and molds can only come from the air, if the liquid is boiled, because the boiling destroys all those that were present in the container or which had been brought there by the liquid. The first question to resolve is therefore: are there germs in the air? Are they there in sufficient numbers to explain the appearance of organized bodies in infusions which have been previously heated? Is it possible to obtain an approximate idea of the number of germs in a given volume of ordinary air? . . .

The procedure which I followed for collecting the suspended dust in the air and examining it under the microscope is very simple. A volume of the air to be examined is filtered through guncotton which is soluble in a mixture of alcohol and ether. The fibers of the guncotton stop the solid particles. The cotton is then treated with the solvent until it is completely dissolved. All of the particles fall to the bottom of the liquid. After they have been washed several times, they are placed on the microscope stage where they are easily examined. . . .

These very simple manipulations provide a means of demonstrating that there exists in ordinary air continually a variable number of bodies. Their sizes range from extremely small up to 0.01 or more of a millimeter. Some are perfect spheres, while others are oval. Their shapes are more or less distinctly outlined. Many are completely translucent, but others are opaque with granules inside. Those which are translucent with distinct shapes resemble the spores of common molds, and could not be told from these by the most skillful microscopist. Among the other forms present, there are those which resemble spherical infusoria and may be their cysts or the globules which are generally regarded as the eggs of these small organisms. But I do not believe it is possible to state with certainty that a particular object is a spore, or more especially the spore of a particular species, or that another object is an egg, and the egg of a certain microzoan. I will limit myself to the statement that these bodies are obviously organized, resembling in all points the germs of the lowest organisms, and so diverse in size and structure that they obviously belong to a large number of species.

By using a solution of iodine, it is possible to show unequivocally that amidst the bodies there are always starch granules. But it is easy to remove all globules of this sort by diluting the dust in ordinary sulfuric acid, which dissolves immediately all of the starch. Without a doubt the sulfuric acid alters and perhaps dissolves other globules, but there still remain a large number. Sometimes it is possible to distinguish more after treatment with sulfuric acid, because the acid dissolves the calcium carbonate and dilutes the other dust particles, so that many of the organized particles are freed from the amorphous debris which had prevented them from being seen well. It is well to make observations immediately after the small bubbles of carbonic acid have been dissipated, and before needles of calcium sulfate precipitate.

A wad of guncotton 1 centimeter long by ½ centimeter in diameter was exposed to a current of air flowing at one liter per minute for 24 hours. After this time, examination revealed 20 to 30 organized bodies deposited per quarter of an hour (15 minutes). There are ordinarily several bodies in the microscope field. It should be noted that the drop of dust mixture placed on the microscope slide is only a fraction of the total mixture obtained. . . .

Chapter III. Experiments with heated air.

We may conclude that there are always in suspension in the air, organized bodies which, from their shape, size and structure, cannot be distinguished from the germs of lower organisms, and without exaggeration, the number of these is quite large. Are there amongst these bodies really those which are capable of germinating? This is a very interesting question which I believe I have been able to solve with certainty. But before presenting the results of experiments which support in particular this argument, it is necessary to determine first if the results of Dr. Schwann on the inactivity of air which has been heated red hot are correct. MM. Pouchet,

Mantegazza, Joly and Musset contest this. Let us see which side is correct, especially since this will be fundamental to our later researches.

In a flask of 250–300 cc., I introduced 100–150 cc. of solution of the following composition: water, 100; sugar, 10; albuminous material and minerals from beer yeast, 0.2–0.7.

The neck of the flask is drawn out and connects with a platinum tube which is heated red hot, as shown in Figure 10. The liquid is boiled for two or three minutes, then it is allowed to cool completely. The flask is filled with ordinary air at atmospheric pressure, but all the air which has entered has been heated red hot first. Then the neck of the flask is sealed under a flame.

The flask so prepared is placed in a constant temperature chamber at 30°. It can be kept this way indefinitely without the liquid within showing the slightest alteration. Its clarity, odor, and slightly acidic character show not the slightest change. Its color darkens slightly after a time, which is undoubtedly due to a direct oxidation of the albuminous material or the sugar.

I may say with the utmost sincerity that I have never had a single experiment which has given me the slightest doubt. Sugared yeast water heated to boiling for two or three minutes, then placed in contact with heated air, does not change in the slightest, even after 18 months at 25° to 30°. At the same time, if the flask is filled with ordinary air, it undergoes an extensive change within a day or two, and is full of bacteria, vibrios, or is covered with mucors.

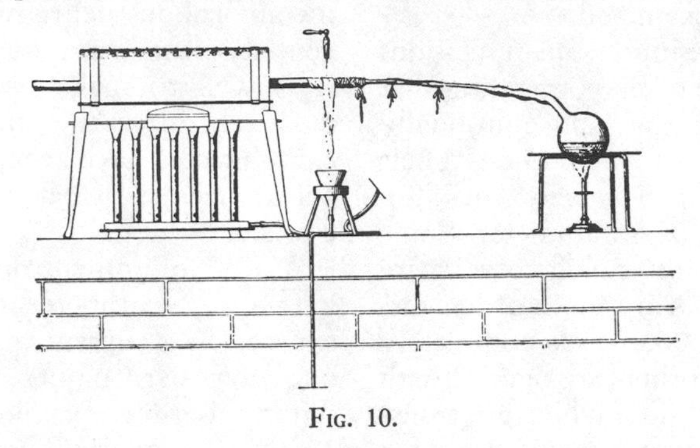

FIG. 10.

The experiment of Dr. Schwann is therefore of irreproachable exactitude. . . .

Chapter VI. Another very simple method to demonstrate that all of the organized bodies in infusions previously heated originate from bodies which exist in suspension in the atmospheric air.

I believe I have rigorously established in the preceding chapters that all organized bodies in infusions which have been previously heated originate only from the solid particles carried by the air and which are constantly being deposited on all objects. In order to remove the slightest doubt in the reader, may I present the results of the following experiments.

In a glass flask I placed one of the following liquids which are extremely alterable through contact with ordinary air: yeast water, sugared yeast water, urine, sugar beet juice, pepper water. Then I drew out the neck of the flask under a flame, so that a number of curves were produced in it, as can be seen in Figure 25. I then boiled

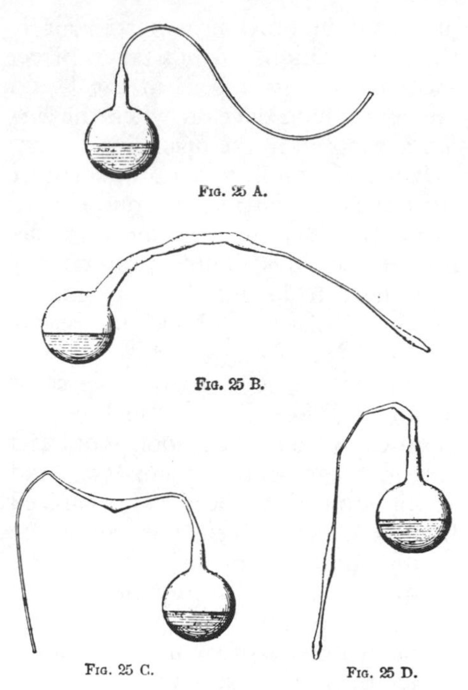

Fig. 25 A.

Fig. 25 B.

Fig. 25 C.

Fig. 25 D.

the liquid for several minutes until steam issued freely through the extremity of the neck. This end remained open without any other precautions. The flasks were then allowed to cool. Any one who is familiar with the delicacy of experiments concerning the so-called "spontaneous" generation will be astounded to observe that the liquid treated in this casual manner remains indefinitely without alteration. The flasks can be handled in any manner, can be transported from one place to another, can be allowed to undergo all the variations in temperature of the different seasons, the liquid does not undergo the slightest alteration. It retains its odor and flavor, and only, in certain cases, undergoes a direct oxidation, purely chemical in nature. *In no case is there the development of organized bodies in the liquid.*

It might seem that atmospheric air, entering with force during the first moments, might come in contact with the liquid in its original crude state. This is true, but it meets a liquid which is still close to the boiling point. The further entrance of air occurs much slower, and when the liquid has cooled to the point where it will not kill the germs, the entrance of air has slowed down enough so that the dust it carries which is able to act on the infusion and cause the development of organized bodies is deposited on the moist walls of the curved tube. At least, I can see no other explanation for these curious results. For, after one or more months in the incubator, if the neck of the flask is removed by a stroke of a file, without otherwise touching the flask, molds and infusoria begin to appear after 24, 36, or 48 hours, just as usual, or as if dust from the air had been inoculated into the flask.

The same experiment can be repeated with milk, except here one must take the precaution of boiling the liquid under pressure at a temperature above 100° and allowing the flask to cool with the reentry of heated air. The flask can be allowed to stand in the open, just as before. The milk undergoes no alteration. I have allowed milk prepared in this manner to incubate for many months at 25 to 30°, without alteration. One notices only a slight thickening of the cream and a direct chemical oxidation.

I do not know any more convincing experiments than these, which can be easily repeated and varied in a thousand ways. . . .

At this moment I have in my laboratory many highly alterable liquids which have remained unchanged for 18 months in open vessels with curved or inclined necks. A number of these were deposited in the office of the Academy of Sciences during the meeting of 6 February 1860, when I had the honor of communicating to them these new results.

The great interest of this method is that it proves without doubt that the origin of life, in infusions which have been boiled, arises uniquely from the solid particles which are suspended in the air. Gas, various fluids, electricity, magnetism, ozone, things known or things unknown—there is absolutely nothing in ordinary atmospheric air which brings about the phenomenon of putrefaction or fermentation in the liquids which we have studied except these solid particles. . . .

The experiments of the preceding chapters can be summarized in the following double proposition:

1. There exist continually in the air organized bodies which cannot be distinguished from true germs of the organisms of infusions.

2. In the case that these bodies, and the amorphous debris associated with them, are inseminated into a liquid which has been subjected to boiling and which would have remained unaltered in previously heated air if the insemination had not taken place, the same beings appear in the liquid as those which develop when the liquid is exposed to the open air.

If we grant this, will a proponent of spontaneous generation continue to maintain his principles, even in the presence of this double proposition? This he can do, but he is forced to reason as follows, and I let the reader be the judge of it:

"There are solid particles present in the air," he will say, "such as calcium carbonate, silica, soot, wool and cotton fibers, starch grains, etc., and at the same time there are organized bodies with a perfect resemblance to the spores of molds and the eggs of infusoria. Well, I prefer to place the origin of the molds and infusoria in the first group of amorphous bodies, rather than in the second."

In my opinion, the inconsistency of such reasoning is self-evident. The entire body of my research has placed the proponents of spontaneous generation in this predicament.

Comment

We should marvel at the logic, the clarity, and above all the simplicity of these experiments. In a few incisive blows Pasteur has ended the controversy concerning spontaneous generation. His experiments were easily reproducible, and this made it simple for others to accept his conclusions. As we shall see from Tyndall's paper, the controversy did not end completely, since Pasteur was working with nutrient fluids which did not allow for the development of heat-resistant bacterial spores. But from this time on, proponents of spontaneous generation found themselves fighting a losing battle.

At the same time that he solved the problem of spontaneous generation, Pasteur founded the science of microbiology. He showed how to sterilize a liquid, and how to keep it sterile. Only when it is possible to culture one organism in the complete absence of other organisms is it possible to determine the unique properties of the one organism under study. We shall see the importance of this requirement later, when reading of the work on infectious disease.

Studies on the biology of the bacilli

1877 · Ferdinand Cohn

Cohn, Ferdinand. 1877. Untersuchungen über Bacterien. IV. Beiträge zur Biologie der Bacillen. *Beiträge zur Biologie der Pflanzen,* Vol. 2, pages 249–276.

1. BACTERIA AND SPONTANEOUS GENERATION

Among the problems remaining to be solved by modern science, there is none of more significance than the question as to whether living beings develop exclusively from germs which have been produced by other beings of the same type, or whether living beings can develop from non-living material (by abiogenesis, archigenesis, spontaneous generation, *generatio spontanea, Urzeugung*). Unjustly most workers in Germany have long considered that this question has been solved in favor of the former alternative. Although since the time of Redi a large number of experiments and observations have been reported showing that a large number of types of animals and plants cannot develop unless their germs (eggs, seeds, or spores) are present, it does not necessarily follow that the development without germs is impossible for all living beings, even those that are the simplest and lowest. Those workers who deny an absolute boundary between inorganic and organic compounds, between living and non-living bodies, and who view life as a function of the identical forces which are operating in the non-living world, do not have the slightest grounds for doubting the possibility that under certain conditions protoplasm could be formed from the atoms of carbon, oxygen, hydrogen, and nitrogen through a certain combination of chemical and physical forces, in the same way that urea can be formed from ammonium carbonate, and that this artificially or spontaneously formed protoplasm could enter into the cycle of nature and become a living Monera, capable of feeding and reproducing. Therefore Pouchet and especially Ch. Bastian have recently performed us a worthy service when they performed experiments, without hypothesizing the impossibility of spontaneous generation, to discover the conditions under which it might be possible for living beings to develop from organic but non-living material. . . .

Although the studies of the adherents of spontaneous generation have followed scientific procedures, there is a criticism of these experiments which the opponents of this doctrine believe has refuted spontaneous generation. The latter have assumed that it is an incontestable fact that living organisms never develop in substances in which no germs of living beings capable of reproducing are present or which enter later. But because of the smallness of the germs, their presence or absence is not directly determined, so generally in these experiments the substances to be used are treated to a temperature which is assumed to be sufficient to kill all germs present. Up until the present, the proper temperature for this has been considered to be that of boiling water. . . . Therefore, the assumption mentioned above reduces to the assumption that substances which have been placed for a time at boiling water temperature develop no living beings.

2. RESISTANCE OF BACTERIA TO BOILING

This assumption is however not justified. Schwann noted that it was not always possible to preserve meat from putrefaction by boiling. Pasteur found that milk was protected from the development of lactic acid bacteria only after it was heated to 110°. Schroeder required 130° to prevent the development of bacteria in meat, egg yolk, and milk; others have needed even higher temperatures. In this regard, of especial interest are the experiments which have been performed in large numbers to effect preservation of meat, vegetables, etc. in hermetically sealed tin cans. It is well known that the production of preserved foods after the method of Appert has become a significant industry in recent times, and it continues to find further applications and new food materials which can be prepared for unlimited storage bacterial-free for export purposes. I have already endeavoured a number of times to learn about the production of such bacterial-free foods. Herr Senator Dr. W. Bremer has been so kind as to inform me that in Lübeck many factories preserve vegetables in tin cans by boiling at 100° (over 80,000 cans in 1873) without once having spoilage, so long as the can was perfectly sealed. A single exception are peas, which were earlier heated to 100°, but in warm years almost half of all the cans spoiled through fermentation in spite of airtight seals. However, if 28% sodium chloride is dissolved and the peas heated to 108°, no spoilage occurs in perfectly sealed cans. The same result was obtained if the peas were heated without salt to 117° by a process invented in France. By this method in Lübeck alone around 50,000 cans of peas are prepared yearly and sent mostly to tropical lands, without a single case of spoilage throughout a number of years. . . .

If indeed Ch. Bastian has drawn the conclusion from this and other experiences that in those substances which still yield living organisms after boiling, this is due to spontaneous generation, and not to the fact that their germs are resistant to 100°, so will his opponents draw the latter conclusion. However, Bastian himself has recently published results which have been attested by as many as 20 reliable witnesses, that through boiling, organic substances are not always disinfected, i.e., rendered incapable of bacterial development and putrefaction. In addition we have other results here that are in agreement with this, namely that by heating over 100° it is eventually possible to sterilize every substance, if the temperature and the length of time for its action are properly chosen. That is, a higher temperature will sterilize in a shorter time, while a lower temperature will sterilize in a longer time. . . .

Has the whole problem therefore been reduced to the fact that if organic substances cannot be sterilized by 100°, they can be by more or less time at temperatures over 100°, if recontamination is prevented? Why is it that 100° is not sufficient to kill bacteria, when much lower temperatures are sufficient for killing other living organisms? . . .

As early as 1872 I attempted to determine directly and experimentally the temperature which would cause bacteria to lose their powers of reproduction. It was shown that when solid substances were used, such as boiled lupines or peas, the results were uncertain, and this was due to the fact that the solid substances were poor conductors of heat and were cooler than the temperature indicated by the thermometer in the water, so that bacteria hidden in cracks and intercellular spaces were protected from the heat. Therefore a small amount of living bacteria (a bacterial drop) was added to a clear fluid (bacterial nutrient) and by comparative studies, which Dr. Horvath undertook at my request, it was found that 60 minutes at 60–62° prevented the bacteria from growing. Dr. Schröter determined in his studies on disinfection the minimum temperature which would render moving bacteria immobile, and probably incapable of further development, therefore without a doubt dead. This temperature was 58°.

But even this temperature exceeds that which is sufficient for killing most other organisms. Although the upper limit for all organisms has not been precisely determined, from the results of most workers this seems to be not higher than 35–50°C., as living protoplasm is coagulated by 43°, while other protein compounds become cloudy at 60° and form a coagulant flock at 70–75°. . . .

3. EXPERIMENTS WITH BOILED HAY INFUSION

Is the contradiction between these studies and those mentioned earlier only due to the fact that in the earlier ones solid nutrients were used, while in the latter exclusively liquid nutrients were used? . . .

However, Bastian had found in his experiments various fluids which were able to develop living organisms after 5–10 minutes boiling, although there was a complete absence of solid substances. In this regard, the experiments of Dr. W. Roberts, presented to the Royal Society of London, are of special interest. These experiments were performed to determine the necessary temperature to sterilize fluids and at the same time to observe the possible development of bacteria by spontaneous generation. Little glass flasks of 30–50 cc. were filled ⅔ full with a fluid. The dried neck of the flask was stoppered with a cotton plug and its long, drawn-out tip was sealed in the flame. Then the flask was allowed to stand in boiling water for longer or shorter times. After cooling, the sealed tip was opened again. This method served to prevent the boiling away of the liquid, as well as the later possibility of chance contamination from the air. Throughout four years, more than a hundred experiments were performed. Decoctions from various organic animal and plant tissues produced by short heating with water, as well as solutions of organic salts and healthy or diabetic urine required 3–4 minutes for sterilization. Infusions produced by slow digestion from meat, fish, grapes, carrots and fruits required 5–10 minutes. Water to which pieces of green vegetables, fish, meat, egg white, or cheese were added, as well as milk, blood and albuminous urine, required 20–40 minutes. The most difficult to sterilize was alkaline hay infusion. In this fluid the development of bacteria was only prevented by one to two hours in a boiling water bath, while in an oil or salt water bath sterilization took only 5–15 minutes.

These results, obtained in rigorous scientific fashion, are worthy of careful confirmation, which I have done many times with hay infusion, assisted by Mr. Robert Hare from Ottawa (Canada), who is working in our plant physiology institute. At the same time I have attempted to observe the organisms which developed in boiled hay infusion. These organisms have until now been poorly characterized as bacteria, and I have observed them under the microscope in order to determine if the nature of their unbelievable resistance to boiling water could reside in some characteristic specific property.

The hay infusion was prepared exactly by the method of Roberts: Hay was placed in a glass cylinder with a little water and digested four hours at 36°, and this extract was diluted with distilled water to a specific gravity of 1006 and filtered twice. It was then completely clear, a pretty gold yellow color, about like Munich beer, and reacted acid. . . .

W. Roberts had found that neutral hay infusion was especially hard to sterilize. 200 cc. of hay infusion was treated with 1.5 cc. of potassium hydroxide liquid, and it then reacted neutral with litmus. . . . This fluid was not clear, but had a turbid opalescence. . . .

If one allows unboiled hay infusion, whether neutral or acid, to stand, after 12–20 hours it becomes turbid and translucent. . . . The most common organism which develops is the thin-rod *Bacterium Termo*. Others which develop are *Micrococcus*, *Ascococcus*, *Sarcina*, *Torula*, *Bacillus*, *Leptothrix*, yeast, and *Penicillium*.

With boiled hay infusion the results are quite different. In this solution changes can likewise take place which are due to the reproduction of microscopic organisms, even after a long time in the boiling water bath, and whether the infusion is neutral or acid. . . .

After 5–15 minutes boiling, organisms develop without exception, while by longer times, the results are variable. Sometimes 20 minutes was necessary to sterilize, other times, 30 minutes, and several times 1.5 to 2 hours. . . .

Before I report on the organisms which developed in boiled infusions, I want to state that there was no possibility of a contamination of the liquids after the boiling period. Tyndall has unjustly held suspicious the cotton plugs used by Roberts to enclose his flasks. Cotton has been confirmed by all workers, even by Tyndall, as a perfect filter against the bacteria floating in the air. . . . I have obtained the same results when I have bent the necks of the flasks in the manner of Pasteur. . . .

4. STUDY ON THE ORGANISMS DEVELOPING IN BOILED HAY INFUSIONS

From the above results I concluded that in disinfection we must deal with the fact that in filtered, perfectly clear fluids, long heating at 100° is not always able to prevent the development of organisms, and that it may take more than an hour of boiling to be certain of sterilizing. These results are in complete confirmation with those of Bastian and Roberts. I now turned to the second part of my problem, which was to determine by microscopic examination what type of organism it was that possessed such an unexpected resistance to the effects of boiling water. Even the first glance indicated that a completely different sequence of development occurred in the boiled hay infusions from that in the unboiled ones. . . . In the boiled hay infusion I found neither yeast, nor

Penicillium, nor *Ascococcus,* nor *Sarcina.* . . .

The first indication of growth in the boiled infusions was the development after about two days of a delicate, iridescent film on the surface of the liquid. Soon thereafter the top layer began to become turbid and assume a slimy-flocculent or scaly character, without the turbidity gradually going all the way to the bottom, as it does in the unboiled infusions. . . .

If one examines the hay infusion under the microscope after 24–48 hours, when the first signs of this iridescent film are noticeable, one finds that already every drop taken from the surface swarms with countless fine, straight, actively motile rods. These rods are about 0.6 microns thick, and vary in length, depending upon their stage of development, from 3, 5, 7, etc. microns. But all of these belong to a single species, *Bacillus subtilis.* The shortest of these rods could be confused with those of the putrefying bacteria *(Bacterium Termo).* However, these latter when at such a size are in the process of dividing and are therefore pinched in the center, while the shortest rods of the hay bacillus show no sign of division. The majority of the *Bacillus* rods are two to four times as long, although occasionally ten or more times as long as the longest putrefying bacterium. These rods are often attached together at angles, and the short members separate easily, and as soon as they become free they swim around rapidly. Longer filaments show a somewhat retarded motion. . . .

5. SPORE FORMATION IN BACILLUS SUBTILIS

Later the *Bacillus* filaments begin to prepare for spore formation. In their homogeneous contents strongly refracting bodies appear. From each of these bodies develops an oblong or shortly cylindrical, strongly refracting, dark-rimmed spore. In the filaments one finds the spores arranged in simple rows (see Figure 4 *). . . .

The process of spore formation can only be observed by careful observations with very strong immersion systems. Although the *Bacillus* filaments seem to be without cross walls even under the strongest magnification, this is in reality not the case. The single members which make up the filament are four times as long as wide. In each member a spore develops, which does not fill the cavity completely, but is separated from the empty cell membrane on each side. The spores are 1.5–2.2 microns long and 0.8 microns wide. . . . In their development they seem to resemble those of Nostocaceae *(Cylindrospermum, Nostoc, Spermosira,* etc.) the most. Depending on whether the *Bacillus* filaments are shorter or longer, out of two or more members, we find the spores in a filament arranged in short chains of two or more. By decomposition of the *Bacillus* filaments, single members become isolated which contain only single spores. When these have completely separated from their mother cell, they show a delicate, jelly-like enclosure (spore membrane) and a strongly refracting interior. . . . With the maturation, release and settling out of the spores, the development of the *Bacillus* is ended, and no further changes take place in the hay infusion. . . .

The spores are viable however. Indeed, it appears that they do not germinate in the same liquid in which

* [The figures for the paper of Cohn are on the same plate with those of the paper of Koch, "The etiology of anthrax, based on the life history of *Bacillus anthracis*" (1876). See page 93.]

they were formed. At any rate, I have never seen in an experimental vessel a development of turbidity which could be attributed to a second generation of *Bacillus* developing from a spore mass. Instead, the hay infusion remains completely unchanged, once it has become clarified, despite the large number of spores present. But when a small portion of the spore mass was placed in a test tube containing a hay infusion that had been completely sterilized by long heating, and which would undoubtedly not have changed after a long time in the warm box, the spores clearly germinated. Even on the next day after inoculation I could often observe a white, slimy film of *Bacillus* on the surface of the liquid, which showed the presence of the filaments with the later production of spore chains again. When I placed a small number of spores which had been present for months in a boiled hay infusion, I was fortunate being able to observe the germination directly. The spores swelled somewhat and extruded at one end a short germ tube. . . . The strongly refracting body of the spore soon disappeared. The germ tube developed immediately into a short *Bacillus* rod, which became motile, separated into parts by cross walls, and lengthened into a filament. Soon in the drop countless short and long bacilli were swarming. . . . When the *Bacillus* spores are inoculated together with germs of *Bacterium Termo*, the spores do not germinate, since the other form grows faster and suppresses the bacilli.

6. CONCLUSIONS

The observations presented here seem to me to contain the solution to the puzzle regarding the development of organisms in boiled organic substances. The following facts were demonstrated: (1) In boiled liquids *Bacterium Termo* does not develop, and so far as is known, no other organisms do except the *Bacillus*. (2) If the bacilli develop in boiled infusions, the cause for this is to be sought in the life history of the organism. We cannot doubt that *Bacillus* spores are present in hay, whether they are already present on the living grass, or whether they first arrive during the mowing and drying of the hay or its later storage. In dry hay these spores are shrunken and very difficult to wet with water. During the digestion they are washed away and get into the hay infusion. But, as long as the spores have not imbibed water and become swollen, they can be heated to 100° without losing their viability, in the same way that has been shown for mold spores and even phanerogam seeds. (Footnote: Pasteur has shown that *Penicillium* spores are not killed when they are heated dry at 121°. . . .) The resistance of the unswollen *Bacillus* spores to at least 15 minutes and even 1–2 hours boiling water is probably due to the fat-like contents of the spore, or perhaps to the layer of air adhering tightly to the spore wall. . . . The longer the boiling proceeds, the fewer the spores which remain alive. . . . Finally, by heating over 100°, all of the spores are killed and the liquid is completely sterilized. . . . (3) In all cases in which organisms have developed in boiled organic substances, I have found only spore-producing bacilli. . . . (7) Amongst the bacilli that have been described, that one which is present in the blood of animals and humans with anthrax has an especially important place, since it is without a doubt of significance for the pathology of this disease. In 1875 I noted that since the bacilli as a rule reproduce by resting spores, it would also be expected that the rods of anthrax would also form spores and that these would be the germs of the infection in apparently rod-free blood. To my great pleasure, I received a letter from Dr. Koch in Wollstein on 22 April. He has been occupied with studies on the anthrax contagium for a long time and has finally been able to discover the complete life history of *Bacillus anthracis.* He was willing to demonstrate this to me at my plant physiological institute and obtain my opinion of his discoveries. Dr. Koch came to Breslau from 30 April to 3 May and with anthrax material he had brought along performed in our institute inoculations into living frogs, mice and rabbits. Through this series of experiments I was given the opportunity to convince myself of the complete correctness of his discoveries on the development of the anthrax bacillus. . . . Herr Dr. Koch reports the results of his experiments at the end of this paper and indicates the highly important conclusions which these studies yield for the nature and spread of the anthrax contagium. I will only remark here that the life history of the anthrax bacillus agrees completely with that of the bacillus of hay infusions. Indeed, the anthrax bacillus does not have a motile stage, but otherwise the similarity with the hay bacillus is so perfect, that the drawings of Koch can serve without change for the clarification of my observations, and some of my drawings could serve as illustrations of the anthrax rods. For those who may wish to challenge this work, let me state clearly that there is no question of any uncertainty in Koch's studies through mistake or contamination. We seem to have here one and the same bacterial form, or, more likely, two species which cannot be differentiated under the microscope. One of these occurs in the human organism as a constant accompaniment of a specific pathological condition, without a doubt the bearer of this contagium, and the other appears outside of the human organism, in indifferent media and with unknown or with completely different ferment actions. Whether we will see in the future a genetic relationship between the hay bacillus and the anthrax bacillus, between the spirochete of marsh water and that of recurrent fever, between the micrococcus colonies of spoiled drinking water or fermenting food and the typhus or diphtheria, or whether we are dealing here with outwardly similar but specifically different species or races, will have to be left to the further development of science to decide. . . .

Comment

It will be seen that by Cohn's time, the technology of sterilization had proceeded effectively even in the absence of a scientific basis. Practical use had been made in the commercial preservation of food of the observations of investigators on spontaneous generation. This is another example of how fundamental research may have practical applications in unsuspected directions.

Cohn clearly saw that not all liquids or solids are sterilized equally easily. Instead of using this information to accept or reject the theory of spontaneous generation, he decided to find out what was responsible for this observation. His experiment with hay infusion was really an enrichment culture. By boiling for a while, he killed all organisms that were sensitive to boiling water. Only heat resistant forms were left, and these could develop and be studied under the microscope. In this way he discovered a new species, *Bacillus subtilis*, and a new process, spore formation. Undoubtedly improved microscopes and the use of immersion lenses aided this discovery. It was easy for him to draw the conclusion that these spores were heat resistant, since spores of other forms, such as molds, had already been reported to be so. We know now that bacterial spores are the most heat resistant of all living organisms. Hence their discovery by Cohn was important in clarifying at last the variable results of workers who had attempted to sterilize materials with boiling water (see Tyndall's work below).

We can also see here in Cohn's work the beginnings of the development of bacteriological techniques, such as the use of cotton for closing flasks and tubes. But probably the most important result of his work is that it was the direct forerunner of Koch's brilliant researches on infectious disease. Koch's first paper, referred to by Cohn, is included in this volume on pages 89–95.

Further researches on the deportment and vital persistence of putrefactive and infective organisms from a physical point of view

1877 • John Tyndall

Further Researches on the Deportment and Vital Persistence of Putrefactive and Infective Organisms from a Physical Point of View. By John Tyndall, D.C.L., L.L.D., F.R.S. Read May 17, 1877. In *Philosophical Transactions of the Royal Society of London*, 1877, Vol. 167, pages 149–206.

. . . WHILE CONTINUING THE CONFLICT and experiencing the defeats recorded in the foregoing pages, a remark of Professor Lister's sometimes occurred to me. To apply the antiseptic treatment * with success, the surgeon must, he holds, be interpenetrated with the conviction that the germ-theory of putrefaction is true. He must not permit occasional failures to produce scepticism, but, on the contrary, must probe his failures, in the belief that his manipulation, and not the germ-theory, is at fault. This may look like operating under a prejudice; but Professor Lister's maxim is nevertheless consistent with sound philosophy and good sense; and if I permitted a bias to influence me in this inquiry, it was one fairly founded on antecedent knowledge, which led me to conclude that the long line of failures above referred to would eventually be traced to my ignorance of the conditions whereby perfect freedom from contamination was to be secured.

I laboured to discover these conditions, and to learn something more regarding the nature of the contamination–its origin, persistence, and manner of action. When these researches began, five minutes boiling, as I have

* [Read Lister's work on pages 83–85 and 86–89.]

frequently stated, sufficed to sterilize the most diversified infusions. Here we have frequently extended the time of boiling to ten and fifteen minutes, and in some cases glanced at above, to immensely longer periods, without producing this result. . . .

STERILIZATION BY DISCONTINUOUS HEATING

It is an undisputed fact that active *Bacteria* are killed by a temperature far below that of boiling water. It is also a fact that a certain period, which I have called the period of latency, is necessary to enable the hardy and resistant germ to pass into that organic condition in which it is so sensitive to heat. There can hardly be a doubt that the nearer the germ approaches the moment when it is to emerge as the finished organism, the more susceptible it is to that influence by which that organism is so readily destroyed. We may learn from experience . . . what is the approximate time required for the Bacterial germ to pass into the *Bacterium.* Say that it is twenty-four hours. Supposing the heat of boiling water to be applied to the germ immediately before its final development, when all parts are plastic:

Here, at all events, we have a theoretical finger-post pointing out a course which experiment may profitably pursue. It is not to be expected that the germs with which our infusions are charged all reach their final development at the same moment. Some are drier and harder than others, and some, therefore, will be rendered plastic and sensitive to heat before others. Hence the following procedure. . . .

On the 1st of February eight pipette-bulbs were charged with two hay-infusions, four bulbs being devoted to each. . . . They were subjected to the temperature of boiling water for a minute; at the same time four other bulbs containing the same infusions were boiled continuously for ten minutes and suspended beside their neighbours. Twelve hours subsequent to their first brief heating the eight bulbs were perfectly brilliant, and while in this condition they were again subjected to the boiling temperature for a minute. On the evening of the same day they were subjected to the boiling temperature for half a minute, and on the following morning the process was repeated. Two additional heatings of the same brief character were resorted to. The result of the whole experiment was that two days after their preparation the four bulbs which had been boiled for ten minutes were found turbid and covered with scum, while two months after their preparation the eight bulbs whose periods of boiling added together amounted to only four minutes were perfectly brilliant and free from scum.

The reason of this procedure is plain. By the first brief application of heat the germs, which are at the moment plastic, are killed; and before any of the remaining germs can develop themselves into *Bacteria* they are subjected to another brief period of heating. This again kills such germs as are sufficiently near their final development. At each subsequent period of heating the number of living germs is diminished, until finally, they are completely destroyed. The infusion, if protected from external contamination, remains for ever afterwards unchanged, although when living *Bacteria,* a sprig of hay, or even the dry dust particles of the laboratory are sown in it, the sterilized liquid shows its power both of enabling the fully developed organism to increase and multiply, and of developing the desiccated Bacterial germ into multitudinous Bacterial life.

Comment

Tyndall experienced great difficulty in sterilizing some liquids, and found his results to be highly variable. Many an investigator observing variable results has discarded a whole line of approach, but Tyndall, having faith in the theoretical basis of his researches, continued his work. The reason Tyndall could not sterilize his fluids reproducibly was because he was working with hay infusions, which are usually teeming with bacterial spores. Pasteur, who worked mainly with fruit juices, did not have this problem. Tyndall postulated the existence of the bacterial spore, and then proceeded on the assumption that it exists. In this way he invented the process of fractional sterilization, which is still used occasionally today under the name Tyndallization.

The distinction between germ and bacteria should also be noted. This distinction was made by most early workers. The germ was that object, usually dormant, which could give rise to the living organism. We generally do not make this distinction today except in the case of the spore-forming organisms.

On the lactic fermentation and its bearings on pathology

1878 • Joseph Lister

Lister, Joseph. 1878. On the lactic fermentation and its bearings on pathology. *Transactions of the Pathological Society of London,* Vol. 29, pages 425–467.

A FEW YEARS AGO IT WOULD HAVE seemed very improbable that the souring of milk should have any bearings upon human disease; but all will now be ready to admit that the study of fermentative changes deservedly occupies a prominent place in the minds of pathologists.

In order that any sure steps may be taken to elucidate the real nature of the various important diseases which may be presumed to be of a fermentative nature, such as the specific fevers or pyaemia, the first essential, as it appears to me, is that we should have clear ideas, based upon positive knowledge, with regard to the more simple forms of fermentation, if I may so speak—more simple because they can be conducted and investigated in our laboratories.

It may be said, indeed, that such information has been already afforded us by the researches of Pasteur and others who have followed in his wake, tending to prove that all true fermentations of organic liquids are due to the development of organisms within them; and I confess that for my own part I am disposed to agree with that view. But this opinion is by no means universal in our profession. We meet with statements by men of very high position, both as physiologists and pathologists, to the effect that in various fermentations—such, for example, as putrefaction—the bacteria which are found to be present may, for aught we know, be mere accidental concomitants, not causes, of the fermentative change. And such being the case, it seemed desirable to obtain, if possible, entirely conclusive evidence upon the subject.

About four months ago I made an attempt of this kind with regard to the lactic fermentation; and I propose on the present occasion to bring forward the results arrived at, and at the same time to afford the members of the Pathological Society an opportunity of seeing with their own eyes samples of the preparations which resulted from that inquiry, and on which my conclusions are based.

First, however, I desire to describe my method of experimenting, which, in its present simplified form, has never been published. It is based, in the first instance, on the fact which experience has now amply demonstrated, that if we have a vessel like this liqueur-glass Fig. 1 (A) in a state of purity, covered with a pure glass cap (B), the capped liqueur-glass being further covered with a glass shade (C), and standing as a matter of convenience on a plate of glass (D), any organic liquid contained in the liqueur-glass, provided it be free from living organisms at the outset, will remain without any organic development occurring in it as long as the arrangement of the glasses is left undisturbed. Or, in other words, although an interchange is constantly taking place between the gases of the atmosphere and those in the liqueur-

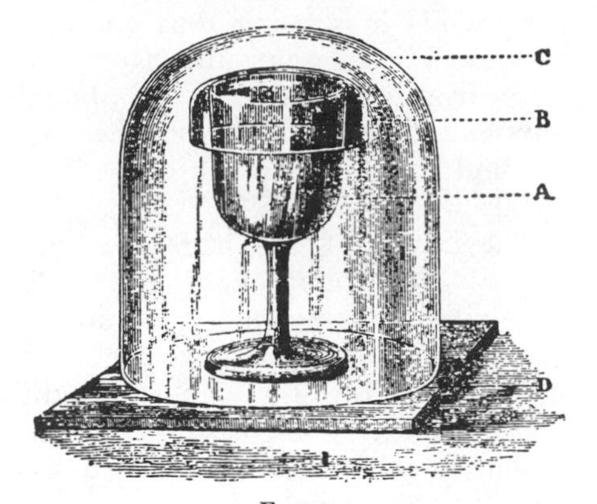

FIG. 1.

glass—for the cap does not fit at all, and the shade is not air-tight—yet the double protection of the glass cap and the glass shade effectually prevents access of the atmospheric dust to the liquid; and if the dust is excluded, organisms do not occur in it.

The glasses are obtained pure by means of heat. I find that exposure to a temperature of 300° Fahr. for two hours is sufficient to deprive all living material of its vitality. But it is not enough that the glasses should be so heated; it is necessary that the air which enters them during cooling should be filtered of its dust. . . .

In the next place, how shall any pure organic liquid be introduced into one of these purified liqueur-glasses without risk of contamination? This we are now able to manage in a comparatively simple manner. The liquid is introduced by means of a flask of [a] . . . form, having a bent spout, large at the commencement and comparatively narrow in its shorter terminal part beyond the bend. . . . And the mouth of the flask being covered with pure cotton-wool, the air that enters the flask during the pouring out of the liquid is filtered of its dust by passing through the cotton. . . .

Now as to the means of preventing the entrace of dust into the liqueur-glass while it is being thus charged. Suppose I were going to charge this glass from this flask. . . . I should remove the cotton cap from the nozzle, and instantly slip the end of the nozzle into the opening which exists in the center of this half of a substantial india-rubber ball, previously steeped in a strong watery solution of carbolic acid. I then remove the shade and take off the glass cap [from the vessel to receive the liquid] and immediately substitute the cap of india-rubber on the nozzle of the flask. . . . Fluid is now poured in . . . and the instant the flask is withdrawn the glass cap is replaced and the shade put on. The hemispherical form of the cap also prevents lateral currents of air from depositing dust on the drop at the nozzle while it is being moved from one glass to another, and a second, a third, or a dozen glasses may be thus charged in succession; and experience shows that this mode of procedure is so secure that the liquid will remain uncontaminated in such a series of vessels until it dries up through atmospheric influence. . . .

. . . suppose we have to deal with a liquid contaminated with organisms, like milk obtained from a dairy, we must purify it by heat. For this purpose I have always found exposure for an hour to a temperature of about 210° Fahr. sufficient. . . .

So much for our method or procedure. I will now go on with the main subject of this communication. I selected the lactic fermentation as one peculiarly favourable for investigation: first, because the effects which it produces in milk are extremely striking and readily recognized—the solidification which takes place being obvious at a glance, and the souring as shown by test paper being also a very conspicuous change; and in the second place because the ferment which occasions these alterations is, in ordinary localities, a very rare ferment; and if it be rare, it is not likely that any defects in our manipulations will lead to its accidental introduction.

It may seem strange that the ferment that leads to the souring of milk should be rare, but such is the fact; in dairies it appears to be universal, but in the world at large it is scarce. If you charge a series of pure liqueur-glasses with boiled milk and take off their caps so as to expose them to the air for about half an hour each, doing this for the various glasses at different times of the day or in different rooms, you will be sure to have organisms develop in all of them, of the nature of filamentous fungi and bacteria; and you will see fermentative changes ensue; but, so far as my experience goes, you will not see the coagulation and souring of the lactic fermentation, nor will you find under the microscope the peculiar organism to which I have given the name of *Bacterium lactis*, . . . which may be seen under one of the microscopes on the table, in milk taken from the cow yesterday.

This organism is a motionless bacterium, that is to say, exhibiting no movement except a slight jogging, occurring most commonly in pairs, but frequently in chains of three, four, or more individuals, each segment being of somewhat rounded form, more or less oval, with the long diameter in the direction of the length of the chain, and often showing, on careful focusing, a line across their central part indicating transverse segmentation. . . . full-sized specimens measuring about 1–20,000th inch. (Footnote: In giving these as the characters of the *Bacterium lactis*, I do not wish to be understood as stating that this species can always be certainly recognized by its morphological features alone.) You always find them, as far as I know, in souring milk from a dairy, and a touch with the point of a needle which has been dipped into such souring milk will induce their development with great rapidity in a glass of boiled milk, together with the characteristic curdling and souring, showing that the boiled milk is thoroughly disposed to the lactic fermentation as soon as the appropriate ferment enters, while that ferment, as already stated, is not likely to occur as the result of exposure to the air of your study or any ordinary situation. . . .

And now, before proceeding further, I desire to correct a mistake into which I fell when investigating this same fermentation some years ago; for, next to the promulgation of new truth, the best thing, I conceive, that a man can do, is the recantation of published error. In the year 1873, I gave . . . an account of the behaviour, as I supposed, of the *Bacterium lactis* in different liquids. I stated that, having obtained souring milk from a dairy, I inoculated a glass of uncontaminated unboiled urine with a small drop, and the result was the development in that liquid of organisms with a very different appearance from the bacteria which I had seen in the souring milk. The latter had the characters of *Bacterium lactis*, as already described. . . . Those in the urine, on the other hand, were broad and extremely long, often coiled up like a spirillum, though motionless. . . . There were, however, what certainly looked like transitional forms between the two. . . .

Here . . . were the facts all wrong. What was the explanation? I had obviously got some accidental contamination of the *Bacterium lactis* with other forms, although in doing the inoculations I had only introduced a very minute portion with the point of a heated needle. So I determined to try if possible to get rid of concomitant bacteria of other kinds, and the way which occurred to me as a possible mode of doing this was to dilute the souring milk with so large a quantity of boiled and therefore pure water as to have on the average, so far as it could be estimated, only one bacterium of any kind to every one of the drops with which a set of glasses of boiled milk should be inoculated. If this could really be done, as the *Bacterium lactis* was certainly in much larger numbers than any other kind, I might hope that some at least of the drops of inoculation might contain it isolated from other species, and that thus I might have the *Bacterium lactis* develop pure and unmixed in the inoculated milk. Accordingly having obtained some souring milk from the dairy, I inoculated a glass of boiled milk from it by dipping the point of a heated needle successively in the two liquids, and when the odour of souring milk was perceptible in the air under the glass shade, I found bacteria present on microscopic examination, and endeavoured to estimate their numbers in proportion to the liquid.

This was done in the following manner. By means of the syringe already described (see Figure 8) one or more hundredths of a minim * could be measured with precise accuracy; and I found that 1/50 minim exactly occupied a circular plate of thin covering glass half an inch in diameter so that when such a drop was placed on a glass slide, and a covering glass of the size mentioned and quite flat was put down upon it, all air was expelled from under the latter, and the rim of fluid that formed round about its margin was so narrow as not to measure a quarter of the diameter of the field of the microscope even when

* [1 minim = 0.062 ml.]

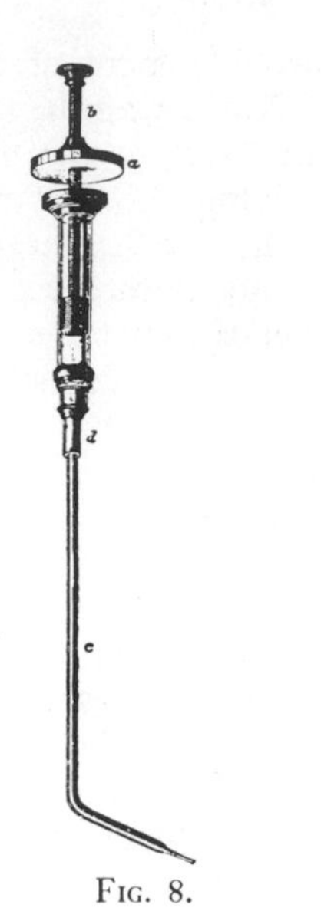

FIG. 8.

the highest magnifying power was used. In other words, 1/50 minim was disposed in a thin uniform layer of the exact size of the covering glass. The micrometer gave the diameter of the field in thousandths of an inch; and the covering glass measured 500 thousandths of an inch across: and the areas of the circles were of course proportioned to the squares of those diameters. All that was needful, therefore, in order to enable me to calculate the number of bacteria in 1/50 minim, was to form a fair estimate of the number of bacteria per field, and this was done by counting the organisms in a considerable number of fields and taking the average.

It so happened that two kinds of bacteria were seen under the microscope, one motionless, with the characters of *Bacterium lactis*, the other much less numerous, with longer segments and in active movement. As a rule, on examining milk which is undergoing the lactic fermentation but is still fluid, the *Bacterium lactis* is alone discoverable; but in this instance we had ocular proof of the admixture of another sort though in much smaller numbers. As the result of the estimate which I made of the number of bacteria present in every 1/50 minim, I found it necessary to dilute the milk with no less than a million parts of boiled water in order that every 1/100 minim should contain on the average a single bacterium. This having been done, 1/100 minim of the infected water was added by means of the small syringe to each of five glasses of pure boiled milk. The result of this inoculation was that only one of the five glasses was affected at all. The contents of four remained permanently fluid and unchanged, and when examined with the microscope after the lapse of thirteen days, showed no bacteria of any kind. The milk of the fifth glass, however, was in the course of the third day after inoculation, converted into a solid mass, and on examination was found to be sharply acid, and under the microscope there were seen among the granular masses of caseine countless motionless bacteria with the ordinary characters of *Bacterium lactis*. . . . No other bacteria, however, were to be discovered; we had got rid of the moving species which was seen to be associated with it in the milk before dilution, and *a fortiori* it might be believed that other species, doubtless present in the original milk, but in too small numbers to be detected, had been avoided. . . .

Now, however, having good reason to believe that I had got the *Bacterium lactis* pure and unmixed, I proceeded to perform the experiment which constitutes the most important feature of this investigation. I have already described the mode of procedure when speaking of the means adopted for isolating the *Bacterium lactis*. Its object now was to obtain, if possible, absolute proof, which would commend the judgment of all, that the

Bacterium lactis is really the cause of the lactic fermentation, and no mere accidental concomitant of the change.

On the 30th of August last, having provided sixteen pure glasses of boiled milk, and having estimated, in the manner already described, the number of bacteria present in every 1/50 minim of a glass of boiled milk which had been inoculated the day before . . . I diluted a drop of this milk with boiled water to the requisite degree, and introduced into each of ten of the sixteen uncontaminated glasses a drop calculated to contain on the average a single bacterium, while five of the rest received each a drop supposed to contain two of the organisms, and the remaining glass was inoculated with a quantity in which, according to the estimate, there would be four bacteria. The result was that within three and a half days the glass into which four bacteria were supposed to have been introduced contained a curdled mass, and the five which had received the drops arranged for two bacteria each had all undergone a similar change. Of the ten inoculated with drops averaging one bacterium each, the majority were at this period still fluid; but some assumed the solid condition in the course of the next twenty-four hours, though at different times. But of this series of ten exactly five, as it so happened, remained permanently fluid.

This was just the sort of occurrence which might have been anticipated if we believed the bacteria to be really the cause of the fermentative change, and supposed that we had succeeded in forming a fair estimate of their numbers. It was to be expected that the bacteria would not be distributed with perfect uniformity in the water with which the milk was diluted; and hence, of the drops containing on the average one bacterium each, some would probably be destitute of the organisms, and the rest have more than one, and in differing numbers, involving slight differences of time in arriving at the stage of the fermentative process which induced coagulation.

But not only were the results of this experiment in harmony with the view that the *Bacterium lactis* was the real fermentative agent; they would, as I believed, afford indisputable evidence of the truth of the theory, provided it should turn out, as former experience made me feel sure would be the case, that every glass which had curdled contained the bacterium, and that every one which remained fluid contained none. . . . I went through the laborious process of examining the contents of all sixteen glasses . . . nine days after the time of the inoculation. All those which had coagulated [showed] . . . under the microscope, in every instance, bacteria with the characters of *Bacterium lactis* . . . but no other organism. And in the case of the remaining five glasses, where the milk was still fluid and unchanged in appearance . . . I could discover under a protracted search no organisms of any kind. I have brought before the Society one of these last glasses (still under the protection of its glass cap and shade) to show that even after the lapse of nearly four months the milk remains fluid and unaltered. . . .

And now let me dwell for a few minutes on the inference to be drawn from these facts. We have seen that boiled water rendered infective by the admixture of a small quantity of milk undergoing the lactic fermentation, having been introduced in drops of equal size into ten glasses of pure boiled milk, five of those glasses underwent the lactic fermentation, characterized by souring and curdling, while five remained altogether unaffected. This proves that the same truth holds regarding the lactic ferment which we have before established with respect to the various ferments that occur in ordinary water, viz. that it is not a material soluble in water, but consists of insoluble particles. For had it been dissolved in the water of inoculation, every equal-sized inoculating drop would have produced the same fermentative effect. Next we have to consider the bearing of our facts on the nature of those insoluble particles, the question being whether they were the bacteria or consisted of some so-called chemical ferment destitute of life, of which the bacteria were a mere accidental concomitant. Let us assume, for the sake of argument, that it is possible for insoluble particles to exist devoid of vitality, yet capable of multiplying like the bacteria. Such a notion is unsupported, I believe, by a tittle of scientific evidence; but, for the sake of argument, let us for a moment assume it. We should then be further obliged to suppose, in order to account for our facts, that these hypothetical particles, though merely accidental accompaniments of the bacteria, were present in precisely the same numbers, a thing which is utterly inconceivable. But we should have to go further and suppose, what is equally inconceivable, that these bodies of different natures, though mere accidental concomitants were not only exactly equally numerous, but invariably accompanied each other in pairs; so that when a bacterium was introduced into one of the glasses, it was always associated with a particle of the hypothetical true ferment, and whenever the bacterium was excluded, the hypothetical ferment likewise failed to enter. Hence, as the only other possible interpretation of our facts involves what is utterly inconceivable, I venture to think that those facts will be admitted by all to afford a conclusive demonstration that the particular species of bacterium which we have been studying is really the cause of this special fermentation. . . .

. . . while we have . . . reason to think it not unlikely that ultramicroscopic fermentative organisms may exist, we have not grounds whatever for believing that bacteria visible under the microscope have ultra-microscopic germs. The sole reason for the frequently expressed opinion . . . to that effect is, I believe, the fact that while ordinary water has been shown to cause the development of bacteria when introduced into organic liquids even in small quantity, yet no bacteria can be discovered in it by aid of the microscope. But we are apt to forget how extremely difficult it is, with the very high magnifying powers which it is needful to use, to discover such minute objects as bacteria unless they are present in large numbers. Now, if we recall the experiments above related of inoculating milk with very minute drops of water, we see that in the sample of ordinary tap-water examined there could not be more than about one particle capable of producing a bacterium in every 1/100 minim. A drop of that size, small as it is, would, if placed between two flat plates of glass for microscopic examination, spread itself over a space about half a square inch in area, and we might search such a drop for an entire day without finding an individual bacterium contained in it. . . . Meanwhile the facts which I have adduced will, I hope, remove the mystery attendant on the notion that water teems with ultra-microscopic or invisible germs of the bacteria which we see of larger or smaller dimensions in organic liquids undergoing fermentative changes.

Comment

Lister has here outlined the first method for the isolation of a pure culture. For its day, it was an elegant method, and one which is still used under certain circumstances. The principle which he used is the basis of the Most Probable Number method for estimating the number of bacteria in a liquid. For the isolation of a pure culture it is not as simple or direct as Koch's method, and it is sufficiently sophisticated that not all people understand it the first time it is described. In its historical context, it is quite important.

It can also be seen how this work is directly descended from all of the earlier work on the nature of fermentation and on spontaneous generation. It had important implications for understanding the nature of infectious disease, as Lister realized. It showed that a specific organism could cause a specific change in a liquid. If this was so, then a specific organism might cause a specific change in a human being, leading to disease and death. This point will be developed further in Koch's papers on pages 96 and 116.

The simple calculation which Lister makes of the probability of seeing bacteria under the microscope in ordinary tap water is one which could have been made much earlier. If it had been made workers would have realized that apparent sterility under the microscope is only a delusion. In fact, if there are fewer than 10^5 bacteria per ml., it is quite difficult to see their evidence by microscopic examination. Anyone would admit that 10^5 bacteria, or 100,000 cells, is quite a large number.

The reader might be interested in speculating on how bacteriology would have developed if the microscope had not been able to resolve objects of the size of single bacterial cells. Would scientists have discovered that there were particulate objects present in various natural substances, capable of multiplying and causing specific changes in these substances? The present work of Lister shows how these particulate objects could have been shown to exist, even if they could not be seen. One generation later, filterable viruses were shown to exist in an analogous way.

Alcoholic fermentation without yeast cells

1897 · Eduard Buchner

Buchner, Eduard. 1897. Alkoholische Gährung ohne Hefezellen. *Berichte der Deutschen Chemischen Gesellschaft*, Vol. 30, pages 117–124.

A SEPARATION OF THE FERMENTING power from living yeast cells has until now never been done. In the following a method is described which can be used to this end.

1000 g. of pure brewer's yeast cells, taken before the addition of potato starch, is mixed carefully with 1000 g. of quartz sand and 250 g. of Kieselguhr, and ground until the mass has become moist and plastic. 100 g. of water is added to the paste and the mixture wrapped up in a filter cloth and gradually placed under a pressure of 4–500 atmospheres. The result is 300 cc. of expressed juice. The residual material is ground again and filtered, and 100 g. of water added. This is then also pressed in the hydraulic press under the same pressure, to yield another 150 cc. expressed juice. From one kg. of yeast can be obtained 500 cc. of juice, which contains 300 cc. of cell contents. To remove a slight turbidity, the pressed juice is finally shaken with 4 g. of Kieselguhr and after several infusions, the first portion is filtered through filter paper.

The expressed juice obtained in this way is a clear, opalescent, yellowish liquid, which has the pleasant smell of yeast. The specific gravity was found to be 1.0416 (17°C.). When it was boiled, an extensive coagulum separated out, so that the liquid almost completely solidified. The formation of insoluble flocks began to occur at 35–40°. Even before this the formation of gas bubbles could be noticed, which were proved to be carbon dioxide. These were formed because the fluid was saturated with carbon dioxide. The juice contained about 10 per cent dry material. In an earlier batch prepared by a poorer technique, from 6.7 per cent dry material, 1.15 per cent ash was present, and 3.7 per cent protein, as calculated from the nitrogen content.

The most interesting property of this juice is that it is able to bring about the fermentation of carbohydrates. When an equal volume of the juice is mixed with a concentrated solution of cane sugar, after only one-fourth to one hour the regular formation of carbon dioxide bubbles begins, and this may last for days. Glucose, fructose and maltose show the same phenomenon. However, no fermentation occurs when saturated solutions of lactose and mannose are added. These two latter sugars are also not fermented by living beer yeast cells. After a number of days of fermentation in the ice box, a mixture of yeast juice and sugar gradually becomes turbid, without any microscopic organisms being visible. At 700 times magnification, a rather large coagulum of proteinaceous substances can be seen, which probably are formed by the acid developing during the fermentation. The saturation of the mixture of yeast juice and sucrose with chloroform has no effect on the fermentation, but causes an earlier formation of a small amount of proteinaceous coagulum. The fermentation is not prevented by the filtration of the yeast juice through a sterilized Berkefeldt-Kieselguhr filter, which is able to retain all of the yeast cells. The mixture of this completely clear filtrate with sterilized cane sugar undergoes fermentation after perhaps a day, even in the ice box, although the beginning of fermentation is somewhat delayed with this sterile filtrate. When a parchment tube was filled with yeast juice and placed in a 37 per cent cane sugar solution, after several hours the surface of the tube was covered with a large number of tiny gas bubbles. Naturally there was also a lively development of gas within the tube, as the result of the inward diffusion of sugar. Further experiments will be necessary to decide if the agent causing the fermentation is able to diffuse through parchment paper, as it seems from this experiment. The fermenting power of the yeast juice is gradually lost with time; after five days in ice water in a half-full flask, a yeast juice was inactive against sucrose. It is remarkable that as opposed to this, the fermentation of yeast juice with cane sugar continues for at least two weeks in the ice box. Perhaps the unfermenting juice is affected by oxygen from the air, which would not be present in the fermenting juice due to the formation of carbon dioxide. It could also be that the easily assimilable sugar is able to maintain the fermenting agent.

Only a few experiments have been carried out so far on the nature of the active material in the yeast juice. When the juice is heated to 40–50°, there is first the development of carbon dioxide, and then gradually the separation of coagulated protein. After an hour this was filtered. The clear filtrate had only weak activity against cane sugar in one experiment, while in another it had none. The active substance seems therefore either to lose its activity at this surprisingly low temperature, or to coagulate and precipitate. In another experiment, 20 cc. of the juice was added to three times this amount of absolute alcohol, and the precipitate which formed was removed under vacuum and dried over sulfuric acid. This yielded 2 g. of dry material which was only very slightly soluble in water. The filtrate from this had no activity against cane sugar. This experiment must be repeated. Especially, it must be attempted to isolate the active substance by the use of ammonium sulfate.

The following conclusions can be drawn concerning the theory of fermentation. First it has been shown that for the introduction of the fermentation process, no apparatus so complicated as the yeast cell is necessary. The bearer of the fermentative power of the expressed juice is more on the contrary a soluble substance, and more than likely it is a protein. This same will be called zymase.

The view that a special proteinaceous substance derived from the yeast cell caused the fermentation, was first announced as the enzyme or ferment theory as early as 1858 by M. Traube, and was later especially defended by F. Hoppe-Seyler. The separation of this kind of an enzyme from the yeast cell had been until now unsuccessful.

The question still remains if zymase can be identified directly with the already known enzymes. As C. v. Nägeli has already emphasized, there are important differences between the action of fermentation and the action of the ordinary enzymes. The latter action is merely hydrolytic, and this action can be reproduced by the simplest chemical agents . . . but the breakdown of sugar into alcohol and carbon dioxide is one of the most complicated of reactions. Carbon-to-carbon bonds are broken in such a way that has not been obtained with other methods. There are also significant differences in the heat stability.

Invertase can be separated from yeast killed by dry heat (one hour at 150°), by water, and can be precipitated with alcohol to obtain a powder easily soluble in water. In the same manner, the fermenting substance cannot be obtained. . . . Therefore it seems reasonable that zymase belongs to the true protein substances, but is more closely related to the living protoplasm of the yeast cell than invertase. . . .

The fermentation of sugar by zymase can take place within the cell, but it is more probable that the protein substances are excreted into the sugar solution, and the fermentation occurs there. The process is then perhaps a physiological act only in so far as it is the living yeast cell which excretes the zymase. . . . It seems evident, as the above observations have shown, that the zymase can diffuse through parchment paper.

In one experiment the gas which

developed was allowed to pass into lime-water, and in this way identified as carbon dioxide. In two other experiments, after three days' fermentation at ice box temperature, the alcohol formed was determined. In one experiment with 50 cc. sucrose solution and 50 cc. of yeast juice, 1.5 g. ethyl alcohol was formed. In another experiment with 150 cc. of sugar solution and 150 cc. yeast juice, 3.3 g. ethyl alcohol was formed. . . . In all cases the alcohol was identified by the iodoform reaction and was finally salted out of the aqueous solution with potassium carbonate. The material collected in this way in the latter experiment showed a boiling point between 79–81° (734 mm.). The distillate was colorless, combustible, and possessed the odor of ethyl alcohol. . . .

It has been shown that the above described method for the expression of yeast juice is also suitable for obtaining the contents of bacterial cells, and experiments on this are in progress, especially with pathogenic bacteria, at the Hygienische Institut in Munich.

Comment

This paper by Buchner is a fitting close to this section on the nature of fermentation and spontaneous generation. The early opinion that fermentation was not a vital process gave way to the opinion that fermentation only occurred when a microorganism developed. Now we see that the whole cell is not necessary for this process, but that a proteinaceous material can be extracted from the cell which carries out the process. It is obvious that Buchner's work marked the beginnings of biochemistry and has led directly to our current vast store of information about enzymes and enzyme reactions. Those workers who believed that all of the properties of a living organism could be explained in terms of chemistry and physics could begin to find their ideas justified by the work of Buchner.

PART II
The Germ Theory of Disease

Contagion, contagious diseases and their treatment

1546 • Girolamo Fracastoro

Fracastoro, Girolamo. 1546. De Contagione, Contagiosis Morbis et eorum Curatione. *Contagion, Contagious Diseases and their Treatment.* Translated by Wilmer C. Wright.

WHAT IS CONTAGION?

I shall now proceed to discuss Contagion, and shall begin with what seem to be its universal principles from which are derived its particular causes. . . . As its name indicates, contagion is an infection that passes from one thing to another. . . . The infection is precisely similar in both the carrier and the receiver of the contagion; we say that contagion has occurred when a certain similar taint has affected them both. So, when persons die of drinking poison, we say perhaps that they were infected, but not that they suffered contagion; and in the case of things that naturally go bad when exposed to the air, such as milk, meat, etc., we say that they have become corrupt, but not that they have suffered contagion, unless indeed the air itself has also become corrupt in a precisely similar way. . . .

Everything that happens, whether actively or passively, affects either the essential substance of bodies or their non-essential parts. When someone has been heated or sullied by something, we do not say, except by a metaphor, that he has suffered contagion; because contagion is a precisely similar infection of the actual substance. Now when a house catches fire from the burning of a neighboring house, are we to call that contagion? No, certainly not, nor in general when the whole thing is destroyed primarily as a whole. The term is more correctly used when infection originates in very small imperceptible particles, and begins with them, as the word "infection" implies; for we use the term "infected," not of a something that is destroyed as a whole, but of a certain kind of destruction that affects its imperceptible particles. By the whole, I mean the actual composite, and by very small, imperceptible particles, I mean the particles of which the composite and mixture (combination) are composed. Now burning acts on the thing as a whole, whereas contagion acts on the component particles, though by them the whole thing itself may presently be corrupted and destroyed.

Contagion, then, seems to be a certain passive affection of elements in combination. But since such combinations can be corrupted and destroyed in two ways, either by the advent of a contrary element, owing to which the combination cannot retain its form, or secondly by the dissolution of the combination, as happens when things have putrefied, we may perhaps hesitate to say whether contagion, when it is carried in by infection of the smallest particles, is produced in the former way or the latter. Moreover, what shall we say is the nature of this infection? Is it a corruption of those particles, or only an alteration? What, in short happens to those particles? Hence it is hard to determine whether every contagion is a kind of putrefaction. All these problems will become clearer if we first investigate the fundamental differences of contagions, and their causes. Meanwhile, if we allow ourselves to sketch a sort of tentative definition of contagion, we shall define it as: A certain precisely similar corruption which develops in the substance of a combination, passes from one thing to another, and is originally caused by infection of the imperceptible particles.

THE FUNDAMENTAL DIFFERENCES IN CONTAGIONS

There are, it seems, three fundamentally different types of contagion: the first infects by direct contact only; the second does the same, but in addition leaves fomes, and this contagion may spread by means of that fomes, for instance scabies, phthisis,* bald spots, elephantiasis and the like. By fomes I mean clothes, wooden objects, and things of that sort, which though not themselves corrupted can, nevertheless, preserve the original germs of the contagion and infect by means of these: thirdly, there is a kind of contagion which is transmitted not only by direct contact or by fomes as intermediary, but also infects at a distance; for example, pestilent fevers, phthisis, certain kinds of ophthalmia, exanthemata of the kind called variolae,† and the like. These different contagions seem to obey a certain law; for those which carry contagion to a distant object infect both by direct contact and by fomes; those that are contagious by means of fomes are equally so by direct contact; not all of them are contagious at a distance, but all are contagious by direct contact. Hence the most simple kind of contagion is that by direct contact only, and it is naturally first in order. . . .

CONTAGION THAT INFECTS BY CONTACT ONLY

An especially good instance of the contagion that infects by contact only is that which occurs in fruits, as when grape infects grape, or apple infects

* [Tuberculosis.]
† [Smallpox.]

apple; so we must try to discover the principle of this infection. It is evident that they are infected because they touch, and that some one fruit decays first, but what is the principle of the infection? Since the first fruit from which all the infection passes to the rest has putrefied, we must suppose that the second has contracted a precisely similar putrefaction, seeing that we defined contagion as a precisely similar infection of one thing by another. Now putrefaction is a sort of dissolution of a combination due to evaporation of the innate warmth and moisture. The principle of that evaporation is always foreign heat, whether that heat be in the air or in the surrounding moisture; hence, in both fruits the principle of contagion will be the same as the principle of putrefaction, namely extraneous heat; but this heat came to the first fruit either from the air or some other source, and we may not yet speak of contagion; but the heat has passed on to the second fruit by means of those imperceptible particles that evaporate from the first fruit, and now there is contagion, since there is a similar infection in both fruits; the heat that evaporates from the first fruit has power to produce in the second fruit what the air produced in the first, and to make it putrefy in a similar way, all the more because there is analogy. Now some of the particles that evaporate from the first fruit are hot and dry, either independently or when in combination, whereas others are hot and moist, either independently or when in combination. Those that are hot and dry are more apt to burn the fruit, whereas those that are hot and moist are more apt to produce putrefaction and less apt to burn. For the moisture softens and relaxes the parts of the fruit that it touches, and makes them easily separable, while the heat lifts them up and separates them. Hence when heat and moisture are produced within and evaporate, the result is dissolution of the combination, and this was our definition of putrefaction. We must therefore suppose that the hot moist particles—moist either independently or in combination —that evaporate from the first fruit, are the principle and germ of the putrefaction that occurs in the second fruit. I use the term "moist in combination," because, in evaporations that occur in putrescent bodies, it nearly always happens that very small particles are intermingled, and thus become the principles of certain generations and of new corruptions; and this combination of hot and moist particles is most apt to convey putrefactions and contagions. We must therefore suppose that it is by means of these principles that contagion occurs in fruits. But in all other bodies also that are in contact and putrefy, if they are analogous with one another, it is reasonable to suppose that the same thing happens, and by means of the same principle. Now the principle is those imperceptible particles, which are hot and sharp when they evaporate, but are moist in combination. In what follows they are called Germs of Contagions.

CONTAGION THAT INFECTS BY FOMES

Now it is not at once obvious that the germs that transmit contagion by means of fomes are produced in the same manner and by the same principle as that above described, for the principle that exists in fomes seems to be of a different nature, inasmuch as, when it has retired into fomes from the body originally infected, it may last there for a very long time without any alteration. Things that have been touched by persons suffering from phthisis or the plague are really amazing examples of this. I have often observed that in them this virus has been preserved for two or three years; whereas particles that evaporate from putrefying bodies never seem to have the power to last as long as that. Nevertheless no one ought on that account to think that the principle of contagion that is in fomes is not the same as the principles that infect by contact only, because the very same particles that evaporate from the body originally infected, after being thus preserved, can produce the same effect as they would have done when they evaporated from the original body. . . . Now a combination is strong and lasting in virtue of two qualities; first it must have the kind of hardness possessed by iron, stones and the like, whose very small, imperceptible particles last for many years; secondly there must be present a certain viscosity, and the mixing process must be thoroughly elaborated. So that even when the germs of contagions are not hard, they may be viscous and elaborated. By an elaborated combination I mean one composed of very small particles well shaken together. . . . Combinations of this sort are produced by evaporations that are closely confined, where what evaporates is not dispersed, but is violently shaken, and hence is very finely and minutely mixed. Now if viscosity be added, the resulting combination is strong and suitable for preservation in fomes. A proof of this is that all germs that infect by means of fomes are without exception viscous and sticky, and only when they have this quality can they occupy fomes. . . .

CONTAGION AT A DISTANCE

Even more surprising and hard to explain are those diseases that cause contagion, not by direct contact only, or by fomes only, but also at a distance. There is a kind of ophthalmia with which the sufferer infects everyone who looks at him. It is well known that pestiferous fevers, phthisis, and many other diseases infect those who live with the sufferer, even though there is no actual contact. It is far from certain what is the nature of these diseases, and how the taint is propagated. We must therefore study these problems with the greatest care, since of this sort are the majority of the diseases that we are investigating. . . .

How the germs of contagions are carried to a distant object and in a circle

Let us first enquire by what sort of movement these germs of contagions are impelled, since it is clear that they are carried far and to persons far distant, a fact that many people think so astonishing. . . . Now the principle of movement in these small bodies, in all directions, is in part independent, in part given by something else. All evaporation independently rises upwards, as may be seen in smoke and many other things, for everyone knows that all evaporation is warm; but the movement may be derived from something else, and then the thrust is sideways and finally downwards. This is due to two main causes, one being the resistance of the air or of floors, or things of that sort, on which fall the particles that are first exhaled; when these particles cannot be carried further, they are thrust sideways by the particles that follow them, and these by others, till the whole surrounding space is filled. The second cause is the air itself, which divides into its smallest and indivisible parts all evaporation that is tenuous and easily soluble. For it is the nature of elements and of all liquids that they seek, so far as is possible, a suitable position; and a position is most suitable when the parts are continuous, or if not continuous, are the least possible distance apart from one another; for thus they are less exposed to violence. . . . Hence the air keeps on dividing the evaporation more and more, until it arrives at those parts which cannot be further divided and separated. Then when this countless division has been made, much of the air is filled and mingled with the evaporation all round and about, as is most evident in the case of smoke. These then are the reasons why the evaporations that occur in contagions are also carried round and about, and occupy a great volume of the air. . . . Thus it is that these germs may infect also those who live with persons infected, and the germs can be preserved for a certain time, not only in fomes but also in the air, though longer in fomes.

But how does it happen that germs whose bulk is so small do not suffer alteration, when thus exposed to the air? That is the first question. What must be the strength of the combination in so small a particle, especially when those particles have not the quality of hardness, that they can last so long in the air? It is those that are viscous and sticky, however small they may be, that can live, if not quite so long as the hard ones, still nearly as long.

The hard particles offer most resistance to alterations, because of three properties: First, in a small bulk they have more substance; secondly, they are colder, on account of their earthy elements; thirdly, by reason of their density, their parts cannot easily be volatilised and rarefied as ought to happen when heat is introduced. . . . hard, but also viscous bodies, defend themselves from many alterations, if only these are moderate; but they cannot endure violent alterations. Hence the germs of all contagions are consumed by fire, and are broken up by very cold water also. . . .

Of such sort are the germs of contagions also, for they are all per se acute, although constituted in viscosity, and they become active when the animal heat vaporises that combination and brings together the similar parts. Now germs of this kind have great power over the humors and spirits, so that they can even cause death in a few hours if they are analogous with the spirits, but about this I shall say more later. These same germs can be shot forth from sore eyes into the eyes of another and carry in a precisely similar infection, and this is not a visual image but a taint in the eye. It is not surprising, if one considers the method by which they attack, that they penetrate into the animal, and some of them very quickly, for they attack and enter from the small pores, veins and arteries into the larger, and from these to others, and often reach to the heart.

One method of penetration is by propagation and, so to speak, progeny. For the original germs which have adhered to the neighboring humors with which they are analogous, generate and propagate other germs precisely like themselves, and these in turn propagate others, until the whole mass and bulk of humors is infected by them. A second method of penetration is by attraction, which works inwards, partly through the breath of inspiration, partly by dilation of the blood vessels. For along with the air that is drawn in, there enter, mixed with it, germs of contagions, and when once these have been introduced, they do not retire as easily by expiration as they entered by inspiration; for they adhere closely to the humors and organs, and some even to the spirits, which retreat from the image of their contrary, and carry their enemy with them even to the heart. . . .

THE ANALOGY OF CONTAGIONS

Contagions have manifold and very surprising analogies (selective properties). For instance, there is a certain pest which attacks trees and crops, but harms no sort of animal; again, there is a pest which attacks certain animals but spares trees and crops. In the animal world, one pest will attack man, another cattle, another horses, and so on. . . . Some pests work promiscuously, so that some persons are infected, others not; some persons can associate with the plague-striken and take no hurt, others cannot. The organs of the body also have their own analogy (affinity), for opthalmia harms no organ save the eyes, while phthisis does not affect the eyes, though they are so delicate, but does affect the lungs. . . .

IS EVERY CONTAGION A KIND OF PUTREFACTION?

Now let us enquire whether every kind of contagion is a kind of putrefaction, and whether every putrefaction is contagious. It seems that every putrefaction is contagious, either absolutely, or at least contagious to a contiguous part, but not every putrefaction is contagious for another body, since, in order that it may act, many factors are required, as I have said. Perhaps one may doubt whether every contagion consists in putrefaction of some sort; since rabies seems to be contagion of a sort, but not putrefaction. Likewise, when wine turns to vinegar, it seems to suffer a sort of contagion by something else, but not to suffer putrefaction. For when it putrefies it has a bad smell and is unfit to drink, whereas vinegar is pleasant to taste and even opposes putrefactions. Yet these cases also must be regarded as putrefactions of a sort. . . .

Now all putrefactions have the power to convey a precisely similar putrefaction, at least to a continuous part; hence, if every contagion is putrefaction, it seems that contagion, simply and generally speaking, might be defined as: A certain precisely similar putrefaction which passes from one thing to another, whether that other be continuous with the original thing, or separated from it. Yet this is not contagion strictly so called, for true contagion occurs between two different bodies. But if we wish to consider, above all and by itself, that contagion which is observed in diseases and does not affect by direct contact only, then we shall define contagion as: A precisely similar putrefaction which passes from one thing to another; its germs have great activity; they are made up of a strong and viscous combination; and they have not only a material but also a spiritual antipathy to the animal organism. This definition will give us the key to all the phenomena that are observed in contagion.

Comment

With this paper we begin to consider the nature of infectious disease. The reader should have several points clearly in mind. An infectious (or contagious) disease is a *process* (not a thing) that occurs in a host as a result of an interaction with a parasite. The disease is not the parasite. The parasite isolated from the host is not the disease, but is merely the potential causer of the disease.

The disease itself is a complex interaction between parasite and host. With a given host and a given parasite, under the usual environmental conditions, a given disease usually arises. This disease is recognized in the host because of certain symptoms which become evident to an observer. In many, if not most cases, two hosts of the same species usually exhibit quite similar symptoms when infected with a given parasite. Because of this, we can recognize the same disease in several or many individual hosts.

It was only these symptoms which were available to early observers. They had no knowledge of microorganisms, and even little knowledge of the nature of the host. But because a number of diseases, such as syphilis, plague, tuberculosis, small pox, usually had characteristic symptoms, it was possible to observe these diseases in populations of individuals. By such observation, it was possible to infer that these diseases were transmitted from one person to another. Remember that the agency of this transfer was unknown. It was merely known that a transfer occurred.

As we see in this paper of Fracastoro, speculation on the nature of contagious, or infectious, diseases began at an early time. Fracastoro's writing is mostly philosophical and he attempts to define the subject in terms of the ideas of his day. Even though his discussion seems crude by current standards, it is a highly interesting discussion, because it attempts to understand a problem which is quite complex, and because it comes as close as it does to hitting the nail on the head. The reader should reflect on what he himself would have invented to explain the observable facts of syphilis or tuberculosis in the year 1546.

This paper was written in Latin. It is interesting that Fracastoro uses a word for the infectious or contagious principle which can best be translated into English as "germ." The reader can see the very ancient origin of this word, and may also find it interesting to compare the usage here with the usage of Schwann, Pasteur, Koch and others in the 19th century. His statement that these germs generate and propagate other germs precisely like themselves is prophetical although not based on direct observation.

Finally, his attempt to find a similarity between putrefaction and contagion is noteworthy, because it was this very similarity which led later workers, such as Henle, Lister, Pasteur, and Koch, to consider seriously the germ theory of disease. It seems quite likely that by their time the speculations that Fracastoro first advanced 400 years before had become so ingrained in theoretical discussions of contagion that it was quite easy to accept them.

Concerning miasmatic, contagious, and miasmatic-contagious diseases

1840 • J. Henle

Henle, J. 1840. Pathologische Untersuchungen. Berlin. August Hirschwald Verlag, 1840. *Von den Miasmen und Contagien und von den Miasmatisch-contagiösen Krankheiten*, pages 1–82.

. . . I WILL NOW INDICATE THE BASES which show that the material of the contagion is not only organic, but also living, and indeed even has a separate existence, which is related to the diseased body in the way that a parasitic organism is related to it. This principle agrees with the old theory of *contagium animatum*, which has been often denied, and which continues to reappear in more precise forms. I would also like to point out a theory which seems on the surface to agree with this, and which has many adherents among the natural scientists in Germany. This theory states that it is not the contagium, but the disease itself which is a parasitic organism, or more ambiguously, is a parasitic life process. It says that the contagium is the germ or seed of this parasitic being with hidden body, through which the contagium itself reproduces. . . . The difference between this theory and our own is that in our theory it is not the disease, but the cause of the disease, which reproduces itself. To offer a crude example, consider the case of a thorn which has been thrust into a finger. One would consider that it is the thorn which causes the inflammation and suppuration. If the thorn is removed, it can be stuck into the finger of another person and cause this same disease a second time. In this case through the transfer of the thorn, it is not the disease, or even a product of the disease, which is being transferred, but the cause itself. And let us suppose that the thorn could reproduce itself in the diseased body, or that every small part of the thorn could turn into a new thorn, then through the transfer of each small part of the thorn, one could cause in other individuals the same disease, inflammation and suppuration. The thorn, not the disease, is the parasite. The diseases are the same, because the causes are the same. The contagium, as we consider it, is therefore not the germ or seed of the disease, but the disease inducer. In the same way, the egg of a tapeworm is not a product of the tapeworm disease . . . or a product of the individual which suffers with a tapeworm, but it is from that parasitic body which reproduces by eggs, and which is responsible, at least in part, for the symptoms of the tapeworm disease. It is not the seed of the disease which is inoculated, but the disease inducer.

The disease inducer or causer reproduces itself in the sick body and at the end of the disease is eliminated from the body. . . .

The facts which prove the separate life of the contagium are the following:

(1) The ability to reproduce by the assimilation of foreign substances is an ability that we recognize only in living beings. No dead, chemical substance, organic or inorganic, can reproduce from the food of another. With dead substances, one can only get compounds which can again be caused to produce the original substances which acted to make these compounds.

Fermentation has been looked upon as a proof of the opposite of this statement, because it is known that during fermentation the ferment itself produces more of itself, because the very smallest amount of ferment is sufficient to keep the fermentation of a large amount of fermentable liquid in operation, and at the same time produce alcohol and carbon dioxide from the sugar. Many believe that this solved the puzzle of how the contagium multiplies, if the contagium is compared with a ferment, which is able to produce more of itself in the blood. Although this comparison seems quite good, this comparison is actually more pertinent to the idea which we are proposing here. This is because Cagniard Latour and Schwann have shown without a doubt, that the fermentation is the decomposition of organic liquids by plantlike beings, lower fungi, which grow on the nitrogenous and sugary nutrients of the liquid, reproduce themselves, and in the process decompose the sugar. . . . Schwann has described these organisms from beer yeast as follows: "partly round, mostly oval globules, which occur partly alone, but most often in rows of 2–8 connected together. . . ."

(2) The action of the contagium may be compared with the fermentation, in that the extent of the effect has no relation to the amount of the ferment used. A needle which is immersed in a solution of a grain of small pox material in a half dram of water is sufficient to cause an infection.* This effect by such a small dose depends on the ability of the agent to reproduce itself, in the same way that such things occur in the fermentation or putrefaction, and is therefore a further proof for the living nature of the contagium.

(3) The exactly constant course in the development of the miasmatic-contagious diseases, and the similarity in the course throughout an epidemic, speaks for a time-wise development of the disease inducer, in the same way that living beings develop.

ORGANIZATION OF THE CONTAGIUM

After it has been shown that the contagium is a material endowed with an independent living existence, which reproduces in the manner of plants or animals, and can multiply by the assimilation of organic substances, and lives as a parasite on the sick body, there still remains the question of how the so far unseen bodies of these parasites are constituted, which give them so clear and destructive life manifestations. It seems to be a rule in the thinking processes of man, that the contagium, if it is to be considered to be something living, must always have ascribed to it a form which comes from the organic world as it is known at the time. Therefore, in the earlier, more naive times, the natural scientists (of whom there are a few today who are still so naive) identified the con-

* [See the paper by Jenner, page 121.]

tagium with insects. And when the microscopic animals were discovered, the infusoria were then blamed, perhaps with more reason, to be contagia and miasma. Now it seems to be more appropriate to think of the contagium with a plantlike body, as one learns more and more about how widespread these organisms are, and how rapidly they multiply, and how endurable they are. . . . [Then are cited a number of works on a disease of silkworms, which seems to show that this disease is due to a fungus, called *Botrytis bassiana*. He cites this work as evidence of the existence of plantlike organisms as disease producers.]

If it was possible with our present-day methods to solve the question of the nature of the contagium through direct observation, then the theoretical discussion which I have advanced as proof would be superfluous and unnecessary. Unfortunately it must be predicted that a rigid proof from positive observations is not yet possible, even if these observations were more favorable to our hypothesis than the current ones. If one finds living, moving animals or distinct plants in the infectious (contagious) material, it is quite possible that these could have developed incidentally when this material was exposed to air. And even if the animals or plants in this contagious material were always present within the body, there would still be the possible objection, and one hard to oppose, that they are only parasitic, although constant elements, which develop in the body fluids and are significant for the diagnosis of the disease, without being the causal material or the seeds of the causal material. In order to prove that they are really the causal material, it would be necessary to isolate the animal seeds and animal fluid, the contagious organism and contagious fluid, and then observe especially the power of each one of these to see if they corresponded. This is an experiment which cannot be performed.

In the same way, a negative result of our observations would not be sufficient to reject our hypothesis. It is not necessary for us to look for excuses, such as that the organisms of the contagium might be too small for our optical methods. But if they are not motile animal beings, but eggs or similar germs of lower plants, then I can think of no way to distinguish these from the cellular bodies which occur in all of the tissues and excretions of the body, if their structure or their development does not give us some independent information. . . . I have examined under the microscope the little spheres of which the lower plants consist, while they were mixed with pus, and compared them with the decomposing bodies in the pus, and have found no microscopical or chemical differences, which would have served to allow further studies on a surer footing. . . .

There is still a second hypothesis to explain the contagium. The first hypothesis was that the contagium contained lower plants or animals. The second possibility is that the contagium consists of animal elements which can be cultured and isolated from the sickness and which can become free-living entities, from which the infection can spread. This last case is improbable in the miasmatic-contagious illnesses of man and higher animals, since we know of no elements which can maintain their energy of life for so long a time after being separated from the body, or after being dried out, to regain their energy of life. More important for the answer to this question is the study of how the contagium initially develops. If the contagium develops initially within the diseased body in the course of an illness which is caused by some other means, then one could not possibly consider it to consist of free-living animal or plant beings. . . . If it would be possible to prove that a contagium can be cultured outside the body, as in the observations of Audouin in the silkworm disease, then such a contagium could only be a plant or animal, even though one which is not yet known.

[Henle's separation of contagious and miasmatic diseases is essentially as follows: There are diseases which are purely miasmatic, such as malaria. In a second group are the diseases which are miasmatic-contagious. These include exanthema, smallpox, measles, scarlet fever, typhus, influenza, dysentery, cholera, plague, and puerperal fever. In the third group are those which develop only through contagium. These include syphilis, mange, glanders and worms of horses, foot-and-mouth disease of sheep, and rabies.]

Comment

By Henle's time, a considerable body of facts were available to use in speculations concerning contagious diseases. The rise of the microscope had revealed a whole new world of invisible things and made Fracastoro's idle speculation about invisible particles seem quite acceptable. But many workers did not have a clear conception of the nature of the disease process. They had not separated in their minds the concept "disease" and the concept "parasite." The word "germ" had acquired a fuzzy meaning. Henle sees through all of this confusion and strikes right to the core of the problem: "The disease inducer or causer reproduces itself in the sick body and at the end of the disease is eliminated from the body. . . ."

The reader should note how Henle was influenced by the contemporary work of Schwann and Cagniard-Latour (see pages 16 and 20) showing that the alcoholic fermentation was caused by a living organism. This is an excellent example of how work in one field of research may influence thought in other related fields. Probably the most important aspect of Henle's work is his statement that it would be necessary to separate a suspected microorganism from the disease process and culture it, thus showing that it had an independent existence. Although he does not carry this one step further, the obvious thing would then be to reintroduce cells of this microorganism into a healthy host and reproduce the disease again. These requirements were formalized by Robert Koch in his paper of 1884 (see page 116). It is interesting that Henle was one of Koch's teachers at Göttingen.

Lecture on the genesis of puerperal fever (childbed fever)

1850 • Ignaz Semmelweis

Semmelweis, Ignaz P. 1850. Vortrag über die Genesis des Puerperalfiebers. *Protokoll der allgemeinen Versammlung der k. k. Gesellschaft der Aerzte zu Wien,* vom 15 May 1850.

HERR DR. SEMMELWEIS, EMERITUS Assistant at the first Obstetrical Clinic, developed his ideas concerning the etiology of puerperal fever (childbed fever). He brought forth numerical data from the records of the clinic to show that, since the year 1839, since which time there have been two separate clinics, one for the instruction of obstetricians and the other in which midwives are responsible, the first of these clinics has had over four times as many deaths in the women due to puerperal fever as in the second clinic. This is irrespective of the fact that the admittance of the women to each of these two clinics alternates every 24 hours, so that it is possible to conclude that the increase in cases of childbed fever in the first clinic is not due to a general epidemic influence but is unique and localized in the first clinic, so that it is probable that endemic influences in the first clinic are responsible. Dr. Semmelweis then listed a number of ordinary endemic causes which might have been responsible, like the crowding of the wards, the saturation through the years of the locality with the miasma of puerperal fever, the more frequent and supposedly cruder examinations of the obstetricians, the frequently common fears of the pregnant women when they discover they have been brought to this clinic for parturition. He indicated that none of these can explain the increased proportion of puerperal fever in the first clinic, and moreover, that many of these presumed causes also exist to a greater extent in the second clinic. These considerations, as well as the pathological-anatomical findings in childbed fever, which show the greatest similarity to a type of pyemia which affects anatomists and surgeons who wound themselves while examining corpses and impregnate into the freshly damaged area the putrefying organic matter of the corpse, have led Herr Dr. Semmelweis to the conclusion that puerperal fever is also due to the uptake of putrefying organic materials from the uterus into the blood of the mother, which causes a similar pyemic process. Further, the continual new introduction of such materials into the first clinic by the assistants and students of this clinic through the frequent examination of corpses may be a basis for the observed facts. Further other methods for the transfer of the putrefying organic materials may exist, such as from the decomposing placental material, or by the direct touching of a healthy woman immediately after touching a sick woman. Following up on this idea, Dr. Semmelweis introduced the practice that every student or other person making an examination should wash his hands carefully in a solution of chloride of lime after the examination of any patient. This was to destroy as much as possible any putrefying organic atoms adhering to the fingers, as well as to remove the odor of these completely. This procedure met with a remarkable success, and for over three years now the death rate, which was previously 8.3% in the first clinic, has sunk to 2.3%, which rate is similar to that in private practice as well as in other obstetrical clinics. These conclusions are supported by the fact that even in the midwife clinic, the death rate increased when there were assistants there making examinations on corpses. Also, before the establishment of the general hospital (1784) until the establishment of a permanent facility for the study of pathological anatomy, there was no extensive epidemic of puerperal fever, and the death rate was below 1%. From this last time until May 1, 1847, that is, until the introduction of the chloride of lime handwashing procedure, during the time that the study of pathological anatomy flourished, the death rate in both clinics was 5.7%. But from the introduction of the handwashing procedure until April of this year, the death rate has reduced to 2.2%. Further grounds for the endemic character of childbed fever are that outside of the maternity hospitals it is not so extensive, it is not influenced by the seasons, and cannot be considered an epidemic disease since it can also arise from traumatic injury. Also, in animals, it only shows itself sporadically, so that it can be produced artificially in the latter.

Puerperal fever, as a phenomenon considered from the above facts, is therefore neither a contagious nor a specific disease, but develops when an animal-organic substance in the process of putrefying, whether from a sick person or from a cadaver, is taken up into the blood of the woman who has just given birth, and causes in her the puerperal (pyemic) disintegration of the blood, with accompanying exudation and metastasis. These substances are transmitted to the female body by the finger of the examiner, or through the use of instruments impregnated with the material, or through the penetration to the uterine cavity of air containing the putrefying materials after parturition. Therefore, the prevention of this disease is possible through the cleansing of the fingers, the utensils, and the air, after which only isolated cases of puerperal fever are found, such as occur after pieces of the placenta or decidua are left behind to putrefy, or through the abrasion and contusion of the opening of the uterus, etc. The place where the absorption takes place has been indicated by the speaker as that area directly above the inner uterine opening, which can be reached by the finger, which during the pregnancy is covered by the amniotic membrane. In this area the mucous membrane is broken, so that it is suitable for absorption. . . . The uterine cavity is most accessible during the first and second deliveries, and it is also in these that examinations are most frequently undertaken. In this way it can be explained that it is not only the mother, but also usually the child which dies, due to the development of the disease because of the large amount of exudation. This also explains why child-bearing women who, because of delayed delivery must spend two or three days in the delivery area, are the ones who most often succumb to this terrible disease.

At the end of this lecture, Dr. Semmelweis indicated that at a future meeting of the society he would clear up the objections which the Doctors Scanzoni and Seyfert have advanced against his earlier opinions on puerperal fever which were presented to the Imperial Academy of Science by Prof. Skoda in the *Prague Quarterly*. Further, Herr President Prof. Rokitansky asked the society, if they were not in general inclined to have a detailed discussion of this question at the next meeting, in view of the great importance of the subject. His proposal was accepted.

Comment

Childbed fever is due to infection with a hemolytic Streptococcus, which usually reproduces rapidly, causing a serious sepsis. It is quite often a fatal disease. It occurs usually in mothers just after they have given birth, hence its name. Because the symptoms are characteristic, it was possible for Semmelweis to make detailed observations on the frequency of the disease. He had the good fortune of having two separate maternity clinics under his supervision. Because the rate of childbed fever was much higher in one clinic than the other, he was encouraged to perform an analysis of these two clinics to see how they differed. After examining a number of possible reasons, he concluded that the clinic with the highest disease incidence was also the clinic in which obstetricians were instructed. He made the inference that the student obstetricians, in the process of making frequent examinations of various women, moving from one to another without previously cleansing their hands, were transmitting whatever it was that was causing the disease. If this inference were valid, it should be possible to reduce the incidence of childbed fever in this clinic by requiring the student obstetricians and physicians making examinations to wash their hands after each observation in a solution of lime chloride. This simple procedure was sufficient to reduce the incidence of childbed fever in this clinic to the same level as existed in the other clinic.

In this work Semmelweis had the bare requirements for a controlled experiment. He had two clinics, one of which differed from the other. Because he had two clinics at his disposal, he could make the observations he did, and reach conclusions which turned out to be valid. In this respect he had a unique opportunity to make a contribution to medicine. It isn't often that a physician is able to institute such a simple experiment. Usually he feels that he is not justified in withholding a suspected cure from any patient who may need it. He has a moral obligation to each patient to give him the best available treatment. Experimentation on humans can only be done when many ethical requirements are met. Semmelweis was able to meet these requirements, and in so doing, was able to show quite clearly that the physicians themselves were the ones who were transmitting the disease. This was a conclusion which many physicians refused to accept, and Semmelweis found himself despised and reviled by many of his colleagues. He lost his job because of his insistence that he was right. It was only some years later that his theories became generally accepted and put into practice by all.

Semmelweis was not a microbiologist and had little conception of the relationships of microorganisms to disease. But his observations were carefully made and accurately documented, and they had an influence on other workers who were more closely concerned with the possibility of living organisms causing disease. In this way, Semmelweis's paper was a milestone in the history of microbiology.

On a new method of treating compound fracture, abscess, and so forth; with observations on the conditions of suppuration

1867 • Joseph Lister

Lister, Joseph. 1867. On a new method of treating compound fracture, abscess, etc. With observations on the conditions of suppuration. *Lancet,* Vol. 1, pages 326, 357, 387, 507; Vol. 2, page 95.

ON COMPOUND FRACTURE

THE FREQUENCY OF DISASTROUS CONSEquences in compound fracture, contrasted with the complete immunity from danger to life or limb in simple fracture, is one of the most striking as well as melancholy facts in surgical practice.

If we inquire how it is that an external wound communicating with the seat of fracture leads to such grave results, we cannot but conclude that it is by inducing, through access of the atmosphere, decomposition of the blood which is effused in greater or less amount around the fragments and among the interstices of the tissues, and, losing by putrefaction its natural bland character, and assuming the properties of an acrid irritant, occasions both local and general disturbance.

We know that blood kept exposed to the air at the temperature of the body, in a vessel of glass or other material chemically inert, soon decomposes; and there is no reason to suppose that the living tissues surrounding a mass of extravasated blood could preserve it from being affected in a similar manner by the atmosphere. On the contrary, it may be ascertained as a matter of observation that, in a compound fracture, twenty-four hours after the accident the coloured serum which oozes from the wound is already distinctly tainted with the odour of decomposition, and during the next two or three days, before suppuration has set in, the smell of the effused fluids becomes more and more offensive.

This state of things is enough to account for all the bad consequences of the injury.

The pernicious influence of decomposing animal matter upon the tissues has probably been underrated, in consequence of the healthy state in which granulating sores remain in spite of a very offensive condition of their discharges. To argue from this, however, that fetid material would be innocuous in a recent wound would be to make a great mistake. The granulations being composed of an imperfect form of tissue, insensible and indisposed to absorption, but with remarkably active cell-development, and perpetually renovated as fast as it is destroyed at the surface, form a most admirable protective layer, or living plaster. But before a raw surface has granulated, an acrid discharge acts with unrestrained effect upon it, exciting the sensory nerves, and causing through them both local inflammation and general fever, and also producing by its caustic action a greater or less extent of sloughs, which must be thrown off by a corresponding suppuration, while there is at the same time a risk of absorption of the poisonous fluids into the circulation.

This view of the cause of the mischief in compound fracture is strikingly corroborated by cases in which the external wound is very small. Here, if the coagulum at the orifice is allowed to dry and form a crust, as was advised by John Hunter, all bad consequences are probably averted, and, the air being excluded, the blood beneath becomes organized and absorbed, exactly as in a simple fracture. But if any accidental circumstance interferes with the satisfactory formation of the scab, the smallness of the wound, instead of being an advantage, is apt to prove injurious, because, while decomposition is permitted, the due escape of foul discharges is prevented. Indeed, so impressed are some surgeons with the evil which may result from this latter cause, that, deviating from the excellent Hunterian practice, they enlarge the orifice with the knife in the first instance and apply fomentations, in order to mitigate the suppuration which they render inevitable.

Turning now to the question how the atmosphere produces decomposition of organic substances, we find that a flood of light has been thrown upon this most important subject by the philosophic researches of M. Pasteur, who has demonstrated by thoroughly convincing evidence that it is not to its oxygen or to any of its gaseous constituents that the air owes this property, but to the minute particles suspended in it, which are the germs of various low forms of life, long since revealed by the microscope, and regarded as merely accidental concomitants of putrescence, but now shown by Pasteur to be its essential cause, resolving the complex organic compounds into substances of simpler chemical constitution, just as the yeast plant converts sugar into alcohol and carbonic acid. . . .

Applying these principles to the treatment of compound fracture, bearing in mind that it is from the vitality of the atmospheric particles that all the mischief arises, it appears that all that is requisite is to dress the wound with some material capable of killing these septic germs, provided that any substance can be found reliable for this purpose, yet not too potent as a caustic.

In the course of the year 1864 I was much struck with an account of the remarkable effects produced by carbolic acid upon the sewage of the town of Carlisle, the admixture of a very small proportion not only preventing all odour from the lands irrigated with the refuse material, but, as it was stated, destroying the entozoa which usually infest cattle fed upon such pastures.

My attention having for several years been much directed to the subject of suppuration, more especially in its relation to decomposition, I saw that such a powerful antiseptic was peculiarly adapted for experiments with a view to elucidating that subject, and while I was engaged in the investigation the applicability of carbolic acid for the treatment of compound fracture naturally occurred to me.

My first attempt of this kind was made in the Glasgow Royal Infirmary in March 1865, in a case of compound fracture of the leg. It proved unsuccessful, in consequence, as I now believe, of improper management; but subsequent trials have more than realized my most sanguine anticipations.

Carbolic acid proved in various ways well adapted for the purpose. It exercises a local sedative influence upon the sensory nerves; and hence is not only almost painless in its immediate action on a raw surface, but speedily renders a wound previously painful entirely free from uneasiness. When employed in compound fracture its caustic properties are mitigated so as to be unobjectionable by admixture with the blood, with which it forms a tenacious mass that hardens into a dense crust, which long retains its antiseptic virtue, and also has other advantages, as will appear from the following cases, which I will relate in the order of their occurrence, premising that, as the treatment has been gradually improved, the earlier ones are not to be taken as patterns.

[Case histories omitted. See original references.]

Comment

Here again we see how a physician can arrive at significant deductions from observations made in the clinic. And we also see the impact of Pasteur's seemingly fundamental work on a practical medical problem. Lister is able to bring his own observations on compound fractures into clear focus, once he is aware of Pasteur's observations on the presence of microorganisms in the air. But it was essential that Lister could go one step further. He had to discover something which could be applied to compound fractures which would be harmful to the microorganisms that were present. It would have been impossible for him to set up experiments in patients to this end, since he would have had to test blindly a large number of available substances, and ethical considerations would have prevented him from doing this. His observation on the effect of carbolic acid on sewage gave him a convenient short-cut. Because he theorized that fermentation, putrefaction of sewage and decomposition of open wounds were all due to similar organisms, it was possible for him to bring these seemingly unrelated observations together and devise a procedure for handling compound fractures successfully. It was then a simple matter for him to extend these procedures to the practice of surgery, as will be seen in the next article.

On the antiseptic principle in the practice of surgery

1867 • Joseph Lister

Lister, Joseph. 1867. On the antiseptic principle in the practice of surgery. *British Medical Journal*, Vol. 2, page 246. Read before the British Medical Association in Dublin, Aug. 9, 1867.

IN THE COURSE OF AN EXTENDED investigation into the nature of inflammation, and the healthy and morbid conditions of the blood in relation to it, I arrived, several years ago, at the conclusion that the essential cause of suppuration in wounds is decomposition, brought about by the influence of the atmosphere upon blood or serum retained within them, and, in the case of contused wounds, upon portions of tissue destroyed by the violence of the injury.

To prevent the occurrence of suppuration, with all its attendant risks, was an object manifestly desirable; but till lately apparently unattainable, since it seemed hopeless to attempt to exclude the oxygen, which was universally regarded as the agent by which putrefaction was effected. But when it had been shown by the researches of Pasteur that the septic property of the atmosphere depended, not on the oxygen or any gaseous constituent, but on minute organisms suspended in it, which owed their energy to their vitality, it occurred to me that decomposition in the injured part might be avoided without excluding the air, by applying as a dressing some material capable of destroying the life of the floating particles.

Upon this principle I have based a practice of which I will now attempt to give a short account.

The material which I have employed is carbolic or phenic acid, a volatile organic compound which appears to exercise a peculiarly destructive influence upon low forms of life, and hence is the most powerful antiseptic with which we are at present acquainted.

The first class of cases to which I applied it was that of compound fractures, in which the effects of decomposition in the injured part were especially striking and pernicious. The results have been such as to establish conclusively the great principle, that all the local inflammatory mischief and general febrile disturbance which follow severe injuries are due to the irritating and poisoning influence of decomposing blood or sloughs. For these evils are entirely avoided by the antiseptic treatment, so that limbs which otherwise would be unhesitatingly condemned to amputation may be retained with confidence of the best results.

In conducting the treatment, the first object must be the destruction of any septic germs which may have been introduced into the wound, either at the moment of the accident or during the time which has since elapsed. This is done by introducing the acid of full strength into all accessible recesses of the wound by means of a piece of rag held in dressing-forceps and dipped in the liquid. This I did not venture to do in the earlier cases; but experience has shown that the compound which carbolic acid forms with the blood, and also any portions of tissue killed by its caustic action, including even parts of the bone, are disposed of by absorption and organization, provided they are afterwards kept from decomposing. We are thus enabled to employ the antiseptic treatment efficiently at a period after the occurrence of the injury at which it would otherwise probably fail. Thus I have now under my care in the Glasgow Infirmary a boy who was admitted with compound fracture of the leg as late as eight and a half hours after the accident, in whom nevertheless all local and constitutional disturbance was avoided by means of carbolic acid, and the bones were firmly united five weeks after his admission.

The next object to be kept in view is to guard effectually against the spreading of decomposition into the wound along the stream of blood and serum which oozes out during the first few days after the accident, when the acid originally applied has been washed out, or dissipated by absorption and evaporation. This part of the treatment has been greatly improved during the last few weeks. The method which I have hitherto published consisted in the application of a piece of lint dipped in the acid, overlapping the sound skin to some extent, and covered with a tin cap, which was daily raised in order to touch the surface of the lint with the antiseptic. This method certainly succeeded well with wounds of moderate size; and, indeed, I may say that in all the many cases of this kind which have been so treated by myself or my house surgeons, not a single failure has occurred. When, however, the wound is very large, the flow of blood and serum is so profuse, especially during the first twenty-four hours, that the antiseptic application cannot prevent the spread of decomposition into the interior unless it overlaps the sound skin for a very considerable distance, and this was inadmissible by the method described above, on account of the extensive sloughing of the surface of the cutis which it would involve. This difficulty has, however, been overcome by employing a paste composed of common whitening (carbonate of lime) mixed with a solution of one part of carbolic acid in four parts of boiled linseed oil, so as to form a firm putty. This application contains the acid in too dilute a form to excoriate the skin, which it may be made to cover to any extent that may be thought desirable, while its substance serves as a reservoir of the antiseptic material. So long as any discharge continues, the paste should be changed daily; and, in order to prevent the chance of mischief occurring during the process, a piece of rag dipped in the solution of carbolic acid in oil is put on next the skin, and maintained there permanently, care being taken to avoid raising it along with the putty. This rag is always kept in an antiseptic condition from contact with the paste above it, and destroys any germs that may fall upon it during the short time that should alone be allowed to pass in the changing of the dressing. The putty should be in a layer about a quarter of an inch thick, and may be advantageously applied rolled out between two pieces of thin calico, which maintain it in the form of a continuous sheet, that may be wrapped in a moment round the whole circumference of a limb, if this be thought desirable, while the putty is prevented by the calico from sticking to the rag which is next the skin. When all discharge has ceased, the use of the paste is discontinued, but the original rag is left adhering to the skin till healing by scabbing is supposed to be complete. I have at present in the hospital a man with severe compound fracture of both bones of the left leg, caused by direct violence, who, after the cessation of the sanious discharge under the use of the paste, without a drop of pus appearing, has been treated for the last two weeks exactly as if the fracture were a simple one. During this time the rag, adhering by means of a crust of inspissated blood collected beneath it, has continued perfectly dry, and it will be left untouched till the usual period for removing the splints in a simple fracture, when we may fairly expect to find a sound cicatrix beneath it. . . .

The next class of cases to which I have applied the antiseptic treatment is that of abscesses. Here, also, the results have been extremely satisfactory, and in beautiful harmony with the pathological principles indicated above. The pyogenic membrane . . . forms pus, not from any inherent disposition to do so, but only because it is subjected to some preternatural stimulation. In an ordinary abscess, whether acute or chronic, before it is opened, the stimulus which maintains the suppuration is derived from the presence of the pus pent up within the cavity. When a free opening is made in the ordinary way, this stimulus is got rid of; but when the atmosphere gaining access to the contents, the potent stimulus of decomposition comes into operation, and pus is generated in greater abundance than before. But when the evacuation is effected on the antiseptic principle, the pyogenic membrane, freed from the influence of the former stimulus without the substitution of a new one, ceases to suppurate . . ., furnishing merely a trifling amount of clear serum, and, whether the opening be dependent or not, rapidly contracts and coalesces. At the same time any constitutional symptoms previously occasioned by the accumulation of the matter are got rid of without the slightest risk of the irritative fever or hectic hitherto so justly dreaded in dealing with large abscesses.

In order that the treatment may be satisfactory, the abscess must be seen before it has opened. Then, except in very rare and peculiar cases, there are no septic organisms in the contents, so that it is needless to introduce carbolic acid into the interior. Indeed, such a proceeding would be objectionable, as it would stimulate the pyogenic membrane to unnecessary suppuration. All that is necessary is to guard against the introduction of living atmospheric germs from without, at the same time that free opportunity is afforded for the escape of discharge from within. . . .

It would carry me far beyond the limited time which, by the rules of the Association, is alone at my disposal, were I to enter into the various applications of the antiseptic principle in the several special departments of surgery.

There is, however, one point more that I cannot but advert to—namely, the influence of this mode of treatment upon the general healthiness of a hospital. Previously to its introduction, the two large wards in which most of my cases of accident and of operation are treated were amongst the unhealthiest in the whole surgical division of the Glasgow Royal Infirmary, in consequence, apparently, of those wards being unfavourably placed with reference to the supply of fresh air; and I have felt ashamed, when recording the results of my practice, to have so often to allude to hospital gangrene or pyaemia. It was interesting, though melancholy, to observe that, whenever all, or nearly all, the beds contained cases with open sores, these grievous complications were pretty sure to show themselves; so that I came to welcome simple fractures, though in themselves of little interest either for myself or the students, because their presence diminished the proportion of open sores among the patients. But since the antiseptic treatment has been brought into full operation, and wounds and abscesses no longer poison the atmosphere with putrid exhalations, my wards, though in other respects under precisely the same circumstances as before, have completely changed their character; so that during the last nine months not a single instance of pyaemia, hospital gangrene, or erysipelas has occurred in them.

As there appears to be no doubt regarding the cause of this change, the importance of the fact can hardly be exaggerated.

Comment

Lister's use of carbolic acid had a tremendous influence on the practice of surgery. It made possible the introduction of many surgical procedures that had previously been impossible. However, carbolic acid is not a good antiseptic, since it is fairly toxic to the host. It has been succeeded by a large number of other agents, and now we have available antibiotics and chemical antiseptics to serve almost every purpose in the practice of medicine. Lister opened up a whole new field for drug research. The importance of his work cannot be too highly stressed.

The etiology of anthrax, based on the life history of *Bacillus anthracis*

1877 · Robert Koch

Koch, Robert, 1877. Untersuchungen über Bakterien V. Die Aetiologie der Milzbrand-Krankheit, begründet auf die Entwicklungsgeschichte des Bacillus Anthracis. *Beiträge zur Biologie der Pflanzen,* Vol. 2, No. 2, pages 277–310.

I. INTRODUCTION

SINCE THE DISCOVERY OF ROD-SHAPED bodies in the blood of animals dying of anthrax, there has been much effort directed to attempts to prove that these rod-shaped bodies are responsible for the transmissability of this disease as well as for the sporadic appearance of it. These studies have sought to determine whether these bodies are the unique contagium of anthrax. Recently Davaine has carried out a number of inoculation experiments with fresh and dried blood containing the rods and has stated decisively that these rods are bacteria, and that only in the presence of these bacteria can a fresh case of anthrax be produced. The lack of proof of the direct transmission of the anthrax disease in man and animals is due to the ability of the bacteria to remain alive for a long time in dry conditions and to be transmitted through the air by insects and the like. It seems here that the mode of transmission of anthrax has been explained.

Nevertheless, these ideas of Davaine have found many opponents. Several workers have obtained experimental anthrax by inoculating blood containing bacteria, but have been unable to show the presence of bacteria in the blood of the diseased animals. Others have been able to induce anthrax by inoculation with blood which could not be shown to contain bacteria, but the diseased animals then had bacteria in their blood. Others have noted that anthrax is not derived solely from a contagium which is transmitted above ground, but that this disease is related in some way with conditions of the soil. . . .

These experiences cannot be explained by the hypothesis of Davaine, and because of this, many people feel that bacteria are of no significance for anthrax.

Since I have had the opportunity several times of examining animals which had died of anthrax, I performed a series of experiments which would clear up the uncertainties in the etiology of anthrax. Through these, I came to the conclusion that the theory of Davaine concerning the transmission of anthrax is only partly correct.

I could show that the rods in the anthrax blood were not so resistant as Davaine had believed. As I will show later, the blood, which contains only rods, keeps its ability to induce anthrax on inoculation only a few weeks in the dry state, and only a few days when moist. How is it possible then for an organism which is so easily destroyed to maintain itself as a dormant contagium for a year in soil and throughout the winter? If bacteria are really the cause of anthrax, then we must hypothesize that they can go through a change in life history and assume a condition which will be resistant to alternate drying and moisture. What is more likely, and what has already been indicated by Prof. Cohn is that the bacteria can form spores which possess the ability to reform bacteria after a long or short resting period.

All of my experiments were designed to discover this developmental stage of the anthrax bacterium. After many unsuccessful experiments, I was finally able to reach this goal, and thus to find a basis for the etiology of anthrax.

Since the life history of the anthrax bacterium offers not only botanical interest but also much light on the heretofore uncertain etiology of the soil-related infectious diseases, I am publishing now the most important results of my experiments, although my work is still in progress.

II. LIFE HISTORY OF *Bacillus anthracis*

It has not been possible for me to observe the multiplication of the bacteria directly in the animal. But it can be inferred that this occurs from the inoculation experiments which follow. I have used the mouse as my experimental animal, as it is simple to use. . . . In most experiments I inoculated them at the base of the tail, where the skin is loose and covered with long hair. . . . I have made a large number of inoculations in this way, using fresh anthrax material, and in every case I have had a positive result, and I believed therefore that the success of the inoculation could be used as an indication of the life or death of the bacilli inoculated. I will show through later experiments that this idea is true.

Partly in order to always have available fresh material, and partly to discover if the bacilli would change into another form after a certain number of generations, I inoculated a number of mice in series, one from the other, each time using a mouse which had just died as a source of the spleenic material. The longest series of mice treated in this way was twenty, which therefore represented that many generations of the bacilli.* In all animals the results were the same. The spleen was markedly swollen and contained a large number of transparent rods which were very similar in appearance and were immotile and without spores. This same type of bacillus could be found also in the blood, but not in so great a number as in the spleen. In these experiments it was shown, therefore, that a small number of bacilli could always develop into a significant mass of individuals of the same type . . . which appeared to reproduce by growing in length and then splitting after they had reached about twice the length of the individual bacilli. These results also indicate that it is highly unlikely that the bacilli would go through some change in form if a longer series of inoculations were made, and therefore it is unlikely that there is ultimately some alternation of generations. . . .

It will take us too far afield to consider whether or not the actual cause of the death of the animals is due to the production of carbon dioxide in the blood through the rapid growth of the bacilli there, or, what seems more probable, that death is due to a metabolic product produced by the parasite through its utilization of proteins as nutrients, and that this metabolic product is poisonous to the animal. . . .†

[To study the life history of the bacilli away from the animal,] a drop of fresh beef serum or aqueous humor from the eye of a cow was placed on a microscope slide. Then a small piece of spleen which contained bacteria and which had been freshly removed from an infected animal was placed in this and a cover glass placed on top. The microscope slide was then placed in a moist chamber to keep the liquid from evaporating, and this was then placed in an incubator. . . .

These preparations were incubated for 15–20 hours at 35–37°. At the end of this time, in the middle of the preparation between the tissue cells could be seen many unaltered bacilli, although in smaller numbers than in fresh preparations. However, away from the tissue in the fluid, one could see bacilli which were 3–8 times longer and showed shallow bends and curvatures (Fig. 2). ‡ The closer to the edge of the cover glass, the longer the filaments, and these finally reached a size which was a hundred or more times the length of the original bacilli (Fig. 3). Many of these long filaments had lost their uniform structure and transparent appearance, and their contents had become finely granulated with the regular appearance of strongly

* [We know now that the number of generations of the bacteria would be much greater than twenty.]

† [Such a metabolic product, known today as a toxin, is usually associated in some way with most infectious diseases, including anthrax.]

‡ [This plate also contains the figures for the paper of Cohn, 1876: "Studies on the biology of the bacilli." See page 49 for the text of this paper.]

light-refracting grains (Fig. 3a). The filaments which lay right at the edge of the cover glass, where the gas exchange with the nutrient fluid was the best, showed the most extensive development. They contained completely formed spores which were imbedded in the substance of the filaments at regular distances and were somewhat oval, strongly light-refracting bodies. In this form the filaments revealed a remarkable appearance, which can best be compared with a string of pearls.

Many filaments had already lost their spores, which can be seen between them as small, free clusters (Fig. 4b). In favorable preparations it is possible to see all of the stages from short bacillus rods, through long, sporulating filaments, to free spores, and this is proof that the latter arises from the former. . . . [Because these spores seemed to form most frequently at the edge, it occurred to Koch that they might not actually come from the Bacillus, but be due to contamination from the air, since his preparations were not pure, and he had observed micrococci and bacterium types from time to time. So he decided that the only way to be sure would be to observe the spore formation actually take place.]

Although I had imagined that such an experiment would be very difficult

Plate—Koch and Cohn

Figs. 1–7. ANTHRAX BACILLUS (*Bacillus Anthracis*). *Fig. 1.* Anthrax bacilli from the blood of a guinea pig. The bacilli appear as transparent rods, occasionally with beginnings of division, or bent. (*a*) white blood cells. (*b*) red blood cells. *Fig. 2.* Anthrax bacilli from the spleen of a mouse, after three hours in a drop of aqueous humor. The bacteria have lengthened into filaments . . . *Fig. 3.* From the same preparation as Fig. 2, after ten hours. The bacilli have grown to long filaments, which often are intertwined. (*a*) in isolated filaments strongly refractile bodies appear regularly spaced. *Fig. 4.* The same culture as Fig. 3, after 24 hours. (*a*) oval spores appearing like beads on a string have developed in the filaments. (*b*) many filaments are undergoing decomposition and releasing free spores, which occur singly or in clumps. *Fig. 5.* Spore germination. Fig. 5a and 5b at different magnifications. The spores elongate into cylindrical bodies, with the refractile area remaining at one pole. This body becomes smaller, breaks up into 2 or more parts, and finally disappears completely. *Fig. 6.* Diagram of the method of culture of anthrax bacilli. The fluid containing the bacteria is in the hollow of a hanging drop slide, covered with a cover glass which is ringed with olive oil to prevent evaporation. The slide is placed on a warm stage heated to body temperature, so that observations can be made during incubation. The bacilli are suspended in a drop of fresh aqueous humor. Even with the naked eye it is possible to see the developing masses of filaments. *Fig. 7.* Appearance of the epithelial layer from the skin of a frog, in which a piece of infected spleen from a mouse had been placed. The layer consists of large, nuclear cells (*a*); in occasional cells many short, occasionally bent or crooked *Bacilli* (*b*) have been taken up, which have developed further within the cells and later escaped; (*c*) disrupted cell; (*g*) free spiral *Bacilli;* (*e*) blood cells of the frog. In addition, unaltered *Bacilli* are visible. *Figs. 8–10.* HAY BACILLUS (*Bacillus subtilis*). *Fig. 8. Bacilli* in long parallel rows of filaments, which had formed an iridescent film on the surface of a boiled hay infusion after 24–48 hours. Between the parallel rows are seen short motile rods, or rods in the process of elongating. *Fig. 9.* Spore formation in the segmenting filaments of the hay *Bacillus* after 3 days. *Fig. 10. Bacillus* filaments encased in slime . . . at the left end of the figure can be seen the beginning of the formation of chains of spores in the filaments. *Fig. 11.* Portion of a mass surrounded by slime, in which the formation of chains of spores is complete, and the single *Bacillus* filaments have become indistinct; the spores however are still arranged in parallel rows and are bound together by the slime. MAGNIFICATIONS, Figs. 1–7, 8 and 10, 650X; Figs. 5b and 9, 1650X; Fig. 11, 900X. [The handwriting on the figure was placed there later by some interested reader.]

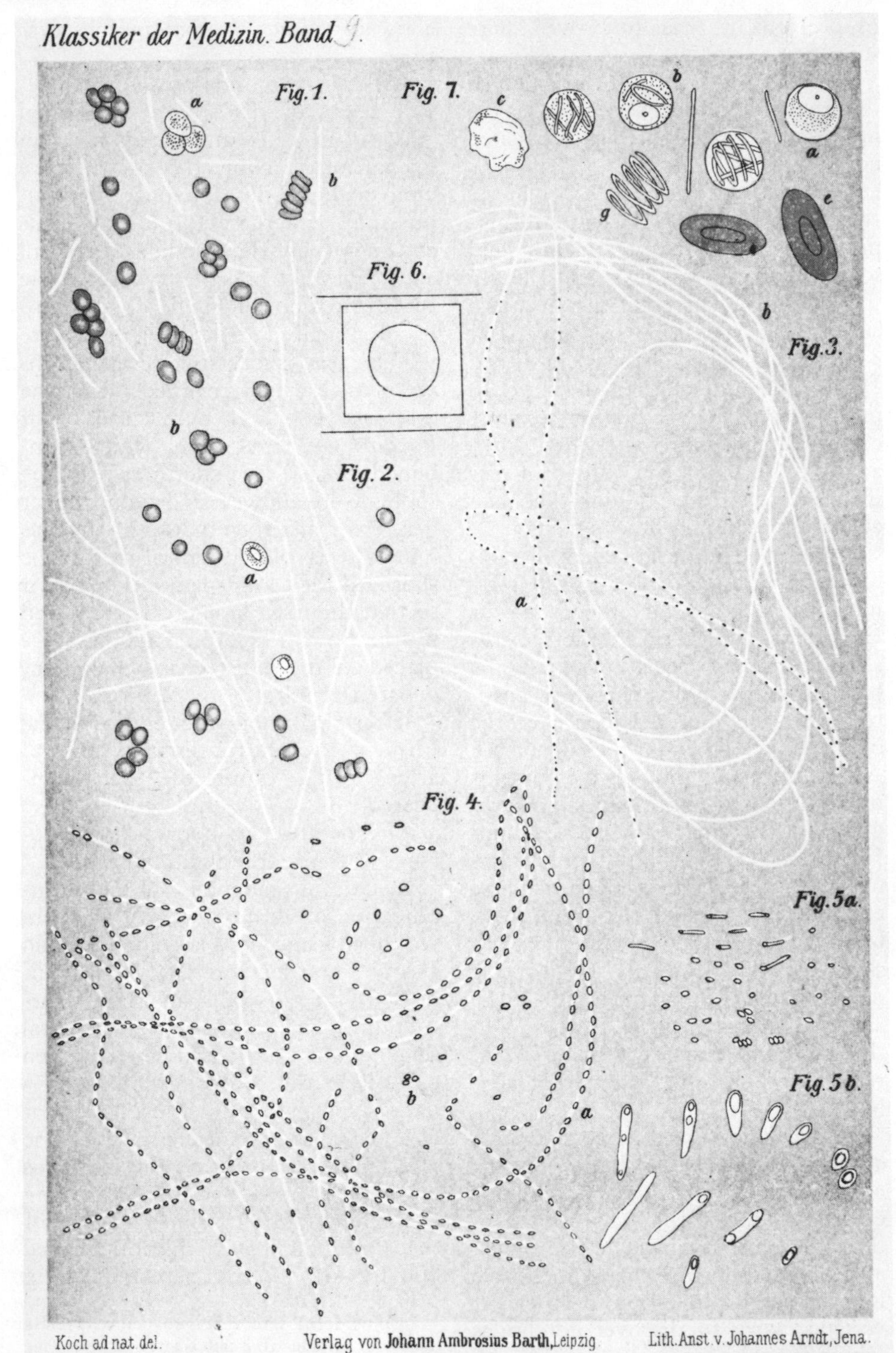

Klassiker der Medizin. Band 9.
Fig. 1.
Fig. 7.
Fig. 6.
Fig. 2.
Fig. 3.
Fig. 4.
Fig. 5a.
Fig. 5b.
a
b
c
e
g
Koch ad nat del.
Verlag von Johann Ambrosius Barth, Leipzig
Lith. Anst. v. Johannes Arndt, Jena.

to perform, it actually proved quite simple. . . .

[The preparations were so arranged that they could be observed under the microscope while being incubated continuously.]

Observations every 10–20 minutes revealed that the bacilli at the beginning were somewhat thicker and seemed to be swollen and hardly showed any changes in the first two hours. Then they began to grow. After 3–4 hours they had already lengthened 10–20 times; they began to curve, to push against each other or to cross each other and make a network. After a few more hours the individual filaments were already so long that they covered several microscope fields. . . .

If the free end of a filament was observed continuously for 15–20 minutes, it was quite easy to observe its lengthening and perceive the remarkable spectacle of actually watching the bacillus grow. It was therefore possible to obtain direct evidence of the further development of these filaments. After only 10–15 minutes the contents of the strongest and most luxuriantly growing filaments were finely granular, and soon the small, refractile grains were cut off in regular sequence. These enlarged in the space of several more hours into the strongly refractile, oval-shaped spores. Gradually the filaments disintegrated, fragmented at the ends, and the spores became free. . . . In this condition the preparations could remain for weeks without changing. . . .

Observations were made to obtain a complete picture of the life history of the *Bacillus anthracis* and to discover whether the spores passed through some intermediate form, such as a swarm spore, or passed directly into a bacillus. In order to do this it was necessary to discover conditions which would permit the spore to develop into the bacillus which would allow for direct microscopic observations.

All efforts to obtain the further development of the spores in distilled water or in well water failed. In serum or aqueous humor, the results were equivocal; bacilli developed without question, and these formed filaments and spores, but their number was small and it was not possible to observe the transformation of single spores into bacilli. Finally I arrived at a procedure which was successful. Preparations were used which revealed under the microscope only a pure culture * of *Bacillus anthracis*, and which contained mostly free spore masses. The spores were allowed to dry on a cover glass . . . and then a drop of aqueous humor was placed on a microscope slide and the cover glass was laid on it so that the mass of spores was wetted by the fluid. These preparations were placed in the moist chamber and incubated at 35°.

After a half hour the remains of the filaments began to disintegrate, and after 1½ to 2 hours they had disappeared.

Already after 3–4 hours the development of the spores could be seen. . . .

By careful examination at high magnification, it appeared that each spore was oval-shaped and was imbedded in a round transparent mass, which appeared like a small, light ring surrounding the spore. The spherical shape of this ring could be easily seen by rolling the spore in various positions. This material first lost its spherical shape, lengthened itself on one side in the direction of the long axis of the spore and became like a long oval. The spore remained in one of the poles of the cylindrical shaped body. Soon the transparent covering

* [This term will receive a more precise meaning in one of Koch's later papers (see page 101). It is not at all certain that he really had a pure culture at this time.]

became longer and filamentous, and at the same time the spore began to lose its strongly refractile characteristics. It became quickly pale and smaller, broke apart into many pieces, and finally completely disappeared. In Fig. 5 is pictured a mass of spores showing the conversion to filaments.

Later I was able to observe in the same preparation and same drop of aqueous humor the appearance of bacilli from the spores and then later a second generation of spore-containing filaments. . . .

It may be assumed that when these spores in some way reach the blood stream of a sensitive animal, a new generation of bacilli will be produced. In order to prove this assumption, the following experiments were performed. . . .

By the inoculation of mice with material rich in spores, or with material with few spores, the interesting fact was discovered, that in the first case, with many spores, the mice died after 24 hours, while in the latter case, the mice did not die from anthrax until after three or four days. I have repeated these experiments many times. Such substances containing spores were dried and allowed to stand for a while. When moistened with water and injected, they had not lost their ability to produce anthrax. . . .

On the other hand I have inoculated mice with spore masses which had come from cultures in glass cells and which I had ascertained by microscopic examination to have been derived from completely pure cultures of *Bacillus anthracis*, and every time the inoculated animals died of anthrax. It follows, therefore, that only a species of Bacillus is able to cause this specific disease, while other schizophytes have no effects or cause entirely different diseases when inoculated. . . .

Comment

This was the first proof that a specific microorganism could cause a specific disease in an animal, and although the proof was not perfect, it was good enough for most people. This work is all the more remarkable when it is recalled that Koch was a mere country doctor, with no formal research training. It is not exactly clear what prompted him to begin work on anthrax, but the choice was quite fortuitous. The disease occurs most often in animals, less frequently in man, and produces characteristic symptoms. The organism can be transferred readily to mice by inoculation, and this made it possible for Koch to study the disease in the laboratory. This is an extremely important point, since laboratory study can be carried out under reproducible conditions, and avoids the complications which beset Semmelweis and Lister. The causal organism of anthrax is a very large bacillus, easily seen under the microscope. In infected animals it occurs in very large numbers in the blood stream, sufficient to be seen by direct examination of blood. The organism has a quite characteristic morphology, making a microscopic identification reasonably certain. In addition, it forms spores. Koch's observation of these spores and of the development of spores from bacilli and bacilli from spores is an important observation in fundamental bacteriology. It confirmed rather nicely the work of Cohn (see page 49) and indicated the medical importance of spore-forming organisms. This was especially noteworthy since Cohn had shown that bacterial spores were highly heat-resistant. Finally, Koch's observations were so thorough and so accurate that his work has remained completely valid up until the present. The success of this work brought Koch to the attention of the German medical world. He soon received the opportunity to move to Berlin, where he was given adequate facilities to further his work. His most important discoveries were yet to come (see pages 96, 101, 109, and 116).

Investigations into the etiology of traumatic infective diseases

1880 • Robert Koch

Koch, Robert. 1880. *Investigations into the etiology of traumatic infective diseases.* Translated by W. Watson Cheyne. London, The New Sydenham Society.

AS AT PRESENT USED, THE TERM "traumatic infective diseases" indicates a group of affections formerly known as traumatic fever, purulent infection, putrid infection, septicaemia, pyaemia, but which were included at a subsequent period (when the view became generally accepted that these diseases were essentially of the same nature) under the title "pyaemic or septicaemic processes." . . .

A considerable number of investigators have advanced the statement that the normal blood and tissues of man and of the lower animals always contain micro-organisms. From this some infer that these organisms are not the cause of the infective disease, but that an abnormal increase in their numbers follows the morbid process, because the fluids of the animal body, when altered by disease, present conditions very favourable for their development. . . . Were it, however, true that bacteria do occur in normal blood, and that the same bacteria,—*e.g.*, micrococci—are found, though in unusual numbers, in organs altered by disease, then the possibility of proving that these micrococci were the cause of the disease would be rendered much more difficult, perhaps indeed quite hopeless. . . .

According to my own experience, the examination of blood, with the view of ascertaining the possible presence of bacteria, is excessively difficult, unless one makes use of the aids to be described afterwards, *viz.*, of staining and suitable illumination. Without the assistance derived from these methods it is in most cases impossible to distingush the bodies, so characteristically described by Reiss, from true micrococci; and I can therefore easily imagine that, according as one wished to find that bacteria were present or absent, the granular constituents of the blood would be regarded as micrococci, or micrococci when present would be regarded as the remains of disintegrated white corpuscles. I have, however, on many occasions examined normal blood and normal tissues by means which prevent the possibility of overlooking bacteria, or of confounding them with granular masses of equal size; and I have never, in a single instance, found organisms. *I have therefore come to the conclusion that bacteria do not occur in the blood, nor in the tissues of the healthy living body either of man or of the lower animals.*

On the other hand, the following objections which have been raised against the assumption that bacteria are the cause of traumatic infective diseases seem to me to be well founded. In order to establish this assertion (that they are the cause of traumatic infective diseases) it would be absolutely necessary *that the presence of bacteria in these diseases be proved without exception*, and further that the conditions as regards their number and distribution be such as to afford a complete explanation of the symptoms. For, if in some cases of a certain form of infective disease bacteria be found, while in others of like nature they are absent . . . then of course nothing remains but to regard the irregular appearance of the bacteria as depending on chance. . . . In other words it is necessary to assume the presence of some other agency. . . .

A third point remains to be urged against the cogency of the facts known as to the occurrence of bacteria. It is this, that morphologically the bacteria found in the most diverse traumatic infective diseases, and also in other infective processes not in any way connected with wounds, are strikingly similar. . . . A number [of cases] may be added in which micrococci, indistinguishable from each other, were found. Such are erysipelas, puerperal fever, mycosis of the navel in newly-born infants, hospital gangrene, intestinal mycosis, endocarditis (with or without acute articular rheumatism), primary infective periostitis, scarlet fever, rinderpest, and pleuropneumonia. It is however, impossible that all these diseases can be produced by one and the same parasite, and we must therefore assume either that the micrococci are in reality always the same, in which case they would be merely associated, as an accidental complication, with the disease enumerated, or that the micrococci—though, on account of their small size, very similar, and indeed apparently the same—are nevertheless different in nature, and consequently capable of giving rise to these diverse results. . . .

If we now look at the facts brought together and the remarks on them, we come to this conclusion, that the frequent discovery of micro-organisms in traumatic infective disease and the experimental investigations made in connection with them render the parasitic nature of these diseases probable; but that a thoroughly satisfactory proof has not yet been furnished, and can only be so when we have succeeded in finding the parasitic micro-organisms in all cases of the disease in question, when we can further demonstrate their presence in such numbers and distribution that all the symptoms of the disease may thus find their explanation, and finally when we have established the existence, for every individual traumatic infective disease, of a micro-organism with well-marked morphological characters.

Is it then possible to fulfil these conditions in any degree? or have we now, as many microscopists assume, reached the limit of the capabilities of our optical appliances?

This question will, indeed, have often enough occurred to every one who has specially devoted himself to the examination of pathogenic bacteria. It has also occupied my attention greatly, and at once forced itself on me when I commenced these general investigations on bacteria, and saw what great advantages might be obtained by a proper use of microscopic aids in recognizing and distinguishing the smallest forms of bacteria with their spores and cilia. . . .

[Koch then proceeds to describe the

methods which he has used for the microscopic examination of blood and tissues. The important points in this are the use of aniline dyes for staining and the use of the Abbé condenser as a light source. He then describes his experiments with various infections in mice and rabbits, in which he has been able to infect these animals by inoculating them, and then examines them for the presence of bacteria. In the diseases studied—progressive destruction of tissue (gangrene) in mice, septicaemia in mice, and spreading abscess, pyaemia, septicaemia, and erysipelas in rabbits—he has been able to show the continual presence of unique types of bacteria whenever he has diseased animals. He mentions also his work with anthrax which showed the same thing.]

CONCLUSIONS

As regards the artificial traumatic infective diseases observed by me, the conditions, which must be established before their parasitic nature can be proved, were completely fulfilled in the case of the first five, but only partially in that of the sixth. For the infection was produced by such small quantities of fluid (blood, serum, pus, etc.) that the result cannot be attributed to a merely chemical poison.

In the materials used for inoculation bacteria were without exception present, and in each disease a different and well-marked form of organism could be demonstrated.

At the same time, the bodies of those animals which died of the artificial traumatic infective diseases contained bacteria in such numbers that the symtoms and the death of the animals were sufficiently explained. Further, the bacteria found were identical with those which were present in the fluid used for inoculation, and a definite form of organisms corresponded in every instance to a distinct disease. . . .

. . . even in the small series of experiments which I was able to carry out, one fact was so prominent that I must regard it as constant, and, as it helps to remove most of the obstacles to the admission of the existence of a *contagium vivum* for traumatic infective diseases, I look on it as the most important result of my work. I refer to the differences which exist between pathogenic bacteria and to the constancy of their characters. A distinct bacteric form corresponds, as we have seen, to each disease, and this form always remains the same, however often the disease is transmitted from one animal to another. Further, when we succeed in reproducing the same disease *de novo* by the injection of putrid substances, only the same bacteric form occurs which was before found to be specific for that disease.

Further, the differences between these bacteria are as great as could be expected between particles which border on the invisible. With regard to these differences, I refer not only to the size and form of the bacteria, but also to the conditions of their growth, which can be best recognized by observing their situation and grouping. I therefore study not only the individual alone, but the whole group of bacteria, and would, for example, consider a micrococcus which in one species of animal occurred only in masses (*i.e.*, in a zooglaea form), as different from another which in the same variety of animal, under the same conditions of life, was only met with as isolated individuals. . . .

As, however, there corresponds to each of the diseases investigated a form of bacterium distinctly characterised by its physiological action, by its conditions of growth, size, and form, which, however often the disease be transmitted from one animal to another, always remains the same and never passes over into other forms, *e.g.*, from the spherical to the rod-shaped, we must in the meantime regard these different forms of pathogenic bacteria as distinct and constant species.

This is, however, an assertion which will be much disputed by botanists, to whose special province this subject really belongs.

Amongst those botanists who have written against the subdivision of bacteria into species, is Nägeli, who says, "I have for ten years examined thousands of different forms of bacteria, and I have not yet seen any absolute necessity for dividing them even into two distinct species."

Brefeld also states that he can only admit the existence of specific forms justifying the formation of distinct species when the whole history of development has been traced by cultivation from spore to spore in the most diverse nutritive fluids.

Although Brefeld's demand is undoubtedly theoretically correct, it cannot be made a *sine qua non* in every investigation on pathogenic bacteria. We should otherwise be compelled to cease our investigations into the etiology of infective diseases till botanists have succeeded in finding out the different species of bacteria by cultivation and development from spore to spore. It might then very easily happen that the endless trouble of pure cultivation would be expended on some form of bacterium which would finally turn out to be scarcely worthy of attention. In practice only the opposite method can work. In the first place certain peculiarities of a particular form of bacterium different from those of other forms, and in the second place its constancy, compel us to separate it from others less known and less interesting, and provisionally to regard it as a species. And now, to verify this provisional supposition, the cultivation from spore to spore may be undertaken. If this succeeds under conditions which shut out all sources of fallacy, and if it furnishes a result corresponding to that obtained by the previous observations, then the conclusions which were drawn from these observations and which led to its being ranked as a distinct species must be regarded as valid.

On this, which as it seems to me is the only correct practical method, I take my stand, and, till the cultivation of bacteria from spore to spore shows that I am wrong, I shall look on pathogenic bacteria as consisting of different species. . . .

I shall bring forward another reason to show the necessity of looking on the pathogenic bacteria which I have described as distinct species. The greatest stress, in investigations on bacteria, is justly laid on the so-called pure cultivations, in which only one definite form of bacterium is present. This evidently arises from the view that if, in a series of cultivations, the same form of bacterium is always obtained, a special significance must attach to this form: it must indeed be accepted as a constant form, or in a word, as a species. Can, then, a series of pure cultivations be carried out without admixture of other bacteria? It can in truth be done, but only under very limited conditions. Only such bacteria can be cultivated pure, with the aids at present at command, which can always be known to be pure, either by their size and easily recognizable form, as the bacillus anthracis, or by the production of a characteristic colouring matter, as the pigment bacteria. When, during a series of cultivations, a strange species of bacteria has by chance got in, as may occasionally happen under any circumstances, it will in these cases be at once observed, and the unsuccessful experiment will be thrown out of the series without the progress of the investigation being thereby necessarily interfered with.

But the case is quite different when attempts are made to carry out cultivations of very small bacteria, which, perhaps, cannot be distinguished at all without staining; how are we then to discover the occurrence of contamination? It is impossible to do so, and therefore all attempts at pure cultivation in apparatus, however skilfully planned and executed, must, as soon as small bacteria with but little characteristic appearances are dealt with, be considered as subject to unavoidable sources of fallacy, and in themselves inconclusive.

But nevertheless a pure cultivation is possible, even in the case of the bacteria which are smallest and most difficult to recognise. This, however, is not conducted in cultivation apparatus, but in the animal body. My experiments demonstrate this. In all these cases of a distinct disease, *e.g.*, of septicaemia of mice, only the small bacilli were present, and no other form of bacterium was ever found with it. . . . In fact, there exists no better cultivation apparatus for pathogenic bacteria than the animal body itself. Only a very limited number of bacteria can grow in the body, and the penetration of organisms into it is so difficult that the uninjured living body may be regarded as completely isolated with respect to other forms of bacteria than those intentionally introduced. It is quite evident, from a careful consideration of the two diseases produced in mice—septicaemia and gangrene of the tissue—that I have succeeded in my experiments in obtaining a pure cultivation.

Comment

This paper shows a stage in the evolution of Koch's thinking. He began to come to grips with the problem of whether individual species of bacteria can cause individual diseases. He made an extensive series of observations of different diseases, especially in experimental animals. He saw how each of these diseases differed from the others he had studied. He also observed carefully the bacteria he found associated with each disease. By new microscopic procedures partly devised by him, he could see that the bacteria associated with one disease were different from those associated with other diseases. It was true that the morphological differences were not large, especially between the various coccus forms, but he was convinced that the minor differences he saw were significant. In a certain sense this was prejudice on Koch's part. He wanted these organisms to be different, because now he believed firmly in the germ theory of disease, and if this theory were correct, each disease should be caused by a different species. Fortunately Koch was right on this point, but there was no a priori reason why he should have been right, and so we must conclude that he was lucky. Things could have been much more complex.

Another point of some interest which arose in this paper was Koch's realization of the great importance of pure cultures for scientific studies. In his work on anthrax this was not such a problem. The anthrax bacillus has a very distinct morphology, and any contaminants could be readily recognized. In the smaller forms, such as the micrococci, this difficulty was not so easily overcome. The botanists and mycologists had already realized the importance of pure cultures in their study of fungi. But fungus cells are quite large, and most fungi form spores. Brefeld was able to produce pure cultures by isolating single spores and letting them grow and produce progeny. It was only in this way that he could be sure that a particular culture contained only cells of a single genetic background. Koch was aware of the reasonableness of this technique, but was also aware that it was then impossible to perform it with cells as small as the bacteria. In the next paper (see below) we will see that Koch proceeded to develop the necessary methods, so that Brefeld's criticism could be answered. In the present paper he begged the question, but it should be noted that there is only one year between this paper and the next. He worked out the methods he needed very quickly.

Methods for the study of pathogenic organisms

1881 • Robert Koch

Koch, Robert. 1881. Zur Untersuchung von pathogenen Organismen. 1881. *Mittheilungen aus dem Kaiserlichen Gesundheitsamte*, Vol. 1, pages 1–48.

. . . After it has been determined that the pathogenic microorganism is present in the animal body, and after it has been shown that the organism can reproduce in the body and be transmitted from one individual to another, the most important experiment remains to be done. This, which is the most interesting part of hygienic studies, is to determine the conditions necessary for the growth and reproduction of the microorganism. As I have mentioned earlier, this problem can only be solved with the help of pure cultures, and I do not believe it is too much to say that the most important point in all studies on infectious diseases is the use of pure cultures.

Since the importance of pure cultures has been known for a long time, it has thus been true that all who have worked in this field of infectious disease have worked the hardest to perfect methods of pure culture. The most recent results have shown that we are not even past the first groping stages in this research. At the most, people have learned to avoid the most obvious errors, and not all have even learned this.

The most important procedures that have been developed for the manipulation of pure cultures can be summarized as follows.

A sterilized container is used which has been closed with mold-proof sterilized cotton, and this is filled with a sterilized nutrient liquid of the proper sort. Then this is inoculated with material containing the microorganism which is wanted in pure culture. After suitable reproduction has taken place in this container, a sterile instrument is used to transfer a little of this to a second container. This process may be repeated a number of times. In short, this procedure is analogous to the inoculation of an experimental animal from a diseased animal to transfer an infectious disease.

Naturally in this procedure one has to make several assumptions, of which the first is that the culture vessel is really sterile. How lightly this sterilization has occasionally been treated can be seen from the controversy between Pasteur and Bastian on spontaneous generation, and the well-known question of the former to the latter: "Flambez-vous vos vases avant de vous en servir?" * which Bastian had to answer in the negative.

Second, one must assume that the sterile cotton is really mold-proof. As Nägeli has shown, this is not always the case.

Third, it must be assumed that the nutrient liquid is both sterile and suitable for the growth of the organism in question. . . .

Fourth, it must be assumed that the substance used as inoculum contains no other microorganisms than the one desired. Even a slight contamination of the inoculum with another species which is faster growing than the organism desired will prevent anyone from ever obtaining a pure culture. Buchner has therefore developed his own method for the preparation of an initial material for his studies on the anthrax bacillus. He inoculates the nutrient medium with such a high dilution of anthrax infected material that, through calculations, it can be assumed that only one bacillus is placed in each culture vessel. Then from the characteristic macroscopic appearance of the developing culture, he concludes that he has obtained a pure culture.† [This method also has difficulties which I shall go into later.] . . .

* ["Do you flame your glassware before using it?"]

† [This is Lister's method; see page 58.]

Fifth, it has to be assumed that during the initial inoculation and also in the subsequent inoculations, that no foreign organism gets into the culture liquid from the air. This is a danger which the experimenter will find difficult to prevent with certainty, even when the protecting cotton plug is exposed to the air for only a very short time. Even if the first, second, and third transfers have been successful, the probability that the culture will get contaminated will increase with the number of transfers. In order to circumvent this eventuality as much as possible, it is customary to prepare a number of replicates, and only use those for further inoculations which appear by macroscopic or microscopic observation to be pure. Unfortunately one cannot even rely on this procedure, because the macroscopic differentiation of several cultures is very uncertain, and even the microscopic examination is fraught with difficulties, since one only knows that the very small drop of culture fluid under the microscope is free of contaminating organisms, and, as well, if the amount of contamination is small, there may be only occasional contaminants amongst the large number of organisms, and this makes them quite easy to miss. Therefore the first initiation of contamination cannot be distinguished either macroscopically or microscopically, and if one by chance uses for further inoculations a culture which is presumed to be pure but which has already become contaminated, and the contaminating organisms are able to overtake the experimental organism, then the pure culture is completely lost. The microscope will reveal in the next generation, without a doubt, that the culture is contaminated, but now it is too late, because it is impossible at this time to rid oneself of the uninvited guest.

All in all the situation with regard to pure culture techniques is quite disappointing. No one who has cultured microorganisms in the ways currently in vogue and has not avoided completely all of the sources of error that I have indicated, can complain if his results are not accepted as fact by his fellow workers. What has been said above should be heeded by the Pasteur school in its noteworthy but blindly zealous researches, since this renders it doubtful that they have obtained in pure culture the organisms of rabies, sheep pox, tuberculosis, and so forth.

As I have emphasized many times before, pure cultures are indispensable for the further development of knowledge in the field of pathogenic organisms and all that is connected with this, and a practical and exact method must, in some way, be developed. The present methods seem to me to offer no hope for a significant improvement. . . .

Therefore I have rejected completely all of the current principles of pure culture technique and have adopted an entirely new way. A simple observation which anyone can repeat has led me to this approach.

If a boiled potato is cut in half and the cut surface is exposed to the air for several hours and then placed in a moist chamber such as a moistened bell jar in order to prevent it from drying, then, depending upon the temperature of the chamber, one will find in the following day or two, on the surface of the potato, a large number of very small droplets, all of which seem to be different from each other. Several of these droplets may be white, others may be yellow, brown, light gray, or reddish, while others appear to be spread out water droplets, or half spheres, or warty. But all of these become larger in time, then appear mycelia of molds, and finally all of the droplets coalesce and the potato soon becomes obviously spoiled. If one examines these droplets under the microscope while they are still isolated, preferably after they have been streaked on a cover glass, heated, and stained, it can be seen that each droplet consists of microorganisms of one particular species. Some reveal large micrococci, others have small micrococci, in a third the cocci will be in chains. Those which have spread considerably usually consist of bacilli of various sizes and arrangements. Many consist of yeast cells, and here and there is a mold mycelium which has come from a germinated spore. There is no doubt where these different organisms have come from. Another potato is peeled with a flamed knife to remove the peeling which contains soil with bacillus spores which have not been killed by the short heating time. This piece of potato is protected from the air by placing it in a glass beaker with a cotton stopper and then incubated and observed. In this potato no droplets develop, no organisms appear, and the potato remains unchanged until it eventually dries up after several weeks. Therefore, the germs from which the drop-like colonies on the first potato developed could only come out of the air. Indeed, often one can see in the center of the droplet a dust particle or piece of thread which was the carrier of the germ. These germs may be dried but still living bacteria, yeast cells, or spores. . . .

What can we conclude from these observations on colonies developing on potatoes? It is possible that two different germs may come to lie close together and develop colonies which quickly coalesce, and it is possible that one dust particle may contain more than one germ and these may develop simultaneously. But these are probably exceptions, and most often each droplet or colony is a pure culture and remains a pure culture until it enlarges to the point that it touches its neighbors. If instead of the potato, a liquid medium of the same surface area were exposed to the air, then undoubtedly the same number and the same kinds of germs would fall as had fallen on the potato, but the development of these germs in the liquid would be different and would follow the manner which has been previously described. The motile bacteria upon dividing would separate from each other. The nonmotile bacteria would probably begin to form tiny colonies, but these would soon be separated by the movements of the motile bacteria. Some of the organisms would sink to the bottom of the liquid, while others would rise to the top. Some of the organisms which would have found places on the potato to grow undisturbed would be choked by the development of other more luxuriantly growing organisms and would never grow. In short, the whole liquid would reveal under the microscope from the beginning a tangled mixture of different shapes and sizes, which no one would mistake for a pure culture. What is the fundamental difference between the nutrient substratum which the potato and the nutrient liquid offer to the microorganisms? It is only that the potato is solid and prevents the various species, even if they are motile, from becoming mixed, while in the liquid medium there is no chance for the different species to remain apart.

How then can we make use of the advantages which a solid nutrient medium offers for the pure culture practice? A number of the colonies which had developed spontaneously on a boiled potato were spread out on other similar potato slices and incubated in the moist chamber. Within one or two days a heavy growth of the seeded microorganism had developed, and these had exactly the same characteristics as those from the original droplet. . . . All of them grew quite quickly from very small colonies of the original potato when transferred to other potatoes and appeared to be perfectly pure cultures. Extra-special precautions to prevent air contamination were not necessary here, since if a germ of another organism fell here and there on the potato, it could only develop where it fell and would slowly spread out but would never endanger the whole culture. As well, any contaminating colonies could be easily distinguished by their appearance, so that a contamination of the culture during the next transfer could be easily avoided. . . . Here therefore was a very simple method for the production of perfect pure cultures, at least for those organisms which could grow on boiled potato, and this number is not small. . . . However, bacteria which had been shown by animal experiments to be pathogenic could not be cultured on potato.

But the principle had been found, and it was only necessary to devise conditions which could be used in all cases. There would be no purpose in outlining all of the experiments which were performed, in order to find a nutrient medium like boiled potato which would suit the pathogenic organisms. I will indicate only the end result of these experiments. In its present form the technique can be used perfectly in the majority of cases where pure cultures are desired, and in time it will undoubtedly be perfected so that all cases will be included.

After I had considered that it would be hardly possible to construct a universal medium which would be equally suitable for all microorganisms, I limited myself to attempting to use the known media and such new ones as I might develop and converting them to a form which would be firm and rigid. The most useful way to obtain this end is to add gelatin to the nutrient liquid. . . . The mixture of nutrient liquid and gelatin, which I will call nutrient gelatin for short, is prepared in the following way: The gelatin is allowed to soak in distilled water and is then dissolved by heating. Both the gelatin and the nutrient liquid are prepared at such concentrations so that when they are mixed in predetermined amounts they will give the desired concentration of gelatin and nutrient in the final medium. I have determined that the best concentration of gelatin for these purposes is 2.5 to 3 percent. . . . One can also dissolve the gelatin directly in the nutrient liquid. Gelatin generally gives a slightly acid reaction, and for this reason it is necessary to neutralize the nutrient gelatin with potassium or sodium carbonate or basic sodium phosphate, if the medium is to be used for the culture of bacteria. The neutralized gelatin is again heated, and since there is usually a precipitate formed either during this heating or the preceding neutralization, the mixture is then filtered. This filtration also removes any impurities that were present in the gelatin. In the meantime a container closed with cotton has been sterilized by heating for a long time at 150°C., and this is then filled with the medium and boiled again. The boiling requires only a short time, since it is only necessary to kill the microorganisms which were already present in the nutrient gelatin, and these are easy to kill. The spores which are present can only be killed by prolonged heating, and this cannot be done, since the gelatin then loses its ability to solidify. For the same reason, it is not possible to sterilize with steam under pressure.* During these manipulations the nutrient gelatin is therefore not sterilized with certainty, but this makes no difference. If the medium were liquid, the spore-forming bacteria would quickly grow and spread throughout the whole liquid, and only reveal themselves through a turbidity on the second or third day. At this time the liquid could no longer be saved, since it would be changed from its original composition, and probably would be full of newly-formed spores. But in the nutrient gelatin the situation is quite different, and here can be seen already the tremendous advantage offered by the solid characteristics of the medium for revealing its content of bacteria. In the next day or two one may see dispersed throughout the transparent, solidified gelatin, a number of very small, translucent little dots, which appear white by reflected light. If one allows the nutrient gelatin to incubate further, then these little dots will soon enlarge into small spheres, and these will continue to increase in circumference, and eventually liquefy the gelatin and convert it into a turbid liquid. These small, white colonies consist of bacilli, which fact can easily be ascertained by a microscopic examination. But if one is aware of this and wishes to sterilize this gelatin, one should not wait until they have achieved such a considerable size, but kill them through boiling of the gelatin when they are just big enough to be seen by the naked eye. Here is a great advantage of the nutrient gelatin, since one cannot overlook the very first beginnings of bacterial development. . . . One discovers quickly if this or that particular nutrient fluid

* [Nutrient gelatin can be steam-sterilized if higher concentrations of gelatin are used.]

when converted into nutrient gelatin is easy or hard to sterilize. Many, as for example alkaline urine or Pasteur's fluid, are easy to sterilize in the form of nutrient gelatin. Others like meat extract or hay infusion are much more difficult; one has to boil them daily for several days. This is because not all of the spores germinate at the same time. Occasional single colonies will develop in the center of the gelatin even days after the last boiling, and their position shows that they were in there from the beginning and did not arrive later. However, if this should be the case, frequent examination of the nutrient gelatin in the first week will allow one to notice these early enough and they can then be killed by another boiling. This frequent examination in the first week should never be omitted.

Because it is so simple and certain to prepare pure cultures using potato slices, I have preferred to prepare the nutrient gelatin in a similar form as a potato slice. It can be poured into flat watch glasses, small glass plates or the like. However, the most useful for the preparation of cultures, and especially for the microscopic examination of these, is to spread the nutrient gelatin as a long, wide drop on a microscope slide, in which form it can be placed under the microscope when so desired. This is done with a previously sterilized pipette, and of course the microscope slide is previously cleaned and sterilized by prolonged heating at 150°C. The drops are about two millimeters thick. The gelatin hardens in a few minutes and the slides are placed on a small glass shelf which will hold two or three slides next to each other. Finally a number of these shelves are placed in layers over each other and placed in a moist chamber. . . . Under such conditions the gelatin drops can be kept two or three weeks before they dry out. The organism to be cultured is seeded by taking a flamed needle or platinum wire, picking up a very small quantity of the liquid or substance containing the organisms, and streaking this in three to six cross lines on the gelatin surface. . . . The expression "inoculation" for this operation seems appropriate. . . .

The bell jar which serves as the moist chamber is sufficient protection from contamination, even though it does not fit tightly. It sometimes happens that foreign organisms may fall on the gelatin during inoculation or manipulation of the slides. But these can only develop at the place on the gelatin where they have fallen and this is usually not on the inoculation streak. It is hardly possible that all of the cultures of an organism will become contaminated so that they cannot be transferred further, and this possibility is even more reduced if the bell jar is not opened often. Within a few days the pure cultures have developed to their maximum extent and can be inoculated further. There is no purpose in allowing the cultures to stand a long time, and this is especially true when the bacteria being cultured are able to liquefy the gelatin, or when sporulation has set in. In these cases a quick transfer is necessary. If it is necessary to keep single cultures for a long time without transfer, then it is necessary to keep them in a container enclosed with cotton. . . .

At low temperatures the development of the cultures proceeds quite slowly, and many organisms require a certain warmth in order to proliferate well. The most luxuriant growth in gelatin cultures has been at 20–25°C., and I have not found any organisms yet which are at all culturable, which could not grow at this temperature. However, if it is necessary to use temperatures over 30°C., where the gelatin is fluid, then one cannot use gelatin or must modify the procedure. . . .*

A very important operation in the pure culture procedure is the procuring of a completely pure material for the first inoculations. This can be easily performed with the help of nutrient gelatin. With the previous methods this problem was almost impossible to solve. If, for example, blood from a septicemic animal was to be used as culture material to obtain a completely pure culture of the septicemia bacteria, previously many precautions of sterile procedure would have to be taken to remove the blood from the animal, and still the desired result would not be obtained. Now it is only necessary to take a flamed needle and remove some blood from the opened heart or a convenient blood vessel and streak it a few times on the nutrient gelatin. There will occur growth in colonies of several types of microorganisms, among which will be a greater or lesser number of pure, characteristically matlike and granular colonies which can be characterized under the microscope as those of the septicemia bacteria. It will be quite easy to culture these further in pure culture. In this case the number of foreign organisms is at a minimum, so that it is quite easy to isolate the pure colonies of the appropriate organism. However, even if this situation were reversed and the sought-for organisms were in the minority, it would still be possible to have success. Although here it would not be as easy, it would be just as certain. It is only necessary to dilute the bacterial mixture considerably and then make a large number of streaks. In such circumstances it is advantageous to inoculate into the still liquid gelatin, in order to spread the various germs over a wide area, and then pour it on the slides and locate the colonies which develop under the microscope. . . .†

I have carried pathogenic and nonpathogenic organisms over a long series of transfers on boiled potato or nutrient gelatin, without ever once observing any noticeable changes in their characteristics. They maintain their morphological as well as their physiological characteristics, so far as one can determine these, without change through months of growth as pure cultures. . . .

In botany and zoology it is a basic rule that all living organisms which have been previously unknown, should be exactly described, named, and tentatively recorded as new species. . . . This tried and approved rule, that all new forms which deviate from each other in significant ways, should be considered as separate from each other, has remarkably been often ignored in studies on bacteria. From the very beginning of bacteriological research, from Hallier to Naegel to Buchner, right up to the present time, there has been a tendency to take all of the different kinds of bacteria and throw them into one pile, and make one or at most several species from them. If it is ever possible to show that one type of bacteria can be converted into another well-known form by merely continued culture, then is the time to consider these demonstrably related forms to be one species. Up until now this proof has not been accomplished, and there is not the slightest basis in bacteriology to deviate from this general maxim of natural science. If at the beginning too many species are assumed, this can be of no dis-

* [In the next paper, page 109, an important modification is presented which allows higher temperatures to be used during incubation.]

† [The first example of a technique known today as "pour plate."]

advantage to the science. But if a priori the utility and necessity of doing research on the different forms of bacteria are denied, making it impossible to acquire knowledge, then a door will be closed on all further research and progress in this field, and this would certainly be a tremendous barrier to the progress in this young and promising subject. . . .

It seems to me indispensable in our studies on bacteria . . . to adhere to the following concept: *All bacteria which maintain the characteristics which differentiate one from another, when they are cultured on the same medium and under the same conditions, through many transfers or many generations, and which seem to be different from each other, should be designated as species, varieties, forms, or other suitable designation.*

Comment

If I had to choose one paper as most significant for the rise of microbiology, this would be it. Koch presented a method for isolating pure cultures that is so simple, reproducible and understandable that it could be performed by anyone. The development of this method led to the isolation and characterization, during the 20 years after 1881, of the causal organisms of all of the major bacterial diseases which affected mankind.

So far as I know, it is not recorded how Koch happened to make his original observations of colonies developing on potato slices. But it could have easily happened that he observed them accidentally while performing other experiments. He already knew the importance of pure cultures (see page 99) and knew that methods for developing them must be worked out. Watching the colonies develop on potatoes, suddenly everything become clear. He had his method, so, as he says ". . . the principle had been found, and it was only necessary to devise conditions which could be used in all cases." He had only to take his known liquid media and devise ways of making them firm and rigid.

Koch saw the advantages that the use of solid media would have for research on infectious disease. These advantages are well outlined in the present paper. But in addition, he saw the implications of his technique for basic bacteriology, for the concept of speciation. It was obvious that different colonial forms developed on the solid media. These colonial forms bred true and could be distinguished from one another by their colony characteristics. They also differed microscopically and in temperature and nutrient requirements. Although Koch was trained as a physician, he realized that these forms met all the requirements that botanists and zoologists set up for the delineation of species. It seemed quite reasonable to him that each form was a separate species, or variety, or other suitable designation. This idea had met with resistance in the past, before it had been possible to culture bacteria on solid media and observe their colonial forms. After the present paper was published, such resistance disappeared quickly, because it was possible for all workers to observe the distinctiveness of various bacterial forms on solid media and convince themselves that they were really separate species. When only observations in liquid media were possible, it was not possible to shed light on the controversy. So Koch's method for solid media cultivation had a tremendous impact on the young science of bacteriology, as well as on the whole field of medicine.

The etiology of tuberculosis

1882 • Robert Koch

Koch, Robert. 1882. Die Ätiologie der Tuberkulose. *Berliner Klinischen Wochenschrift*, No. 15, April 10, 1882, pages 221–230. First presented at a meeting of the Physiological Society of Berlin, March 24, 1882.

THE DISCOVERY OF VILLEMIN THAT tuberculosis can be transmitted to animals has been confirmed a number of times, but has also been opposed on seemingly good grounds, so that up until recently it has not been possible to state for certain whether tuberculosis is an infectious disease or not. Since then, Cohnheim and Salomonsen, and later Baumgarten, have achieved success by inoculation in the anterior chamber of the eye, and Tappeiner has been successful with inhalation. These studies have shown without a doubt that tuberculosis must be counted amongst the infectious diseases of mankind.

If the importance of a disease for mankind is measured from the number of fatalities which are due to it, then tuberculosis must be considered much more important than those most feared infectious diseases, plague, cholera, and the like. Statistics have shown that 1/7 of all humans die of tuberculosis. . . .

The nature of tuberculosis has been studied by many, but has led to no successful results. The staining methods which have been so useful in the demonstration of pathogenic microorganisms have been unsuccessful here. In addition, the experiments which have been devised for the isolation and culture of the tubercle virus * have also failed, so that Cohnheim has had to state in the newest edition of his lectures on general pathology, that "the direct demonstration of the tubercle virus is still an unsolved problem."

In my own studies on tuberculosis I began by using the known methods, without success. But several casual observations have induced me to forego these methods and to strike out in a new direction, which has finally led me to positive results.

The goal of the study must first be the demonstration of a foreign parasitic structure in the body which can possibly be indicted as the causal agent. This proof was possible through a certain staining procedure which has allowed the discovery of characteristic, although previously undescribed bacteria, in organs which have been altered by tuberculosis. . . .

The material for study was prepared in the usual manner for the study of pathogenic bacteria. It was either spread out on cover slips, dried, and heated, or cut into pieces after dehydration with alcohol. The cover

* [The word "virus" as used here means "infective agent."]

slips or pieces were placed in a dye solution which contained 200 cc. distilled water with 1 cc. of a concentrated alcoholic solution of methylene blue. They were shaken and then 0.2 cc. of 10% potassium hydroxide added. This mixture should not give a precipitate after standing for days. The material to be stained should remain in this solution for 20–24 hours. By heating this solution at 40°C. in a water bath, this time can be shortened to ½ to 1 hour. The cover slips are then immersed in a freshly filtered aqueous solution of vesuvin for 1–2 minutes, and then rinsed in distilled water. When the cover slips are removed from the methylene blue, the adhering film is dark-blue and strongly overstained, but the treatment with vesuvin removes the blue color and the films seem light brown in color. Under the microscope the structures of the animal tissues, such as the nucleus and its breakdown products, are brown, while the tubercle bacteria are a beautiful blue. Indeed, all other types of bacteria except the bacterium of leprosy assume a brown color. The color contrast between the brown colored tissues and the blue tubercle bacteria is so striking, that the latter, although often present in very small numbers, are quite easy to find and to recognize.

The tissue slices are handled differently. They are removed from the methylene blue solution and placed in the filtered vesuvin solution for 15–20 minutes and then rinsed in distilled water until the blue color has disappeared and a more or less strong brown tint remains. After this, they can be dehydrated with alcohol, cleared in clove oil and can be immediately examined under the microscope in this fluid or first placed in Canada balsam. In these preparations the tissue components are brown, and the tubercle bacteria are a most distinct brown.

Further, the bacteria are not stained exclusively with methylene blue, but can take up other aniline dyes with the exception of brown dyes, when they are treated at the same time with alkali. However, the staining is not so clear as with methylene blue. Further, it can be shown that the potassium hydroxide solution can be replaced with sodium or ammonium hydroxide, which shows that it is not the potassium which is especially important, but the strongly alkaline properties of the solution which are necessary. . . .

The bacteria visualized by this technique show many distinct characteristics. They are rod-shaped and belong therefore to the group of Bacilli. They are very thin and are only one-fourth to one-half as long as the diameter of a red blood cell, but can occasionally reach a length as long as the diameter of a red cell. They possess a form and size which is surprisingly like that of the leprosy bacillus. . . . In all locations where the tuberculosis process has recently developed and is progressing most rapidly, these bacilli can be found in large numbers. They ordinarily form small groups of cells which are pressed together and arranged in bundles, and frequently are lying within tissue cells. They present in places a picture similar to that in tissue which contains leprosy bacilli. Many times the bacteria occur in large numbers outside of cells as well. Especially at the edges of large, cheesy masses, the bacilli occur almost exclusively in large numbers free of the tissue cells.

As soon as the peak of the tubercle eruption has passed, the bacilli become rarer, but occur still in small groups or singly at the edge of the tubercle mass, with many lightly stained and almost invisible bacilli, which are probably in the process of dying or are already dead. Finally they can disappear completely, but this complete disappearance occurs only rarely, and then only in such sites where the tuberculosis process has stopped completely. . . .

Because of the quite regular occurrence of the tubercle bacilli, it must seem surprising that they have never been seen before. This can be explained, however, by the fact that the bacilli are extremely small structures, and are generally in such small numbers, that they would elude the most attentive observer without the use of a special staining reaction. Even when they are present in large numbers, they are generally mixed with finely granular detritus in such a way that they are completely hidden, so that even here their discovery would be extremely difficult. . . .

On the basis of my extensive observations, I consider it as proven that in all tuberculous conditions of man and animals there exists a characteristic bacterium which I have designated as the tubercle bacillus, which has specific properties which allow it to be distinguished from all other microorganisms. From this correlation between the presence of tuberculous conditions and bacilli, it does not necessarily follow that these phenomena are causally related. However, a high degree of probability for this causal relationship might be inferred from the observation that the bacilli are generally most frequent when the tuberculous process is developing or progressing, and that they disappear when the disease becomes quiescent.

In order to prove that tuberculosis is brought about through the penetration of the bacilli, and is a definite parasitic disease brought about by the growth and reproduction of these same bacilli, the bacilli must be isolated from the body, and cultured so long in pure culture, that they are freed from any diseased production of the animal organism which may still be adhering to the bacilli. After this, the isolated bacilli must bring about the transfer of the disease to other animals, and cause the same disease picture which can be brought about through the inoculation of healthy animals with naturally developing tubercle materials.

The many preliminary experiments which helped to solve this problem will be passed over, and only the final method will be described. The principle of this method is based on the use of a solid, transparent medium, which can remain solid even at incubator temperature. The advantage of a solid medium for bacteriological research in the production of pure cultures has been discussed by me in an earlier paper.* This same procedure has led to the solution of the difficult problem of the pure culture of the tubercle bacillus and is further proof of the value of this method.

Serum from cow or sheep blood, which is obtained as pure as possible, is placed in cotton-plugged test tubes and heated every day for six days, one hour per day at 58°C. Through this procedure it has been possible in most cases to completely sterilize the serum. This serum is then heated for a number of hours at 65°C., until it has solidified completely. The serum appears after this treatment as an amber-yellow, perfectly transparent or lightly opalescent, solid gelatinous mass. When this is placed for a number of days in the incubator, no bacterial colonies develop. . . . In order to obtain a large surface for the culture, the serum is allowed to harden while the test tubes are in a slanted position. . . .

* [See page 101.]

On this solidified blood serum, the tuberculous materials are placed in the following manner.

The simplest way, and one which is almost always successful, is by the use of an animal which has just died of tuberculosis, or by the use of an animal suffering from tuberculosis which is killed for this purpose. First the skin of the breast and abdomen is laid to the side with a flamed instrument. Then the ribs are cut in the middle with a flamed scissors and forceps, and a portion of the ribs are removed without at the same time opening the abdominal cavity. The lungs are then to a great extent uncovered. The instruments used here are now discarded and freshly sterilized ones taken up. Single tubercles or particles about the size of a millet seed are quickly cut out of the lung tissue and immediately carried over to the surface of solidified serum in a test tube, with the use of a flamed platinum wire. Naturally the cotton plug should only be exposed to the air for the shortest possible time. In this way, a number of test tubes, perhaps 5–10, are inoculated with tuberculous material. Such a large number are prepared because even with the most careful manipulations, not all test tubes can remain free of accidental contamination. . . .

These test tubes are now placed in an incubator and are kept there for a long time at 37–38°C. In the first week, no noticeable changes take place. Indeed, if bacteria develop in the first days, either around the inoculum or away from it, these usually white, gray, or yellowish droplets, which often bring about the liquefaction of the serum, are due to contamination, and the experiment is a failure.

The growth of the tubercle bacilli can first be seen by the naked eye in the second week after seeding, ordinarily after the 10th day. They appear as very small dots, dry and scale-like. This growth arises from the material inoculated, and if the tubercle has been spread around extensively on the surface, then a large amount of growth ensues, while if the tubercles have remained in small patches, then the bacterial growth is less extensive. If there are only very few bacilli in the inoculum, then it is hardly possible to free the bacilli from the tissue and have them growing directly on the nutrient medium. . . . With the help of low magnification, 30–40 power, the colonies of the bacilli can already be seen at the end of the first week. . . .

The growth of the culture ceases after several weeks, and a further increase probably does not occur because the bacilli have lost their own power of movement,* and only spread because of the slow reproduction of the bacilli, being pushed forward on the surface, and because of the slow growth of the bacilli, this spread can only occur to a small extent. In order to keep such a culture going, it must be brought onto a new medium 10–14 days after the first inoculation. This is done by removing several of the small scales with a flamed platinum wire, and transferring them to a fresh, sterilized serum slant, where the scales are broken up and spread out as much as possible. Further scaly, dry masses then develop which coalesce and cover more or less of the surface of the serum, depending upon the extent of the seeding. In this way the culture can be continued.

The tubercle bacilli can also be cultured on other nutrient substrates, if the latter possess similar properties to the solidified serum. They are able to grow on a solidified gel which remains solid at incubator temperature,

* [The tubercle bacillus is not motile.]

prepared by adding agar-agar * to a meat infusion or peptone medium. However, on this medium the bacilli form only irregular small crumbs, which are not nearly so characteristic as the growths on blood serum.

Originally I cultivated the tubercle bacilli only from lung tubercles of guinea pigs which had been infected with tubercular material. Therefore the cultures from various sources had first to pass through the intervening stage of the guinea pig before they were obtained in pure cultures. In this way there was a possibility for error, in the same way as in the transfer of a culture from one test tube to another. This might occur through the accidental inoculation of other bacteria into the animal, or through the appearance in the guinea pig of spontaneous tuberculosis. In order to avoid such errors, special precautions are necessary, which can be deduced from observations on the behavior of this spontaneous tuberculosis.

From hundreds of guinea pigs that have been purchased and have occasionally been dissected and examined, I have never found a single case of tuberculosis. Spontaneous tuberculosis develops only occasionally and never before a time of three or four months after the other animals in the room have been infected with tuberculosis. In animals which have become sick from sponstaneous tuberculosis, the bronchial glands become quite swollen and full of pus, and in most cases the lungs show a large, cheesy mass with extensive decomposition in the center, so that it occasionally resembles the similar processes in the human lung. . . . Animals that have been inoculated with tuberculosis show a completely different picture. The place of inoculation of the animals is in the abdomen close to the inguinal gland. This first becomes swollen and gives an early and unmistakable indication that the inoculation has been a success. Since a larger amount of infectious material is present at the beginning, the infection progresses much faster than the spontaneous infection, and in tissue sections of these animals, the spleen and liver show more extensive changes from the tuberculosis than the lungs. Therefore it is not at all difficult to differentiate the artificially induced tuberculosis from the spontaneous tuberculosis in experimental animals. From a consideration of these facts, it can be concluded that the development of tuberculosis in an experimental animal is due to the action of inoculated material, when a number of guinea pigs are purchased and inoculated at the same time in the same way with the same material, and kept separated from other animals in their own cage, and when they show the development of the characteristic tuberculosis symptoms of inoculated animals in a short period of time.

In this way, a substance can be tested for its virulence by inoculating four to six guinea pigs with it, after making use of all precautions, such as previously disinfecting the site of inoculation, using sterile instruments, etc. The results are uniformly the same. In all animals which are inoculated with fresh masses containing tubercle bacilli, the small inoculation site has almost always coalesced on the next day, then remains unaltered for about eight days, then forms a little nodule which may enlarge without breaking open, although it most often changes into a flat, dry abscess. After about two weeks, the inguinal glands and axillary glands on the side where the inoculation has occurred enlarge

* [Koch did not at this time seem to be aware of the superiority of agar as a solidifying agent.]

until they are the size of peas. From then on the animals become progressively weaker and die after four to six weeks, or are killed in order to exclude the later development of spontaneous tuberculosis. In the organs of all of these animals, and most especially in the spleen and liver, the recognizable changes due to tuberculosis occur. That these changes in the guinea pigs are only due to the inoculation of material containing the tubercle bacilli, can be see from experiments in which inoculation was performed with scrofulous glands or fungus masses from joints, in which no tubercle bacilli could be found. In these cases, not a single animal became sick, while in the animals inoculated with bacilli-containing material, the inoculated animals always showed an extensive infection with tuberculosis after four weeks.

Cultures of tubercle bacilli were prepared from guinea pigs which had been inoculated with tubercles from the lungs of apes, with material from the brain and lungs of humans that had died of miliary tuberculosis,* with cheesy masses from phthisistic lungs, and with nodules from lungs and from the peritoneum of cows affected with bovine tuberculosis. In all these cases, the disease processes occurred in exactly the same way, and the cultures of bacilli obtained from these could not be differentiated in the slightest way. In all, 15 pure cultures were made of tubercle bacilli, four from guinea pigs infected with ape tuberculosis, four with bovine tuberculosis, and seven with human tuberculosis.

In order to answer the objection that the nature of the bacilli was changed through the preliminary inoculation into guinea pigs, so that they became more similar, experiments were set up to cultivate tubercle bacilli directly from spontaneous cases in man and animals.

This was successful a number of times, and pure cultures have been obtained from the lungs of two people with miliary tuberculosis, as well as one with cheesy pneumonia, twice from the contents of small cavities in phthisistic lungs, once from cheese-like mesenteric glands, twice from freshly removed scrofulous glands, twice from lungs of cows with bovine tuberculosis, and three times from the lungs of guinea pigs that had suffered spontaneous tuberculosis. All of these cultures were quite similar and also resembled those that had been isolated through the preliminary guinea pig inoculation, so that the identity of the bacilli occurring in the various tuberculous processes cannot be doubted. . . .

Up until now my studies have shown that a characteristic bacillus is always associated with tuberculosis, and that these bacilli can be obtained from tuberculous organs and isolated in pure culture. It now remained to prove the most important question, namely, that the isolated bacilli were able to bring about the typical tuberculosis disease process when inoculated again into animals. . . .

The results of a number of inoculation experiments with bacillus cultures inoculated into a large number of animals, and inoculated in different ways, all have led to the same results. Simple injections subcutaneously, or into the peritoneal cavity, or into the anterior chamber of the eye, or directly into the blood stream, have all produced tuberculosis with only one exception. Further, the infection was not limited to only isolated nodules,

* [An acute, systemic form of the disease.]

but depending upon the size of the inoculum, large numbers of tubercles were produced. . . .

A confusion with spontaneous tuberculosis, or an accidental infection with tubercle virus in the experimental animals, is excluded for the following reasons: (1) Spontaneous tuberculosis or accidental infection cannot develop in so short a time into the extensive eruption of tubercles experienced here. (2) The control animals, which were handled in exactly the same way as the inoculated animals, remained healthy. (3) The typical picture of miliary tuberculosis does not occur when guinea pigs or rabbits are injected with other substances. . . .

All of these facts taken together lead to the conclusion that the bacilli which are present in the tuberculous substances not only accompany the tuberculosis process, but are the cause of it. In the bacillus we have, therefore, the actual tubercle virus.

Comment

The scientific world quickly recognized the importance of this work of Koch, and it was widely acclaimed. We must consider it his masterpiece and the culmination of all the work he had done before. We can see the evolution of his work clearly through the last four papers. This evolution is all the more remarkable when we remember that in 1876, only six years previously, Koch published his first work on anthrax. In those six years he developed a series of new techniques, and it was these techniques which enabled him to discover the tubercle bacillus.

Several properties of the tubercle bacillus make it an organism that is extremely difficult to work with, and it is remarkable that Koch achieved such quick success in his experiments. The organism is extremely tiny, being at any rate a tenth the size of the anthrax bacillus. It is very difficult to stain successfully, due to a waxy layer on its cell surface. Further, it is a very slow-growing organism and requires several weeks for good growth on solid media. Thus Koch had to be extremely persistent in his work. If he had thrown out his cultures after one week, he would have been unsuccessful. It was necessary to have patience and a faith that tuberculosis was an infectious disease.

Koch was also fortunate that the strain of tubercle bacillus that is pathogenic for humans can be transferred so readily to guinea pigs. Without an experimental animal which showed characteristic symptoms upon inoculation with tuberculous material, his work would have been much harder. He might have cultured the organism successfully, but the actual proof that this organism was the causal agent for tuberculosis would have been much more difficult. It should be noted that in this paper he does not have a final proof that the organism he has isolated in pure culture is really the cause of human tuberculosis. This could only be done by making inoculations in humans. Since this cannot be done, we can only infer that the isolated organism causes the human disease. Such a dilemma is always with the investigator of human diseases. He must learn to live with it

The etiology of tuberculosis [Koch's postulates.]

1884 • Robert Koch

Koch, Robert. 1884. Die Aetiologie der Tuberkulose. *Mittheilungen aus dem Kaiserlichen Gesundheitsamte,* Vol. 2, pages 1–88.

IT WAS FIRST NECESSARY TO DETERMINE if characteristic elements occurred in the diseased parts of the body, which do not belong to the constituents of the body, and which have not arisen from body constituents. When such foreign structures have been demonstrated, it is further necessary to ascertain if these are organized and if they show any of the characteristics of independent organisms, such as motility, growth, reproduction, and fructification. It is further necessary to determine the relationships of these structures to their surroundings, the behavior of the neighboring tissue substances, the distribution of the foreign substances in the body, their appearance in the various stages of the disease, and such similar circumstances, which would allow one to conclude with greater or lesser probability that there is a causal relationship between these structures and disease. The facts obtained in this study may possibly be sufficient proof of the causal relationship, that only the most sceptical can raise the objection that the discovered microorganism is not the cause but only an accompaniment of the disease. However, many times this objection has a certain validity, and then it is necessary to obtain a perfect proof to satisfy oneself that the parasite and the disease are not only correlated, but actually causally related, and that the parasite is the actual direct cause of the disease. This can only be done by completely separating the parasite from the diseased organism, and from all of the products of the disease which could be subscribed to a disease-inducing influence, and then introducing the isolated parasite into healthy organisms and induce the disease anew with all its characteristic symptoms and properties. An example will clarify the above statements. If the blood of an animal dying of anthrax is examined, one finds in it a large number of regular, rod-shaped, colorless, immotile structures. It is not directly evident that these rods are plant-like, since in fact they were first taken by many to be non-living, crystalline structures. Only when it was possible to watch these structures grow, form spores, and then form new rods from the spores, could it be concluded with certainty that they were living and belonged to the class of lower plants. Further, if the rod-containing blood of an animal which had died of anthrax was inoculated in an extremely minute quantity into another animal, this second animal always died of anthrax and its blood contained the characteristic rods, the so-called anthrax bacilli. However, this has not proved that through the inoculation of the rods, the disease was transmitted, because not only the rods were inoculated, but also the other formed and unformed elements of the blood. In order to decide whether it is the bacilli or some other substance of the anthrax blood which causes anthrax, the bacilli must be isolated from the blood and inoculated by themselves. The most certain way of isolating the bacilli is through continued pure culture. For this purpose, a small amount of blood containing bacilli is placed on a solid medium on which the bacilli are able to grow, such as nutrient gelatin, or boiled potato. On these they begin to reproduce quickly and soon are present in large numbers, while the other substances of the blood, the red and white cells and the blood serum, remain unchanged. After two or three days when the bacilli have formed a dense mass of sporulating filaments, a very small amount of this white mass is taken and streaked again on nutrient gelatin or boiled potato. The bacilli reproduce again in the same way as before and form again a dense white covering on the potato, and already in this second culture there are no traces under the microscope of the other elements of the blood. In the same way the culture is transferred a number of times. After the third or fourth transfer, one can consider the bacilli to be completely free from the original blood substances that were inoculated with them. Now if the culture is transferred twenty or fifty times, then it can be assumed with complete certainty that the bacilli no longer are associated with even the slightest amount of disease products from the body. They cannot even have captured these products within their own cells, since the original seeded bacteria are long since gone and the many generations of offspring have taken their nutrients for the production of their necessary substances from the potato itself. The anthrax bacilli in pure culture in this way have therefore no relationships with the first organisms that came out of the blood, or with the disease products which belong to the metabolism of the animal. In spite of this, they are able to induce fatal anthrax as soon as they are inoculated into a healthy animal. The inoculated animal dies just as fast, and with the same symptoms, as one inoculated with fresh anthrax blood, or one which succumbs to spontaneous anthrax. Also, in its blood appear the characteristic anthrax bacilli in countless numbers. From these facts no other conclusion can be drawn than that the anthrax bacilli are the actual cause of this disease, and not merely an attendant phenomenon or symptom. . . . These conclusions are so certain, that no one will dispute them, and the anthrax bacillus will be looked upon by the scientific world as the causal agent of ordinary, typical anthrax infection in both our domestic animals and in man himself.

The process outlined above, which has been successful in proving the parasitic nature of anthrax, and which has led to inescapable conclusions, has been used as the basis for my studies on the etiology of tuberculosis. These studies first concerned themselves with the demonstration of the pathogenic organism, then with its isolation, and finally with its reinoculation.

Comment

This paper presents for the first time the so-called "Koch's postulates." It should be noted that they are published in 1884, after Koch had completed his major work. They were therefore not guideposts for Koch's research, but formalizations of certain assumptions that he had adopted unconsciously in his earlier work. Furthermore, it is apparent that Koch would not have developed these postulates if the nature of the diseases he had worked with, especially tuberculosis, had not forced him to do so. In his first work on anthrax there is no evidence of the organized concept we see in the present paper. Therefore, although he uses anthrax as the example to illustrate his present ideas, we cannot assume that he originally got the ideas while working on anthrax. Rather these ideas were forced upon him during his later work, and especially while working on tuberculosis.

A method for staining the tubercle bacillus

1882 • Paul Ehrlich

Ehrlich, P. 1882. Aus dem Verein für innere Medicin zu Berlin. Sitzung vom 1. Mai 1882. *Deutsche medizinische Wochenschrift*, Vol. 8, pages 269–270.

ONLY A FEW WEEKS AGO HERR Regierungs-Rath Dr. Koch reported on his highly significant work on the etiology and on the bacillus of tuberculosis. There is now a general duty to evaluate this extension of our knowledge in its relationships to diagnosis and therapy. I have been working in this direction and believed that this subject must be reexamined in its diagnostic significance. The main results of my experiments concerning a particular modification of the method will be presented here. It seems to me that these results indicate a certain simplification in the diagnostic procedures and at the same time give clues to certain characteristics of the tubercle bacillus.

As is known, the method of Herr Reg.-Rath Koch consists of staining dried preparations in a weakly alkaline solution of methylene blue. After 24 hours they are then treated with a solution of vesuvin. Then the preparation becomes brown, and under the microscope all of the substances are strongly brown, while the bacillus alone remains an intense blue. The principle of the method is that the methylene blue solution must remain alkaline. I have begun with this requirement for alkaline conditions and have attempted to substitute another alkali for that used by Koch. I have found that aniline is a suitable substitute.

Permit me now to describe my

method in a few passages. I have worked almost exclusively with dried preparations prepared from sputum, but I have also made control experiments which show that the method is also useful for tissue sections.

With a preparation needle I remove a small particle from the sputum and press it flat between two cover slips. Cover glasses of 0.10 to 0.12 mm. are the most suitable. Under these conditions it is quite easy to spread the drop of sputum equally thin throughout. The two cover slips are then pulled apart and one has two thin sheets, which are allowed to dry in the air. These preparations are not yet ready. It is expedient to fix the proteins. I ordinarily do this by keeping the preparations for one hour at 100–110°C. However, in actual practice, a more suitable procedure, which I have seen in operation at the Imperial Office of Health, is to take the dried preparations with forceps and pass them three times through the flame of a Bunsen burner.

For the staining I use water-saturated aniline oil. This is prepared by shaking water with an excess of aniline oil and filtering through a moist filter. To the water-clear liquid obtained is added dropwise an alcoholic-saturated solution of fuchsin or methyl violet until a distinct opalescence develops which indicates that the dye is saturated. The preparations are immersed in this liquid and within one-fourth to one-half hour they have become intensely colored. The differentiation of the tubercle bacilli proceeds only very slowly with vesuvin, so that it is necessary to use acid. I have used strong, even heroic, concentrations of acid. The most effective is to mix one volume officinal nitric acid and two volumes of water. Within a few seconds one can see the preparation fade under the acid treatment, yellow clouds are given off, and the preparation becomes white. If one would examine the preparation at this stage, one would see that everything had been decolorized except the bacteria, and they had remained intensely colored. Such a preparation could be examined at this stage, but the technical difficulties are considerable, since it is difficult to focus on the bacilli. It is better to stain the background in a contrasting color, such as yellow, if the preparation is violet, or blue, if the preparation is red.

May I briefly point out the advantages of this method. The aniline acts more gently on the tissues than the alkali. In the latter, the slime especially is easily dissolved. A further advantage is the rapidity of the technique. While the Koch procedure takes 24 hours, in this procedure three-fourths to one hour is sufficient to make a preparation. Even more important is that the preparations are intensely colored, and the bacilli seem to be significantly larger than in Koch's procedure. Also considering that the background is lighter, this makes it possible to view the bacilli at a lower magnification. I believe also that more bacilli appear on the preparations in my technique, which may be important for the statistics of sputum examinations.

I have been able to stain the bacilli in all basic aniline dyes, as well as bismarck brown, so it is apparent that the substance of the bacillus is not different in its staining characteristics from that of other bacterial species. However, the tubercle bacillus differs in its staining from all other fungi, and this is apparently due to the presence of an outer layer which possesses characteristic and specific properties. The first of these is that the outer layer is permeable to dyes only under the influence of alkali.

A further point which I would

like to mention is the question of what information on the nature of the bacillus is revealed by the staining characteristics. Since I have found that the outer layer is completely impermeable to the action of strong mineral acid, it would seem that this condition would have a practical interest regarding the light it may throw on the question of sterilization or disinfection. It may be that all disinfecting agents which are acidic will be without effect on this bacillus, and one will have to be limited to alkaline agents.

I will now report the results of my examinations using this method. All of the cases which I examined were frank cases of Phthisis pulmonum. I examined in all 26 cases and in all of them the bacillus could be demonstrated. In the examination and preparation of these sputums, I took no special cares and made no special selection. In almost all cases it was sufficient to examine a single preparation, and in most only a single microscopic field. Only in one case was it necessary to examine both preparations.

Naturally I have convinced myself through control experiments with other lung diseases, that there were no bacilli present. As further proof I can indicate a certain case. I had asked a friend to send me phthisistic sputums. I obtained one sputum in which I could find no bacilli. After questioning, I discovered that this sputum had been sent to me mistakenly from a man who did not have phthisis, but a perforated empyema.

The further question of what prognostic significance can be derived from this discovery cannot be answered without further experiments. I have found in certain acute cases a large number of bacteria, while in other more chronic cases, I have found a smaller number. On the other hand I have found a large number of bacilli in cases which are only progressing slowly.

Comment

This paper presents the first staining procedure for the tubercle bacillus which makes use of its acid-fast characteristics. The Ziehl-Nielsen method currently used differs in detail but not in principle from Ehrlich's method. This paper also illustrates an early stage in Ehrlich's scientific development. He attempts to consider what properties of the tubercle bacillus are revealed by this remarkable acid-fastness and considers briefly the practical implications for disinfection. We will see later (page 176) how he uses such considerations to develop the whole field of chemotherapy.

PART III
Immunology

An inquiry into the causes and effects of the variolae vaccinae, a disease discovered in some of the western counties of England, particularly Gloucestershire, and known by the name of The Cow Pox

1798 • Edward Jenner

Jenner, Edward, M.D., F.R.S. 1798. An inquiry into the causes and effects of the variolae vaccinae, a disease discovered in some of the western counties of England, particularly Gloucestershire, and known by the name of The Cow Pox. Abridged from a facsimile edition published in 1923 by R. Lier and Co., Milan.

THERE IS A DISEASE TO WHICH THE Horse, from his state of domestication, is frequently subject. The Farriers have termed it *the Grease*. It is an inflammation and swelling in the heel, from which issues matter possessing properties of a very peculiar kind, which seems capable of generating a disease in the Human Body (after it has undergone the modification which I shall presently speak of), which bears so strong a resemblance to the Small Pox, that I think it highly probable it may be the source of that disease.

In this Dairy Country a great number of Cows are kept, and the office of milking is performed indiscriminately by Men and Maid Servants. One of the former having been appointed to apply dressings to the heels of a Horse affected with *the Grease*, and not paying due attention to cleanliness, incautiously bears his part in milking the Cows, with some particles of the infectious matter * adhering to his fingers. When this is the case, it com-

* [Considering the early date of this paper, Jenner analyzes the disease in extremely modern terms.]

monly happens that a disease is communicated to the Cows, and from the Cows to the Dairy-maids, which spreads through the farm until most of the cattle and domestics feel its unpleasant consequences. This disease has obtained the name of the Cow Pox. It appears on the nipples of the Cows in the form of irregular pustules. . . . These pustules, unless a timely remedy be applied, frequently degenerate into phagedenic ulcers, which prove extremely troublesome. The animals become indisposed, and the secretion of milk is much lessened. Inflamed spots now begin to appear on different parts of the hands of the domestics employed in milking, and sometimes on the wrists. . . .

Thus the disease makes its progress from the Horse to the Nipple of the Cow, and from the Cow to the Human subject.

Morbid matter of various kinds, when absorbed into the system, may produce effects in some degree similar; but what renders the Cow-pox virus so extremely singular, is, that the person who has been thus affected is for ever after secure from the infection of the Small Pox; neither exposure to the variolous effluvia, nor the insertion of the matter into the skin, producing this distemper.

In support of so extraordinary a fact, I shall lay before my Reader a great number of instances.

Case I. Joseph Merret, now an Under Gardener to the Earl of Berkeley, lived as a Servant with a Farmer near this place in the year 1770, and occasionally assisted in milking his master's cows. Several horses belonging to the farm began to have sore heels, which Merret frequently attended. The cows soon became affected with the Cow Pox, and soon after several sores appeared on his hands. Swelling and stiffness in each axilla followed, and he was so much indisposed for several days as to be incapable of pursuing his ordinary employment. Previously to the appearance of the distemper among the cows there was no fresh cow brought into the farm, nor any servant employed who was affected with the Cow Pox.

In April, 1795, a general inoculation * taking place here, Merret was inoculated with his family; so that a period of twenty-five years had elapsed from his having the Cow Pox to this time. However, though the variolous matter was repeatedly inserted into his arm, I found it impracticable to infect him with it; an efflorescence only, taking on an erysipelatous look about the centre, appearing on the skin near the punctured parts. During the whole time that his family had the Small Pox, one of whom had it very full, he remained in the house with them, but received no injury from exposure to the contagion.

It is necessary to observe, that the utmost care was taken to ascertain, with the most scrupulous precision, that no one whose case is here adduced had gone through the Small Pox previous to these attempts to produce that disease.

Had these experiments been conducted in a large city, or in a populous neighbourhood, some doubts might have been entertained; but here, where population is thin, and where

* [Even in this early time, people were inoculated for smallpox by using scabs from another individual suffering with the disease. If just the right amount of material was used, the inoculant suffered a mild smallpox infection, recovered, and was forever immune. Frequently, however, the inoculations were a failure, and the person either did not get any infection and was thus not immune or got too severe an infection and was seriously incapacitated or died. Therefore, in Jenner's time, inoculation for smallpox was a rather controversial procedure.]

such an event as a person's having had the Small Pox is always faithfully recorded, no risk of inaccuracy in this particular can arise. . . . [Then follow 15 other cases which follow this same pattern.]

Case 17. The more accurately to observe the progress of the infection, I selected a healthy boy, about eight years old, for the purpose of inoculation for the Cow Pox. The matter was taken from a sore on the hand of a dairymaid, who was infected by her master's cows, and it was inserted, on the 14th of May, 1796, into the arm of the boy by means of two superficial incisions, barely penetrating the cutis, each about half an inch long.

On the seventh day he complained of uneasiness in the axilla, and on the ninth he became a little chilly, lost his appetite, and had a slight headache. During the whole of this day he was perceptibly indisposed, and spent the night with some degree of restlessness, but on the day following he was perfectly well.

The appearance of the incisions in their progress to a state of maturation were much the same as when produced in a similar manner by variolous matter. The only difference which I perceived was, in the state of the limpid fluid arising from the action of the virus, which assumed rather a darker hue . . . but the whole died away (leaving on the inoculated parts scabs and subsequent eschars) without giving me or my patient the least trouble.

In order to ascertain whether the boy, after feeling so slight an affection of the system from the Cow-pox Virus, was secure from the contagion of the Small-pox, he was inoculated on the 1st of July following with variolous matter, immediately taken from a pustule. Several slight punctures and incisions were made on both his arms, and the matter was carefully inserted, but no disease followed. The same appearances were observable on the arms as we commonly see when a patient has had variolous matter applied, after having either the Cow-pox or the Small-pox. Several months afterwards, he was again inoculated with variolous matter, but no sensible effect was produced on the constitution. . . .

[The material from this boy was used to inoculate another person, with the same results. The material was passed from one person to another through five passages from the cow.]

These experiments afforded me much satisfaction, they proved that the matter in passing from one human subject to another, through five gradations, lost none of its original properties, J. Barge being the fifth who received the infection successively from William Summers, the boy to whom it was communicated from the cow.

I shall now conclude this Inquiry with some general observations on the subject and on some others which are interwoven with it. . . .

They who are not in the habit of conducting experiments may not be aware of the coincidence of circumstances necessary for their being managed so as to prove perfectly decisive; nor how often men engaged in professional pursuits are liable to interruptions which disappoint them almost at the instant of their being accomplished: however, I feel no room for hesitation respecting the common origin of the disease, being well convinced that it never appears among the cows (except it can be traced to a cow introduced among the general herd which had been previously infected, or to an infected servant), unless they have been milked by some one who, at the same time, has the care of a horse affected with diseased heels. . . .

The active quality of the virus from the horses' heels is greatly increased after it has acted on the nipples of the cow, as it rarely happens that the horse affects his dresser with sores, and as rarely that a milk-maid escapes the infection when she milks infected cows. . . .

It is singular to observe that the Cow-pox virus, although it renders the constitution unsusceptible of the variolous, should, nevertheless, leave it unchanged with respect to its own action. I have already produced an instance to point out this, and shall now corroborate it with another.

Elizabeth Wynne, who had the Cow-pox in the year 1759, was inoculated with variolous matter, without effect, in the year 1797, and again caught the Cow-pox in the year 1798. . . .

It is curious also to observe, that the virus, which with respect to its effects is undetermined and uncertain previously to its passing from the horse through the medium of the cow, should then not only become more active, but should invariably and completely possess those specific properties which induce in the human constitution symptoms similar to those of the variolous fever, and effect in it that peculiar change which for ever renders it unsusceptible of the variolous contagion.

May it not, then be reasonably conjectured, that the source of the Small-pox is morbid matter of a peculiar kind, generated by a disease in the horse, and that accidental circumstances may have again and again arisen, still working new changes upon it, until it has acquired the contagious and malignant form under which we now commonly see it making its devastations amongsth us? And, from a consideration of the change which the infectious matter undergoes from producing a disease on the cow, may we not conceive that many contagious diseases, now prevalent among us, may owe their present appearance not to a simple, but to a compound origin? For example, is it difficult to imagine that the measles, the scarlet fever, and the ulcerous sore throat with a spotted skin, have all sprung from the same source,* assuming some variety in their forms according to the nature of their new combinations? The same question will apply respecting the origin of many other contagious diseases, which bear a strong analogy to each other. . . .

At what period the Cow-pox was first noticed here is not upon record. Our oldest farmers were not unacquainted with it in their earliest days, when it appeared among their farms without any deviation from the phaenomena which it now exhibits. Its connection with the Small-pox seems to have been unknown to them. Probably the general introduction of inoculation first occasioned the discovery.

Its rise in this country may not have been of very remote date, as the practice of milking cows might formerly have been in the hands of women only; which I believe is the case now in some other dairy countries, and, consequently that the cows might not in former times have been exposed to the contagious matter brought by the men servants from the heels of horses. . . .

Should it be asked whether this investigation is a matter of mere curiosity, or whether it tends to any beneficial purpose? I should answer, that notwithstanding the happy effects of Inoculation,† with all the improvements which the practice has received since its first introduction into this

* [This idea is now known to be wrong.]

† [Meaning inoculation with smallpox itself.]

country, it not very unfrequently produces deformity of the skin, and sometimes, under the best management, proves fatal.

These circumstances must naturally create in every instance some degree of painful solicitude for its consequences. But as I have never known fatal effects arise from the Cow-pox, even when impressed in the most unfavourable manner . . . and as it clearly appears that this disease leaves the constitution in a state of perfect security from the infection of the Small-pox, may we not infer that a mode of Inoculation may be introduced preferable to that at present adopted, especially among those families, which, for previous circumstances we may judge to be predisposed to have the disease unfavourably? . . .

Thus far have I proceeded in an inquiry, founded, as it must appear, on the basis of experiment; in which, however, conjecture has been occasionally admitted in order to present to persons well situated for such discussions, objects for a more minute investigation. In the mean time I shall myself continue to prosecute this inquiry, encouraged by the hope of its becoming essentially beneficial to mankind.

Comment

Smallpox is a disease that has been known since antiquity. Its symptoms are characteristic, and because of this, it was easy to observe its transfer and to determine that it was infectious. We know today that it is caused by a large filterable virus, although we use the word virus in a different sense than did Jenner.

The process of inoculation had been practiced for centuries in the Far East. It was introduced into England from Turkey by Lady Mary Montagu, the wife of the British ambassador to Turkey. Although the process first was greatly resisted, it eventually became established in England. But because complications often arose, inoculation was not without its dangers.

Today we would consider cowpox virus to be a mutant of smallpox virus which has lost some of its virulence for man. It still retains the ability to induce the production of virus-neutralizing antibodies in man. It is this fact which makes the process of inoculation with cowpox possible. This process was called vaccination (vacca = cow), and this word has later been applied to all artificial immunization procedures.

Jenner's observations are quite acute. He was able to bring together a number of diverse facts, and in so doing, arrive at a theory which seemed reasonable. He was fortunate that he could readily test his theory and clearly show that he was right. It is interesting that Jenner attempted to publish in the Transactions of the Royal Society the results of his first case. When the work was rejected, he then collected twenty-three cases and published his book. Because of the simplicity of the vaccination procedure and its high degree of safety, it has eventually become accepted as a common medical procedure, although Jenner met much resistance when his book was first published.

The attenuation of the causal agent of fowl cholera

The original French article appears in the Appendix, pages 269–272.

1880 • Louis Pasteur

Pasteur, Louis. 1880. De l'atténuation du virus du choléra des poules. *Comptes rendus de l'Academie des sciences,* 26 October 1880, Vol. 91, pages 673–680.

I WOULD LIKE TO REITERATE THE following results, which I have previously had the honor of presenting to the academy:

1. Fowl cholera is a virulent disease of the first order.

2. The virus * consists of a microscopic parasite which multiplies readily in culture away from the animal body. From this it is possible to obtain the virus in a state of purity and demonstrate irrefutably that it is the sole cause of the disease and death.

3. The virus may vary in its virulence. At times the disease is followed by death, while at other times, after causing disease symptoms of variable intensity, recovery occurs.

4. These differences in the virulence of the virus are not merely the result of natural variations, as the experimenter can alter them at his will.

5. As is generally the case for all virulent diseases, fowl cholera does not recur, or rather the recurrence is of such a degree that it is inverse in intensity with that of the earlier infection, and it is always possible to extend the resistance so far that inoculation with the most virulent virus does not produce any effect.

6. Without wishing to make a definite assertion on the relationship between the small pox and the cow pox viruses, it seems from the above facts that in fowl cholera, there exists a state of the virus relative to the most virulent virus, which acts in the same way as cow pox virus does in relation to small pox virus. Cow pox virus brings about a benign illness, cow pox, which immunizes against a very serious illness, small pox. In the same way, the fowl cholera virus can occur in a state of virulence that is sufficiently attenuated, so that it induces the disease but does not bring about death, and in such a way that after recovery, the animal can undergo an inoculation with the most virulent virus. Nevertheless, the difference between small pox and fowl cholera is considerable, in certain respects, and it is not amiss to remark that, with respect to an understanding of the principles, studies on fowl cholera will probably be more helpful. Whereas there is still a dispute about the relationships between small pox and cow

*[The causal agent of fowl cholera is not a filterable virus, but a bacterium, now known as *Pasteurella multocida*.]

pox, we know for certain that the attenuated virus of fowl cholera is derived directly from the most virulent virus of this disease, so that their natures are fundamentally the same.

I will now demonstrate the truth of this main assertion, which is at the base of the preceding proposition, namely, that there are variable states in the virulence of the virus of fowl cholera. This result may seem strange when one thinks that the virus of this disease is a microscopic organism which can be handled in a state of perfect purity, in the same way as the yeast of beer or the mycoderma of vinegar. And further, if one considers this strange fact of variation in virulence clearly, one cannot keep from recognizing that this phenomenon is probably common to diverse species of organisms causing infectious diseases. But where is the common factor in these various infectious diseases? In order to cite only one example, is it not usually considered that the variations in the severity of epidemics of variola, sometimes serious, while at other times benign, are due to differences in environmental conditions, to the climate, or to the resistance of the individuals? Is it not equally seen that major epidemics gradually die out and then reappear and die out again?

The idea of the existence of variable degrees of virulence of the same virus is not made in order to surprise the physician or laymen, but because it may be of tremendous importance if it could be established scientifically. In the present case, the mystery is certainly connected with the fact that the virus, being a microscopic parasite, reveals variation in its virulence that are at the control of the observer. It is this point which I would like to establish rigorously.

Let us begin with a virus of fowl cholera which is in the most virulent state possible. Previously I developed a curious method for obtaining a virus with this virulence. This was done by collecting the virus from a chicken which was about to die, not from an acute but from a chronic infection. I have observed that the cholera exists occasionally in this latter form. These cases are rare, but a number of these were obtained. In these cases, the chicken, after becoming very sick, grows more and more emaciated but does not die for weeks or months. When death occurs, it is because the parasite, up until this time localized in certain organs, passes into the blood and reproduces there. One can then observe that the virulence of the virus cultured from the blood of an animal which has taken a long time to die is considerably higher than the virulence of the virus which was used to inoculate this animal, so that it is able to kill all animals that are injected with it.

Let us now make successive cultures of this virus, always in a state of purity, in a medium of chicken broth, by taking the inoculum for each culture from the preceding culture, and then testing each culture for its virulence. This observation shows that the virulence does not change significantly from one culture to the next. In other words, if we admit that two virulences are identical when, upon injecting the same number of animals under the same conditions, and the proportion which die in the same length of time is the same, then we have established that the virulence of our successive cultures is identical.

I made no statement above about the length of time between one culture and the next, or the possible influence of this duration upon the virulence of successive cultures. Let us now turn our attention to this point, and consider the smallest interval which appears to bring about changes. An interval from one to eight days between cultures does not effect the virulence of the successive cultures. An interval of 15 days shows the same results. After an interval of one month, six weeks, or two months, one does not observe any more change. Nevertheless, as the interval is increased, it is possible to perceive at times certain signs of a decrease in virulence, like a weakening of the inoculating virus. For example, the rapidity of death seems to diminish, although the proportion of deaths does not. In various series of inoculated animals, one sees chickens which languish, are very sick, often become very lame because the parasite, in its progression, has passed into the muscles and has attacked those of the legs. The pericardium may become elongated; abscesses appear around the eyes; and finally the virus has lost its fulminating character. Let us then go to longer intervals than these, before we transfer the culture again. Intervals of three, four, five, eight months, or more can be used, before the virus is again tested for virulence. Each time the picture is changed considerably. Differences in virulence are found which are considerable, while previously the changes were either insignificant or were only revealed in a very equivocal manner.

When such intervals between culturing are used, one finds that instead of identical virulence, the virulence is decreased, so that out of ten chickens inoculated, only nine, eight, seven, six, five, four, three, two, or one out of ten die, and at times not a one dies, so that the sickness may develop in all of the inoculated chickens, but they all recover. In other words, by a simple change in the way in which the parasite is cultured, merely by lengthening the time between transfers, we have obtained a method for decreasing progressively the virulence of the virus, until finally we have a virus which is a true vaccine, in that it does not kill, but induces a benign illness which immunizes against a fatal illness.

It would not be expected that in all attempts at attenuation, the procedure would occur with a reproducible and mathematical regularity. Some cultures which have not been transferred in five or six months show a considerable virulence, while others of the same origin may already be quite attenuated after three or four months. We will later explain this anomaly, which is only apparent rather than real. Often there is a steep jump from a good virulence down to a state in which the microscopic parasite has died, and this may take place in a very short interval. In passing it from one culture to the next, one is surprised by the absence of growth of the culture. The death of the parasite is furthermore a usual thing if one allows the culture to stand a sufficient length of time. . . .

In the course of these changes, what happens to the microscopic organism? Does it change its form or appearance while changing its virulence in such a profound way? I would not like to insist that there does not exist a certain relationship between the morphology of the parasite and its virulence, but I must admit that such a relationship is almost impossible to perceive, and if it exists, it is missed even when using the microscope, because of the very small size of the virus. Cultures of all degrees of virulence seem similar. If it seems at times that small changes have occurred, it may only be accidental, and in later observations these changes may disappear or appear in an inverse relationship to earlier changes.

It is possible to say that if one takes a number of cultures of varying virulences, and starts new cultures with them and continues to make successive transfers at the same time interval for all of the cultures, the virulence of each culture is maintained at the same level as it originally was. For example, an attenuated virus which only kills once out of ten inoculations maintains this virulence in successive cultures, if the interval of transfer is not too long. Although probably in line with the preceding observations, it is equally interesting that an interval between transfers which is sufficient to bring about the death of an attenuated virus may not necessarily bring about the death of a highly virulent virus, although it serves to attenuate it.

We must now consider the important question of what is the cause of the decrease in virulence.

The cultures of the parasite are necessarily kept in contact with air, because the virus is an aerobic organism and cannot develop when kept in the absence of air. It is therefore natural to ask initially if it would not be in the contact with air that the attenuating influence exists. Would it not be possible that the tiny organism which is the virus, while developing in the presence of oxygen in the culture medium, undergoes several modifications which become permanent, and still remain when the organism is removed from the influence of the oxygen? It is also possible to ask if there might not be some other principle in the atmosphere, chemical or fluid, which might intervene in the accomplishment of this phenomenon.

Our first hypothesis that it is oxygen in the air which brings about the change can easily be put to experimental test. This can be done by observing whether the phenomenon is suppressed in the absence of oxygen.

To this end the following experiment was performed. A convenient quantity of chicken broth was seeded with the most virulent virus we have, and glass tubes were filled with this seeded broth to two-thirds, of their volume, three-fourths, and so on, and then the tubes were closed under a flame. With the small amount of air remaining in the tubes, the virus began to grow, and this could be determined by observing the turbidity of the tube develop. During the growth in the tube, gradually all of the oxygen in the tube was used up. Then the turbidity fell to the bottom and the virus was deposited on the walls of the tube and the liquid became clear. It took two or three days for this to occur. The tiny organisms are thereafter without oxygen and they will remain in this state for as long as the tube is not opened. What has happened to their virulence? In order to avoid mistakes in this study, we prepared a large number of parallel tubes, and at the same time an equal number of flasks of the same culture, but freely exposed to contact with the air. We have already stated what will happen to cultures that are exposed to air. They become progressively less virulent. This phenomenon was found again in this experiment in the cultures exposed to the air. Let us speak only of those cultures in the closed tubes, cultures without contact with air. Let us open tubes at various intervals, one month, two months, three months, four months, and so on, and transfer them into new broth tubes and then test their virulence. . . . Remarkably, the experiment shows that the virulence under these conditions is always the same as that which was used to inoculate the original closed tubes. Meanwhile the cultures which have been exposed to air are either dead or possess a very weak virulence.

Our problem is therefore solved: it is the oxygen which brings about the attenuation in virulence.

Truly, there is more here than an isolated fact. We are in possession of a general principle. One may hope that there may be an action inherent in atmospheric oxygen, a natural force always present, which will enable us to perform similar attenuations on other viruses. In any case, we have here a circumstance worthy of interest for the possible large generality regarding a method for the attenuation of virulence, which may derive from some influence of the cosmic order of things. May we not assume here that it is perhaps this influence which is responsible, in the present as in the past, for the eventual limitation of the large epidemics? . . .

Comment

The problem of virulence is not as simple as Pasteur visualized. As we have said before, a disease results from the interaction of a host with a parasite. If either the host or the parasite changes in some way, this may alter the severity of the disease. We know that individual hosts may vary in their susceptibility to a given parasite. Thus when 10 chickens are inoculated with a standard number of parasites and only 5 die, this means that 5 of the chickens were more resistant than the other 5. This variation in resistance of individual hosts may be genetically or environmentally controlled. If it is genetically controlled, a surviving host may be able to pass this resistance on to its offspring, and a resistant breed may arise.

In the same way, the virulence of a parasite may be genetically or environmentally determined. Again, if it is genetically determined, a lack of virulence may be passed on to the offspring of the parasite, and a nonvirulent strain may arise.

Pasteur was not aware of genetic concepts in the way we are today. We know that the virulence of a parasite is to a great extent genetically determined. Thus we can guess at what took place in Pasteur's experiments. If he continued to inoculate chickens with cultures freshly isolated from diseased chickens, he would continue to select virulent mutants or strains, which would be the only ones which could grow in chickens. Nonvirulent mutants would not grow and would be lost. However, if he carried his parasite in artificial culture media, the virulent organisms would no longer have any selective advantage over the nonvirulent ones. Probably a mixture of virulent and nonvirulent cells would be present. If he allowed a culture to grow to its maximum extent and then continued to incubate it, many complex changes would take place. Because all of the food in the culture has been used up, many of the cells would die, due to autolytic processes. If the virulent cells had more tendency to die than the nonvirulent ones, we would eventually have only nonvirulent cells left. This is what Pasteur found. It is of course not possible to know if this is the exact mechanism of the phenomenon reported by Pasteur, but it is a most probable one. His work on the effects of oxygen is puzzling. Although oxygen may play a role in speeding up autolytic processes, it probably does not have as direct an action as Pasteur thought.

The process of isolating a nonvirulent strain from a virulent culture has been called attenuation since Pasteur first used the word. Now that we know the genetic basis of virulence, this word no longer has much meaning, since it seems to imply some process of continuous variation from virulent to nonvirulent.

The present paper demonstrates the highly important concept that the virulence of a parasite may vary. Since the disease Pasteur was studying was one which could be produced in experimental animals, it was possible to analyze the situation in some detail. It was obvious to him that attenuation was not an isolated incident in fowl cholera but probably a general phenomenon. The similarity to the smallpox-cowpox story was quite evident to him (see page 121). The practical applications of his observations also did not escape him and are presented in the next paper.

On a vaccine for fowl cholera and anthrax

1881 • Louis Pasteur

Pasteur, Louis. 1881. Sur les virus-vaccins du choléra des poules et du charbon. *Comptes rendus des travaux du Congrès international des directeurs des stations agronomiques, session de Versailles,* June 1881, pages 151–162.

[IN TESTING CULTURES FOR VIRULENCE, a number of animals survived after injections with a nonvirulent culture.]

. . . In examining the animals which have not died during this study on virulence, it can be seen that they have been sick from the same disease which has killed the other animals, since one can find in their muscles the same tiny organism in large numbers. If we consider that in general acute diseases do not ordinarily recur, it may be asked if this disease, which exists without killing the animals, would not be able to hinder the recovery from the fatal disease. But, if we inoculate these chickens which have not died, with a virus which is highly virulent to fresh chickens, one which has not been in contact with air, and which is able to kill 100 times out of 100, we find that it no longer kills, and that it never is able to develop. Evidently, we have discovered a virus-vaccine for fowl cholera, and one may even say, we have discovered many virus-vaccines, since we have a number of degrees of attenuation of the virus. Indeed, if one studies the virulence of a culture every 15 days for a number of months, it is easy to show that there is a gradual diminution in virulence, and each virus can be considered as a vaccine for the virus above it. Here is therefore the idea of a vaccine, well known in medicine since Jenner, but which reveals itself in very new conditions, in a microscopic organism. In the vaccine of Jenner, no organism has been recognized so far, so I reserve this question until it has been elucidated. . . .

Comment

This work developed logically from that of the preceding paper (page 126). It had been a common observation in medicine that when a person recovered from a disease, he quite often was immune to any future attack. Therefore Pasteur was probably not surprised when he found a similar phenomenon in his experiments on fowl cholera. The important discovery was that this immunity had been induced by injection with bacterial cultures which had been rendered nonvirulent by laboratory manipulation. This immediately opened up a whole field of preventive medicine for study. It led to some of the most important triumphs of medicine over disease. It also provided the foundation for the science of immunology.

Pasteur was involved in work on vaccines for the rest of his career. In a number of diseases, he was able to develop vaccines successfully. The actual mechanism of immunity remained a puzzle for a long time and is still subject to debate concerning its details. However, two mechanisms that are of great importance were discovered in the next few years after Pasteur's work and are presented in the next two papers below and on page 138.

A disease of Daphnia caused by a yeast. A contribution to the theory of phagocytes as agents for attack on disease-causing organisms

1884 • Elias Metschnikoff

Metschnikoff, E. 1884. Ueber eine Sprosspilzkrankheit der Daphnien. Beitrag zur Lehre über den Kampf der Phagocyten gegen Krankheitserreger. *Archiv f. pathologische Anatomie und Physiologie und f. klinische Medicin*, Vol. 96, pages 177–195.

THE COMMON WATER FLEA OR DAPHNIA seems quite suitable for studies on pathological processes and may be able to throw some light on many general questions in medicine. Although these crustaceans, because of their small body size and delicate nature, have proven to be quite unsuitable for the production of all kinds of artificial diseases, they offer many advantages for the study of those disease phenomena with which they become afflicted artificially. Because they are relatively small and fairly transparent animals, they can be observed without damaging them for many hours at a time, and also repeatedly from day to day. . . .

The disease which I wish to describe in the following lines is a disease due to a budding fungus, or loosely, a yeast. So far as I know, this disease has not been described earlier and was even unknown to me two years ago when I described another disease of Daphnia. I first found it last fall in an aquarium in which Vallisneria and Daphnia were almost the only flora and fauna. I noticed many Daphnia which seemed to be ill, and under the microscope could see that this was a different disease than I had seen earlier. The whole body cavity up to the last antenna was filled with a massive accumulation of fungus cells, which I demonstrated to be different stages of a single species of fungus. I have named this fungus *Monospora bicuspidata*. . . .

From the characteristics of this parasite, it seems to be very similar to the ordinary yeasts, although it is not possible to ascertain its definitive place in the system of fungi, since we know from the recent work of Brefeld that yeastlike stages occur in many different fungi (Ustilago, Tremella, etc.). . . .

I have observed all of the stages of the Monospora in the abdominal cavity of sick Daphnia. [See Plates IX and X] In the early period of the disease, one sees predominantly the budding conidia, while in the later stages the ascospores prevail. In spite of many experiments, I have not yet been able to cultivate this fungus on artificial media. I have tried various nutrient media such as acidified meat extract, orange juice, and so on.

In the individuals dying of the disease, a large number of spores in asci are produced, which are consumed by healthy individuals. Although the asci do not rupture in water, the spores which occur in the intestinal canal are mostly free of the asci, which I believe is due to the action of the digestive juices of the Daphnia on the asci. As a result of peristalsis, the spores penetrate the intestinal wall, so that they are partly in the intestine and partly in the body cavity. The most favorable spores for observation are those which are partly in the body cavity, but with most of the spore in the intestinal wall and intestinal cavity. Hardly has a piece of the spore penetrated into the body cavity, than one or more blood corpuscles attach to it, in order to begin the battle against the intruder. (Footnote: The blood corpuscles of Daphnia, like most vertebrates, are colorless, amoeboid cells which are adapted to the uptake of solid particles. They circulate in a system of cavities and are kept in circulation by a tubelike heart. Daphnia are completely lacking in blood vessels, except for a short outlet tube which several authors have called the aorta.) The blood cells fasten so tight to the spore that they are only seldom broken free by the blood stream. In this case, they are replaced by new blood cells, so that in most cases the spore is more or less completely surrounded by them. Often the spores penetrate completely into the body cavity, in which case they are even more likely to fall prey to the blood corpuscles. The number of blood cells which collect around one spore varies considerably. When many spores are in the body cavity at the same time, such a large number of blood cells surround them, that the whole area appears highly inflamed, so far as one can speak of inflammation in a vessel-less animal. The blood cells collected around the spore do not always maintain their individuality, and may unite occasionally into a more or less extensive plasmodium (a so-called giant cell). . . .

In the intestinal contents or excrement, one finds that the majority of the spores are intact, which indicates that they are unaffected by the diges-

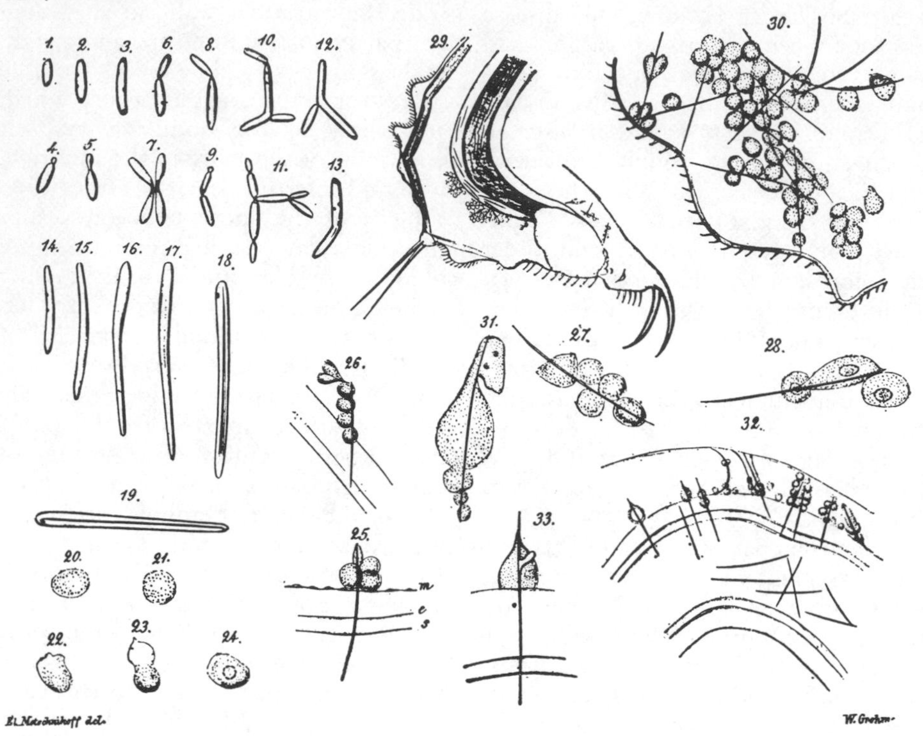

Plate IX

Figs. 1–14. Conidia of Monospora in various configurations. *Figs. 15, 16.* Elongated conidia, just before spore formation. *Figs. 17–19.* Formation of the ascospore. *Figs. 20–23.* Blood cells of Daphnia magna, drawn from life. *Fig. 24.* A blood cell treated with acetic acid. *Fig. 25.* A spore that has penetrated the intestinal wall, surrounded by four blood cells. *(m)* Muscle layer of the intestine, *(e)* Epithelial layer, *(s)* Layer of rods. *Fig. 26.* Another spore, as in Fig. 25. *Fig. 27.* A spore surrounded by blood cells from the body cavity of a Daphnia. *Fig. 28.* Another spore after treatment with acetic acid. *Fig. 29.* The abdomen of an infected Daphnia, with many spores in the body cavity surrounded by blood cells. Many spores are also seen in the intestinal wall and in the intestinal cavity. *Fig. 30.* Area of the abdomen of another Daphnia with intense accumulation of phagocytes around the spores. *Fig. 31.* Blood cells that have coalesced around a spore. *Fig. 32.* An area from the anterior portion of the body, with many free spores and engulfed spores. *Fig. 33.* A germinating spore and an adherent blood cell.

tive juices. Those that are surrounded by blood cells behave completely otherwise. After a spore has lain for a time in the middle of a number of these cells, it begins to undergo quite regular changes. First it thickens, turns light yellow in color, and its contours become jagged. Then it swells in several places to various sizes, assuming round or irregular shaped balls, which become brownish yellow. Meanwhile the rest of the spore, which is still rod-shaped, seems lighter and yellower. Still later the whole spore comes apart into irregular, brownish yellow, dark brown and almost black grains, some large and some small. The connection of these particles to

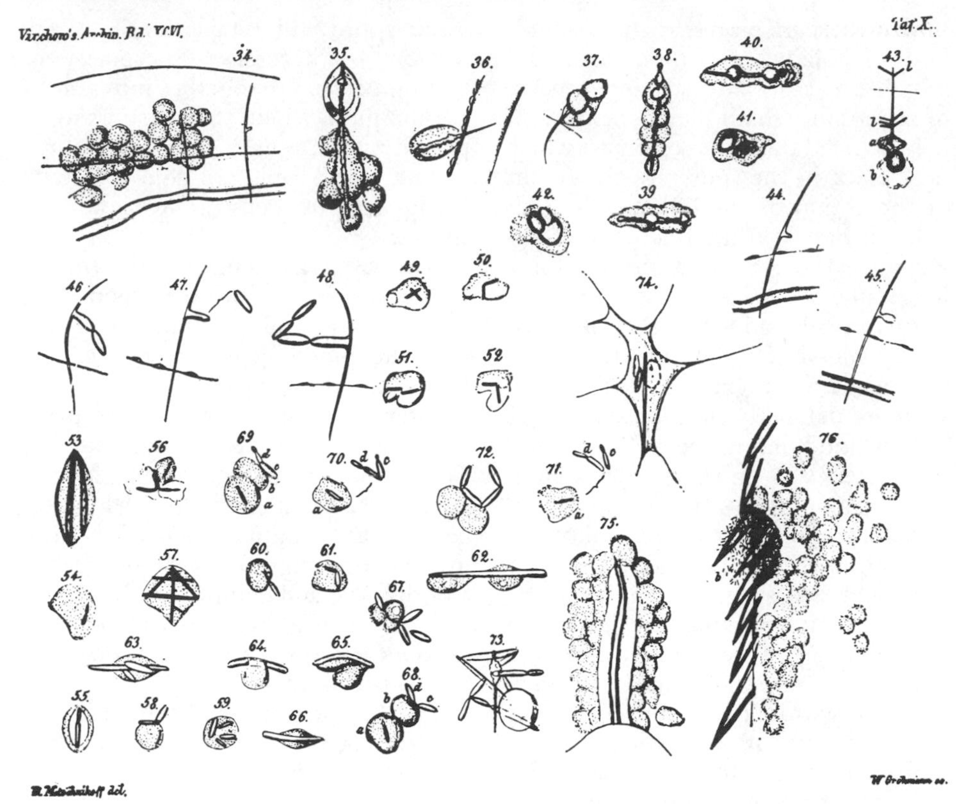

Plate X

Fig. 34. An area from the posterior portion of another Daphnia. *Figs. 35–42.* Spores in various stages of alteration due to the action of blood cells. *Fig. 43.* A spore partially penetrating the wall: *(a)* small wall opening; *(b)* the lower portion of the spore, engulfed by a blood cell and markedly altered; *(l)* young Leptothrix, which has settled on the free portion of the spore. *Figs. 44–48.* Various stages of spore germination and conidia formation. *Figs. 49–52.* A single blood cell in four different configurations. *Figs. 53–57.* Various blood cells and conidia. *Fig. 58.* A blood cell adjacent to two conidia. *Fig. 59.* The same cell as in Fig. 58, one-half hour later. *Figs. 60–66.* Various blood cells and conidia. *Fig. 67.* A ruptured blood cell, from which the conidia have escaped. *Fig. 68.* Two blood cells, in one of which *(a)* a germinating conidium rests, while in the other *(b)* two conidia *(c, d)* remain in contact outside. *Fig. 69.* The same picture as Fig. 68, one-half hour later. The conidium *(d)* has begun to form a bud. *Fig. 70.* The same as Fig. 69, without blood cell *(b)*, which has in the meantime moved away, two hours later than Fig. 69. Conidium *(c)* is beginning to bud. *Fig. 71.* The same picture, one and one-half hours after Fig. 70. *Fig. 72.* Two blood cells adjacent to four conidia. *Fig. 73.* A group of conidia which have brought about the dissolution of a blood cell that had engulfed a spore. All that remains is an empty shell and fine debris. *Fig. 74.* A fibrous phagocyte containing three fungus cells. *Fig. 75.* An injured layer of tissue with many blood cells attached. *Fig. 76.* A disrupted area of another Daphnia, also with many blood cells.

the earlier delicate spore can only be determined through knowledge of the whole process of transformation of one into the other. In the meantime the blood corpuscles have united into a fine-grained, pale plasmodium, which still has the ability to move by amoeboid motion. Occasionally one can find in certain places in the Daphnia body, whole heaps of these plasmodia, which are especially striking because of the grains which they contain. . . .

I believe that the changes which take place in the spore are the results of the action of the blood cells. This belief is based on the following observations. When a spore remains for a long time with half of it in the intestinal wall and only half ingested by the blood cell, only this latter part undergoes the regressive changes and becomes definitely decomposed, while the portion lying in the wall maintains completely its normal appearance. Such examples are too frequent to make one doubt of their generality. . . .

From what has been said above, it is evident that spores which reach the body cavity are attacked by blood cells, and—probably through some sort of secretion—are killed and destroyed. In other words, the blood corpuscles have the role of protecting the organism from infectious materials. This does not always occur. In cases in which a large number of spores reach the body cavity, or for some other reason one or more spores remain unaffected by the blood cells, the disease may break out. . . .

Because the process described here can be observed much more favorably than the battle of phagocytes against bacteria, I will make a few additional comments on the observations. In order to obtain certain results, one must observe one organism for many hours. Then one can see that the blood cells really ingest the spores. Sometimes this process occurs very quickly, but other times it is a very slow process. . . . The number of spores which one blood cell can ingest varies; ordinarily one finds only two spores in each cell, but occasionally one can find three, four, or more spores.

The blood cell which ingests a parasitic spore still retains its ability to move. . . . Occasionally spore-containing cells unite together into a small plasmodium, which then harbors more parasites than usual. . . .

The blood cells are able to attack living fungus cells as well as the spores. . . .

Although it is true that the fungus cells are destroyed by the blood cells, on the other hand, it can not be denied that the blood cells can also be affected by the parasites. Several times I observed a blood cell full of parasites rupture before my eyes, setting the fungus cells free again. Also I could see a number of times that blood cells in the neighborhood of a large number of fungus cells would gradually dissolve and completely disappear. This indicates that the fungus cells produce a substance which is deleterious to blood cells. . . .

The farther along the disease has progressed, the more blood cells that are dissolved, so that at the time when the Daphnia contains a significant number of ripe spores, it reveals only a few or no blood cells.

Aside from the blood cells, only the isolated connective tissue cells play a similar role as phagocytes (eating cells). They behave in the same way as the blood cells in the ingestion of fungus cells. In the same way, they are dissolved by the fungus cells, so that in the later stages of the disease, all of the phagocytes of the animal body disappear. Other tissue elements do not suffer such a remarkable change. One may see a large number of fungus cells develop on the heart muscle, but the heart continues to contract regularly. . . .

When the Daphnia has once become sick and has produced fungus cells in it, it generally dies without a chance of recovery. In the last period of the disease, so many spores have been formed that the body takes on a diffuse milk-white color. . . . The whole disease takes two weeks to proceed. A young isolated Daphnia which had just begun to form fungus cells from spores died 16 days later. . . .

From the above it can be seen that the infection and illness of our Daphnias is a battle between two living beings—the fungus and the phagocytes. The first consist of one-celled lower plants, while the latter represent the lowest tissue element and have a great similiarity with the simple organisms (Amoeba, Rhizopodium, and so on). The phagocytes, which have retained the primitive property of taking up solid food, can act because of this as destroyers of parasites. They seem therefore, as the bearers of nature's healing power, which has been known to exist for a long time, and which Virchow first placed in the tissue elements. The whole course of the Daphnia disease fits in with the basic thoughts of this master of cellular pathology, all the more so since the main role has here been found to be an independent cellular element. . . .

As I remarked at the beginning, the yeast disease of Daphnia is of special interest in so far as it helps us to understand the pathological processes of the higher animals. It strengthens the statement that the white blood corpuscles and other phagocytes of vertebrates eat disease producers, and particularly the schizomycetes, and in this way are of considerable service to the organism. Although this conclusion had been drawn from the sum total of our knowledge of this subject, there has not been one conclusive example of the whole process of ingestion and digestion of fungus cells by phagocytes. Therefore we can criticize the conclusions which have been drawn concerning the presence of bacteria in whole blood cells. For example, R. Koch concluded, from his observations of various quantities of septicemia bacteria in the white cells of mice, that the bacteria could "penetrate the white blood cells and reproduce there." The process of penetration and reproduction could not be directly observed by him. . . . It seems to me more probable, that in this case also, the parasites were eaten by the blood cells. . . .

Because of the paucity of knowledge of this subject, it can be concluded that the pathological results obtained through studies on lower animals can be viewed as a new support for certain basic ideas of cellular pathology.

Comment

This paper shows how observations on lower animals can be of value in developing concepts of medical importance. There had been a long controversy, which still continues to some extent, as to whether the tissue or fluid elements of the body were responsible for natural resistance to disease. As we know today, both are important, but in Metschnikoff's time, there was no certain knowledge. As he mentions, bacterial cells had been seen inside white cells in the body, but no one had seen how they got inside. Did they force their way in as invaders? If so, then the white cells were not defensive cells, but merely favored host cells of the parasite. Were the bacteria actively eaten by the white cells? If so, then the white cells might be a line of defense of the animal body.

In studying a fungus disease in Daphnia, Metschnikoff had an ideal system to examine this question. The animal is simple and transparent, and the fungus is large and easily seen without staining. As he describes, he could watch the whole process in some detail. This work gave considerable encouragement to those who felt that the tissues of the body acted in disease resistance, and it was later possible to watch phagocytes of vertebrates ingest bacteria and verify that a similar process took place in higher animals.

Remember that there is no mention here of artificial immunity. Phagocytosis may be viewed as a mechanism of resistance in both immunized and nonimmunized animals, but it is only the latter that Metschnikoff studied in Daphnia.

The mechanism of immunity in animals to diphtheria and tetanus

1890 • Emil von Behring and Shibasaburo Kitasato

von Behring, Emil, and Kitasato, Shibasaburo. 1890. Ueber das Zustandekommen der Diphtherie-Immunität und der Tetanus-Immunität bei Thieren. *Deutsche Medizinische Wochenschrift,* Vol. 16, pages 1113–1114.

IN THE STUDIES WHICH WE HAVE BEEN carrying out for some time on diphtheria (von Behring) and tetanus (Kitasato), we have also considered questions of therapy and immunization. In both infectious diseases, we have been able to cure infected animals, as well as to pretreat healthy animals so that later they will not succumb to diphtheria or tetanus.

In what way the therapy and immunization have been obtained will only be stated here in enough detail to demonstrate the truth of the following sentence: "The immunity of rabbits and mice, which have been immunized against tetanus, depends on the ability of the cell-free blood fluid to render harmless the toxic substance which the tetanus bacillus produces."

This explanation of immunity has not been considered in any of the works on the immunity question which have appeared in recent years.

Aside from the studies on phagocytosis, which seek to explain immunity in terms of the vital activities of the cells, others have considered the bactericidal action of the blood and the acclimatization of the animal body to the toxin.

When one of these explanations has been found not acceptable, then it has been believed that this exclusion of one explanation is an argument for the other. Thus Bouchard stated: "Let us no longer speak of the action of the leucocytes or the adaptation of the nerve cells to the bacterial toxin: this is pure rhetoric." and "It is actually the bactericidal action which is responsible for vaccination or acquired immunity."

This positive statement derives from that which Roger expressed as follows: "Vaccination induces in the organism chemical modifications which make the fluids and tissues less favorable to the growth of the microbe which has been used to immunize the animal."

However, one of us (von Behring) could determine in his studies on rats immune to diphtheria and on immunized guinea pigs, that none of the theories mentioned above could explain the immunity of these animals, and he realized that it was necessary to look for another principle to explain these phenomena. After many negative experiments, it was discovered that the blood of immune animals had the ability to neutralize the diphtheria toxin, and this discovery revealed the reason for the insensitivity of these animals to diphtheria. But it was only by applying this concept to tetanus that we were able to achieve results which, so far as we can tell, are completely conclusive.

The experiments to be outlined below show:

1. The blood of rabbits immune to tetanus has the ability to neutralize or destroy the tetanus toxin.

2. This property exists also in extravascular blood and in cell-free serum.

3. This property is so stable that it remains effective even in the body of other animals, so that it is possible, through blood or serum transfusions, to achieve an outstanding therapeutic effect.

4. The property which destroys tetanus toxin does not exist in the blood of animals which are not immune to tetanus, and when one incorporates tetanus toxin into nonimmune animals, the toxin can be still demonstrated in the blood and other body fluids of the animal, even after its death.

As proof of these statements, we present the following extensive series of experiments:

A rabbit was immunized against tetanus by a method which will be reported in detail later. The degree of immunity of this animal was such that it would stand a dose of 10 cc. of a bacteria-containing culture of virulent tetanus bacilli, while a normal rabbit would always die from a dose of 0.5 cc. Every rabbit remained completely healthy after this injection.

This was not only true of the infection with living tetanus bacilli, but also by injection with tetanus toxin, for each immune rabbit would tolerate without symptoms a dose of toxin 20 times that which would kill normal rabbits.

From these rabbits, carotid blood was removed.

Before coagulation, the fluid blood was injected into the abdominal cavity of mice, 0.2 cc. and 0.5 cc. After 24 hours these two mice were inoculated along with two control mice with virulent tetanus bacilli. The inoculation was strong enough so that the control mice became sick after 20 hours and died after 36 hours. However, both treated mice remained perfectly healthy.

Most of the blood from the immunized rabbits was allowed to stand until it had coagulated and the serum had formed.

Six mice were injected with this serum in the abdominal cavity. The infection 24 hours later had no effect on these, while the control mice died from tetanus after 48 hours.

The serum could also be used for therapeutic treatment, in which the mice were infected first, and then the serum was injected intraperitoneally afterward.

Also we have done experiments with the serum, which show the enormous toxin destroying activity of it.

A ten day old tetanus culture was filtered to render it free of bacteria. 0.00005 cc. of this was sufficient to cause death of mice after 4–6 days, and 0.0001 cc. caused death after less than 2 days.

Now we mixed 5 cc. of serum from rabbits that were immune to tetanus with 1 cc. of this toxin-containing culture and allowed the serum to act on the toxin for 24 hours. We injected 0.2 cc. of this mixture into each of four mice. This would correspond to 0.033 cc. of culture fluid, or more than 300 times the lethal dose for mice. All four mice remained perfectly healthy. The control mice, however, died 36 hours after an injection of 0.0001 cc. of culture.

The mice from all of the experiments that had received either serum alone or serum with toxin were rendered permanently immune, so far as one can tell. Repeated injections at a later time with virulent tetanus bacilli caused not a trace of illness in them. . . .

Naturally we have performed every experiment with control blood and serum from non-immune rabbits. Such blood and serum are neither therapeutic nor active against tetanus toxin.

The same is true of serum from cows, calves, horses, and sheep. . . .

In earlier times blood transfusions were considered to be effective . . . methods for the treatment of diseases. Recently it has been believed that physiological saline can exert the same effects. The results of our experiments remind us forcibly of these words: "Blut ist ein ganz besonderer Saft," ["Blood is a very unusual fluid."]

Comment

The science of serology can be said to have begun with this paper. It presented the first evidence that there were substances formed in the serum in response to infection which were able to neutralize foreign materials. It was probably only with tetanus that the facts could first have been demonstrated so clearly.

All of the symptoms of tetanus are due to a toxin that is elaborated by the causal organism. This toxin had been discovered earlier by Kitasato. It is produced in artificial culture media. When this toxin is injected into mice, they die rapidly, and a reproducible curve relating dose and number of deaths can be obtained. By a method not mentioned by the authors, it was possible to immunize rabbits against tetanus. We know now that this immunity is due to the production of antibodies by the rabbit in response to the toxin. These antibodies are able to neutralize the toxin and completely prevent the symptoms of the disease. Because a rabbit is a much larger animal than a mouse, it is possible to inject into the mouse serum from immunized rabbits in sufficiently large amounts so that there is sufficient antibody present in the mouse to neutralize all of the tetanus toxin injected. In this way the mouse is passively immunized. We know now that this immunity is not permanent, but gradually wears off. This is because the rabbit antibodies are gradually destroyed by the mouse. von Behring and Kitasato missed this point, for a number of rather complex reasons which need not concern us here.

Note that the substance in the serum of immunized rabbits is highly specific, neutralizing only tetanus toxin and none other. This point is quite important, since it demonstrates that antibodies are highly specific. This specificity is at the base of a number of immunological procedures.

Note also that this discovery opens up the possibility of specific therapy for diseases through the injection of immune serum. It also opens up the possibility of specific prevention of diseases through the induction of specific antibody production in a potential host. This is one of the most important contributions that has been made to medicine by microbiology.

Studies on the mechanism of immunity to diphtheria in animals

1890 · Emil von Behring

von Behring, Emil. 1890. Untersuchungen über das Zustandekommen der Diphtheria-Immunität bei Thieren. *Deutsche Medizinische Wochenschrift*, Vol. 16, pages 1145–1148.

In no. 49 of this journal, Kitasato and I reported on experiments which show that the immunity to tetanus of experimental animals resides in the ability of the blood to render harmless the toxic products of the tetanus bacillus.

The same mechanism was advanced in that paper for diphtheria immunity, without actually reporting experiments which supported this idea. It is the purpose of this paper to present this data.

Löffler, as well as Roux and Yersin, have shown that there are animals which are naturally immune to diphtheria. I have found myself that mice and rats fall into this category, and that these animals can be injected with cultures which have no effects on their health, while equal volumes of these cultures are fatal to guinea pigs, rabbits, and sheep.

A broth culture of diphtheria bacteria isolated from the membrane of a child which had died in January of this year from diphtheria can kill a guinea pig in 3–4 days after an injection of only 0.05 cc. In rabbits, 0.3 cc. subcutaneously kills after 2–4 days. 2.0 cc. kills a mature sheep after 50 hours. From the same culture, rats can receive 2 cc. and mice 0.3 cc. without showing any disease symptoms.

It is also possible to take animals which originally are sensitive to diphtheria and make them immune, and this can be done in different ways.

1. One of these immunization methods, which I can state from my own work is very reliable, has been described by Prof. C. Fränkel. Guinea pigs are inoculated with cultures that have been sterilized, and after 10–14 days of such inoculations, they are insensitive to doses which would ordinarily kill them.

2. I have also immunized guinea pigs in the following way: I take 4 week old cultures and add iodine trichloride in concentrations of 1/500 and allow this to act for 16 hours. Then I inject two guinea pigs with 2 cc. of this treated culture in the abdominal cavity.

After 3 weeks I inject these guinea pigs with 0.2 cc. of a diphtheria culture, which has grown for four days in broth containing 1/5500 iodine trichloride. The control animals die after 7 days; the treated animals remain alive.

After another 14 days the two treated animals receive an injection of a fully virulent culture, which is sufficient to kill normal guinea pigs in 36 hours.

In both of these methods, it is the metabolic products which are produced by the diphtheria bacilli, which induce the immunity. . . .

4. Another method which has not been reported previously is based on the metabolic products of the bacteria.

It consists of first infecting the animals and then alleviating the deleterious effects through therapeutic treatment.

This method is similar to that which occurs in humans when they recover from an infection. Of the many chemical agents which could be effective, the best seems to be iodine trichloride. Eight guinea pigs which had been infected subcutaneously with 0.3 cc. of culture were divided into two groups. The two controls died after 24 hours. The other four animals were treated immediately after the infection with 2 cc. of a solution of iodine trichloride (two smaller animals received a 1% solution, two larger received 2%) at the point of infection. These remained alive. With two other animals the treatment was begun 6 hours after infection, and one animal died after four days while the other remained alive. All animals received new injections of iodine trichloride on the next three days. When the treatment was delayed more than six hours after infection, I have had no success. Also, I have been unsuccessful in immunizing animals when they receive so light an infection with the bacteria that the controls do not die until four days later. . . .

I have animals treated with iodine trichloride, and Dr. Boer has animals treated with gold sodium chloride, which have recovered from the infection, and these guinea pigs are able to stand injections of fully virulent diphtheria bacteria which would kill control animals in 36 hours. . . .

5. Another method for the induction of complete resistance in guinea pigs against infections, and in rabbits against the lethal effects of culture fluids, is not related to the metabolic products of the diphtheria bacillus. This is through the use of hydrogen peroxide. . . .

Hydrogen peroxide is an excellent disinfectant in certain cases, and I originally used it with the idea that it might have a therapeutic effect in diphtheria. However, animals which were treated after infection with hydrogen peroxide became diseased much faster than the controls. Hydrogen peroxide also seemed to increase the virulence of the culture when added to it.

However, when I treated the animals with hydrogen peroxide several days before infection, I found that they had achieved a more or less pronounced degree of immunity. . . .

Of the five different immunization methods which have been mentioned, the first four are basically those methods which we have been acquainted with by Pasteur. The fifth, which produces immunity through completely heterogeneous material of a simple chemical nature, has no analogy in previous work. . . .

In my opinion, all five methods of immunizing against diphtheria could not be useful in humans, at least in the form they have been described here.

However, from the scientific viewpoint, and especially for the solution to the problem of the mechanism of diphtheria immunity, these methods may serve a very useful purpose.

No matter how the immunity is induced, and I include here also natural immunity, diphtheria-immune animals have certain characteristics in common, which differentiate them from non-immune animals.

First, all living immune animals are protected not only against the infection with living diphtheria bacilli, but also against the harmful effects of that toxic substance which is produced by the diphtheria bacilli in culture and in the animal body. . . .

For my purposes, it was not necessary to separate or purify the diphtheria toxin or toxins from the culture fluid. Old filtered cultures furnished me with suitable material.

I prepared cultures in alkaline broth and found that after 10 weeks these contained so much toxic substance, that after filtering to make them bacterial-free, 1 cc. of this broth injected into middle-sized guinea pigs induced typical symptoms of diphtheria toxification, that did not completely disappear until after three to four weeks. Three to four cc. was sufficient to kill a large guinea pig in three to eight days. Regularly necrosis of the skin appeared, and not only at the injection site, but quite distant from it, most frequently on the abdomen.

All guinea pigs which had a solid immunity, so that they could stand repeated infections of bacteria without any diseased symptoms, could also stand injections of 3 to 5 cc. of this toxic broth without showing any toxic symptoms or only a localized reaction. . . .

These observations and considerations led me nearer to the question of whether the cause of the toxin resistance was possibly due not to a characteristic of the living cellular part of the animal, but to a special property of the cell-free blood.

In order to answer this question, I took rats which had received intraperitoneally a large amount of diphtheria toxin and removed, three hours later, their blood and injected it or the serum from it intraperitoneally into guinea pigs. There was no sign of toxic symptoms, while blood from diphtheria-sensitive animals, which had received an equal amount of diphtheria toxin, was injected in equal amount (4 cc.) into guinea pigs, and although it did not kill them, it made them distinctly sick.

A further fact in support of this argument is that the extravascular blood from animals immune to diphtheria possesses the ability to render diphtheria toxin harmless. The intensity of this effect as well as the therapeutic possibilities of blood from immunized animals will be taken up in a further communication. . . .

After I had made these observations regarding the mechanism of immunity to diphtheria, Herr Kitasato and I performed similar experiments with tetanus. In the earlier communication we were able to offer uncontestable proof that the toxin-destroying property of the blood of tetanus-immune animals is a sufficient explanation for tetanus immunity.

In tetanus we had the happy situation that we were able to obtain large quantities of blood and serum from immune rabbits and then inject this into very small mice. This made it possible to demonstrate completely and conclusively the therapeutic consequences of this treatment. Perhaps the therapeutic possibilities of the transfusion of blood from tetanus-immune rabbits was not sharply enough expressed in our earlier communication. If not, I would like to mention especially here that mice that have received blood from tetanus-immune rabbits are not merely immunized, and are not merely protected from some future attack of the disease. Even when the mice have been infected in several extremities with tetanus, and would, from past experience, die within a few hours, it is even then possible with high certainty to produce a cure by serum injection. And this cure takes place so rapidly that even after a few days there are no signs of the illness left.

The possibility for the cure of very acute diseases can therefore no longer be denied.

Comment

This work is an extension of von Behring and Kitasato's paper, just preceding. It presents observations on diphtheria which seem to indicate that a similar situation exits here as in tetanus. Diphtheria was more important medically than tetanus, since it was at one time one of the great killers of children. This statement is no longer true largely because of the work of von Behring.

Like tetanus, diphtheria is also a disease in which all of the symptoms are due to the action of a toxin which is elaborated by the causal organism. Once successful results had been obtained with tetanus, it was natural to attempt to extend this work to diphtheria. von Behring has apparently done so in this paper, although the data are not completely convincing. But it is true that, with the help of Ehrlich and others, he was able to develop this discovery into a practical therapeutic method for diphtheria. Within the next ten years, serum containing antibodies against diphtheria (diphtheria antitoxin) was available commercially and was being used widely to cure children of the disease. Also available later were vaccines which could be used to induce formation of diphtheria antitoxins in potential hosts, and thus eventually eliminate diphtheria as an important disease. Today cases of diphtheria are so rare that they receive widespread attention when they occur.

Leucocytes and the active property of serum from vaccinated animals

1895 • Jules Bordet

Bordet, Jules. 1895. Les leucocytes et les propriétés actives du sérum chez les vaccinés. *Annales de l'Institut Pasteur*, Vol. 9, pages 462–506.

IN CONSEQUENCE OF THE EXTENSIVE researches of recent years, the theory of phagocytosis seems capable of explaining entirely the phenomenon of immunity.* The intervention of phagocytes constitutes a regular process in the defense by animals against parasites, and their cooperation is a powerful aid in the recovery of infected organisms. There have been cases, nevertheless, where the importance of phagocytes has been considered as a secondary factor, and the essential thing for the destruction of the infec-

* [For an example where this is not so, see the preceding paper (page 141).]

tious agent, following an ancient idea, has been attributed to a bactericidal substance dissolved in the fluids. This viewpoint has been especially defended with regard to the infections that are produced by microbes of the vibrio group, and it is the case of cholera in particular which has been cited most often in the battle between the doctrine of phagocytosis and the so-called theory of the bactericidal nature of the fluids. Indeed, in this special case, there exists an evident correlation between the appearance of an antiseptic power in the serum and the appearance of immunity in the animal. This bactericidal power, which is very weak in the unimmunized animal, becomes very marked in the vaccinated animal. And if it could be proven that the microbicidal material can be found during the life of the animal, uniformly distributed in the fluid, the cooperation of the phagocytes would not seem to be particularly necessary for the destruction of the infectious agent. But this point has not been precisely demonstrated. The researches of various workers, and M. Metschnikoff in particular, seem to show that the fluids of the body do not have the bactericidal power in the living animal which the serum possesses *in vitro*.

Nevertheless, it remains to be asked if, in the living organism, the active materials are not formed by fixed cells, and, for example, if it is not the leucocytes which contain the bactericidal material, it is possible that the production of this substance is a special manifestation of their protective activity, and results from an adaptation undergone by these elements under the influence of the vaccination. If such may be the origin of this material, the appearance of bactericidal power in the serum of a vaccinated animal may only be an indication of an actual increase in phagocytosis. . . .

Specificity of the bactericidal power.

Does the bactericidal power which has been found in the serum of a vaccinated animal act uniquely against the strain of vibrio which was used to vaccinate the animal? Or is it, on the contrary, more general, acting on most of the microbes belonging to the vibrio group? M. Pfeiffer indicates that the bactericidal power of serum from a vaccinated animal is specific. We have performed several experiments on this subject. We have immunized guinea pigs or rabbits completely with various strains of vibrios. We have collected the serum from the animals and have assayed the bactericidal power of each serum against each of the vibrios. [These results are presented in Table 1.]

[Similar results were obtained with other strains and in guinea pigs.]

We do not believe it necessary to extend these experiments further, first because these results are in agreement with those of others, further because we have devised a procedure for determining whether an immune serum is bactericidal against a given vibrio, which is much easier than doing plate counts in gelatin. This procedure, which we shall stress later, consists of revealing bactericidal action, in the case where it exists, by a granular transformation of the vibrios (Pfeiffer phenomenon), which the vibrios undergo when they are mixed with an immune serum alone, or an immune serum mixed previously with a certain amount of fresh non-immune serum.*

We can see in Table 1 that the bactericidal power of a serum is always the strongest against the vibrio identical with that which was used to vac-

* [This serum-induced granular transformation (Pfeiffer phenomenon) is probably lysis of the cells.]

TABLE 1

Serum of rabbit vaccinated with *Vibrio cholera*, strain Massaouah *

Hour of counting	*V. Metchnikovii*	*V. cholera* Massaouah	*V. cholera* Calcutta	Vibrio of Finkler	*V. cholera* Constantinople	*V. cholera* Canino
4:30 P.M.	16,800	13,800	7,800	13,500	17,400	12,400
6:15 P.M.	16,200	8	0	70	16,800	30
11:15 A.M.	very high	0	0	120	very high	very high

* The experimental procedure here was as follows. A known number of cells of each strain of Vibrio were inoculated into the immune serum and viable counts made at intervals by pour plate procedures. The values are the viable counts per unit volume. In certain strains the viable count dropped quickly to zero, while other strains were unaffected and grew to a large number of cells.

cinate the animal which provided the serum, and for those vibrios which are in the same pathogenic species as it. The microbicidal effect is without a doubt quite specialized. . . .

The organism has at its disposal a means of defense against microbes against which it has been vaccinated, which is well adapted in its end, and which is directed exclusively against the species of vibrio with which it has undergone repeated contact. In addition, the unvaccinated animal possesses a bactericidal power which is able to destroy actively diverse types of vibrios, although this action is truly much less powerful than the specific action of the immune serum. . . . The bactericidal substance of the non-immune animal and that of the immune animal have in common the fact that they are destroyed by heating to about 60°. Sera which are heated to this temperature become excellent culture media. But these two sera can be easily distinguished from each other in that the one, weakly active, from non-immune animals, is non-specific, while the other, highly active, from immune animals, possesses a high degree of specificity. In this way the two substances can be easily differentiated. Moreover, the bactericidal substance in the serum of vaccinated animals differs from the immunizing substance present in the same serum. The work of MM. Fraenkel and Sobernheim has shown that the immunizing substance is resistant to prolonged heating at 70°, while the bactericidal substance is destroyed under these same conditions. Serum which has been heated to 70° has the same ability as unheated serum to immunize animals when injected into them. Even though it possesses no antiseptic action, it produces in the injected animals a bactericidal condition in the fluids. It appears therefore that there may be at least three substances in serum which are active and play roles in immunity: two distinct bactericidal substances and an immunizing substance. In reality, it does not seem to be so complicated. It may be that the bactericidal substance of the vaccinated and unvaccinated animals may be identical, even though the one is highly active and specific, while the other does not possess these characteristics. Although weakly active in non-immune serum, this material is strongly active in the immune serum, under the influence of an immunizing substance which, when the two are together, the latter incites the bactericidal power and at the same time brings about the specificity.

Indeed, it is only necessary to add to non-immune serum which is weakly bactericidal, a small quantity of anti-cholera serum, unheated or previously heated to 60° (this latter removes its toxic properties for the vibrio), to induce in it a very marked bactericidal power. Thus two fluids, weakly bactericidal when separated, form together a mixture which is strongly antiseptic. This antiseptic power, as we will see later, is specific. It is only manifested against the species of vibrio which was used to vaccinate the animal from which the immune serum was taken. The bactericidal power can be observed just as easily whether the serums contain cellular elements or have been deprived of these elements. . . .

EXPERIMENT

Fresh blood was taken from a guinea pig. The serum was allowed to separate, and only the fluid part free of cellular elements was taken. In the tubes were placed the following: (1) 12 drops of serum from a goat highly vaccinated against *Vibrio cholera* Prusse orientale; this serum had been heated to 58° for one hour. (2) 12 drops of fresh guinea pig serum. (3) 12 drops of a liquid composed of 8 drops of fresh guinea pig serum and 4 drops of goat serum heated to 58°. (4) 12 drops of a mixture as in 3, but containing the portion of the guinea pig serum which contained the cellular elements. These tubes were seeded with a culture of *Vibrio cholera* Prusse orientale that was 24 hours old, which had been diluted in 10 cc. of 0.6% NaCl. Plate counts were made on gelatin. [These results are presented in Table 2.]

Therefore, a very small amount of immune serum suffices to convert fresh serum into a strong bactericide. This power can be determined by doing plate counts, but it can also be determined by observing the transformation of the vibrios into granulations (Pfeiffer phenomenon), which occurs very rapidly when a drop containing the vibrios is mixed with the serums. . . .

TABLE 2

Counts of *Vibrio cholera* Prusse orientale in various cultures

Hour of counting	*Heated immune serum of goat*	*Fluid serum of non-immune guinea pig*	*8 drops guinea pig, 4 drops goat serum*	*8 drops guinea pig with cells, 4 drops goat serum*
6:00 P.M.	8,640	9,600	10,200	9,120
7:30 P.M.	4,320	2,160	0	0
10:00 P.M.	6,480	3,600	0	0
10:00 A.M.	very high	very high	0	0

Comment

Although it illustrates a number of immunological points, this paper is presented here because it presents the discovery of a substance which we now call complement. Although this paper may seem confusing, it is amazing that Bordet was able to clarify the situation as well as he could, considering the puzzling phenomena which he was observing.

Complement is a complex substance present in non-immunized animals which is able to act in cooperation with a specific antibody to bring about a number of immunological reactions. The two reactions reported here are bactericidal action and bacteriolysis. Because all sera, both immune and non-immune, contain complement, it is only possible to demonstrate it by some special procedure. If we use an immune serum, it contains both complement and specific antibody. Since the action we are studying requires the cooperation of both of these substances, we will never know that both exist unless we can remove one. Fortunately complement is readily destroyed by mild heating, leaving specific antibody unaffected. The heated immune serum is not able to kill bacteria. Its killing power can be restored by adding to it some unheated serum from a non-immune animal. This latter serum will not kill bacteria since it lacks specific antibody. Only the mixture of the two sera will be effective. This is what Bordet showed.

The number of immunological reactions involving complement continues to be extended even today. Even so, its mechanism of action remains an unsolved problem.

Much of immunology falls outside the domain of microbiology. I have tried to present here certain of the key papers in the development of immunological concepts and show that these discoveries were made because of studies on infectious diseases caused by microorganisms. Although even today this is one of the main areas of immunological study, it is by no means the only area.

PART IV
Virology

Report of the commission for research on the foot-and-mouth disease

1898 • Friedrich Loeffler and P. Frosch

Loeffler, Friedrich, and Frosch, . 1898. Berichte der Kommission zur Erforschung der Maul-und Klauenseuche bei dem Institut für Infektionskrankheiten in Berlin. *Zentralblatt für Bakteriologie, Parasitenkunde und Infektionskrankheiten.* Part I, Vol. 23, pages 371–391.

THE UNDERSIGNED COMMISSION IS honored to present to your excellency the most recent results of our studies on foot-and-mouth disease.

1. STUDIES ON THE ETIOLOGY OF THE DISEASE

Thanks to the use of the telegraph to inform us of fresh outbreaks of foot-and-mouth disease, which your excellency placed at our disposal through contacts with many local officials, it has been possible to obtain extensive quantities of fresh material for these studies. For this purpose, locations were selected which would permit the collection of material and its return to us on the same day, so that immediate processing of fresh material could be done. . . . Especially suitable for the etiological studies were the contents of freshly removed vesicles in the mouth and udder of sick animals, because from these places the contents of the vesicles could be obtained free from contamination at the surface. Since the feet were mostly soiled with feces it was very difficult to remove material from this area that was uncontaminated, in spite of careful disinfection of the surface layers.

Because we discovered that bacteria penetrated into the vesicles from outside after several days, we decided to always use freshly developed vesicles for these studies. Such completely fresh vesicles are relatively scarce. Even after an outbreak of the disease in a large herd, at most only one or two animals offered such vesicles to us.

The contents of the vesicle were removed after previous treatment of the vesicle with absolute alcohol. A sterilized glass capillary was plunged into the interior of the vesicle to remove the material. In this way the contents of the vesicles from 12 animals from different locations could be studied closer. The examinations were carried out with the help of bacteriological methods (hanging drop, staining, culture). As culture media, the following were used: ordinary broth, acid and alkaline broth, peptone-broth, glucose-broth, liquid and solidified serum, milk, nutrient agar, and nutrient gelatin, and with atmospheres of air, hydrogen, hydrogen sulfide, and carbon dioxide.

The results of all of these studies were completely the same. Staining and examination in hanging drop revealed no bacteria. The culture media inoculated with fluid from the vesicles remained absolutely free of any bacterial development, in most cases for weeks. In those cases where bacteria did occur, it could always be determined on the first observation that they were merely isolated colonies which had arisen from germs which had accidentally entered from the air. It could be shown at the same time that these sterile fluids contained the causal factor of foot-and-mouth disease, because when this material was inoculated on the mucous membranes of the upper and lower lips of calves and heifers, they always became sick from the disease with typical symptoms in two or three days after inoculation. . . . It can be shown that we are not concerned with the action of a germ-free toxic substance produced in the vesicles from the fact that when animals are inoculated and come down with the disease, it is transferred to other animals which are housed in the same stall. From these experiments it seems certain that any species of bacteria which is able to grow on the usual culture media cannot be the etiological agent of foot-and-mouth disease. . . .

Through its efforts to find a practical method for the immunization of animals, the commission has attempted to use not only blood and lymph mixtures from immune animals, but has also attempted to introduce the blood of animals that were acutely sick with and without additions of immune blood, as well as the use of lymph which has been freed from all its cellular elements by filtration. These experiments have not as yet led to satisfactory results. However, the filtration experiments have offered extremely important results in another area. In these experiments the lymph was diluted with 39 parts of water, then inoculated with an easily culturable and identifiable bacterial species which had been cultured from lymph—*Bacillus fluorescens*—and then filtered two to three times through a sterilized Kieselguhr candle. The addition of the bacteria served to demonstrate that the filtrate was really bacterial-free, since large inoculums of this were then placed on nutrient media and examined for growth after incubation. If colonies of this bacterium did not appear on the seeded media, then it was assumed that the filtration had succeeded, and that all of the bacteria that had been previously present in the lymph were retained by the filter candle. Filtrates tested in this manner were always bacterial free. A series of calves were inoculated intravenously with measured amounts of these filtrates corresponding to 1/10 to 1/40 cc. of pure lymph, in order to ascertain if there was a dissolved substance in the lymph which would aid in the production of immunity.

The results of these injections were quite surprising. The animals inoculated with the filtrates died in the same time as had the control animals which had received corresponding amounts of unfiltered lymph. Their deaths occurred with all of the typical symptoms of the disease, such as high fever, and vesicles on the mouth and hoofs. We had the impression that the activity of the lymph was not influenced by the filtration. In order to confirm this important result, these experiments were repeated in a large number of calves and hogs. The results with the use of fresh lymph were always the same. The animals treated with the filtrate became sick just as did the controls treated with unfiltered lymph, and always in a completely typical way.

How could these surprising results be explained? For the explanation there were two possibilities: 1. The bacterial free lymph contains a soluble, extremely active toxin. 2. The causal agent of the disease, which up to now has not been found, was so small that the pores of filters which were able to hold back the smallest known bacteria were able to allow this agent to pass through. If we were dealing with a soluble poison, then it must be one with an astounding activity. With a sample of filtrate corresponding to 1/30 cc. lymph, the disease could be produced in two days in calves of around 200 kg. weight. . . .

If we assume that in the lymph of the vesicles of foot-and-mouth disease there is an active substance which is 1/500 of the lymph, in 1/30 cc. or 1/30 g. of lymph there would be 1/15,000 g. active substance. This amount is sufficient to cause disease in a 200 kg. calf. If an equal distribution of the substance occurred through all parts of the body of the calf, then one part of toxin is distributed in $15{,}000 \times 200{,}000$, or 3,000,000,000, parts of calf and is still active. Knorr has been able to produce in a tetanus culture a toxin which is toxic in 1/1,000,000 parts of rabbit, 1/150,000,000 parts for white mouse, and 1/1,000,000,000 parts for guinea pigs. The toxicity of the foot-and-mouth disease toxin would then be considerably higher than that of the highly toxic tetanus toxin. We would have in the foot-and-mouth disease toxin a poison of completely astounding activity.

However, it has been possible for us to take the lymph from the vesicles of an animal which had been infected with filtered lymph, and use only 1/50 cc. of this to induce a similar disease with typical symptoms in two days in a 30 kg. hog. The toxin has here experienced a further dilution of $1/50 \times 500 \times 3{,}000$ or 1/750,000,000, so that the dilution of the original lymph is $1/2.25 \times 10^{17}$. A toxin which is this active would be truly unbelievable.

In these calculations we have made the assumption that the toxin is equally distributed throughout the whole animal. We are, however, not compelled to make such an assumption. Although in tetanus it is quite easy to demonstrate the presence of the toxin in the blood and organs of the animal, such a demonstration is not so easy in animals suffering from foot-and-mouth disease. In order to transmit the disease to healthy animals by blood of diseased ones, it is necessary to use 50–100 cc. It therefore seems quite possible that the greatest part of the toxin injected into such animals is accumulated in the vesicles which form. If we assume that an animal which has been injected with filtered lymph undergoes typical symptoms of disease after the injection of 1/30 cc. of filtered lymph, and if we further assume that all of the vesicles on the four hooves, in the mouth and on the tongue together contain about 5 cc. of fluid, which is just about right, then the original 1/30 cc. of lymph is now distributed in 5 cc. of new lymph in the vesicles, for a dilution of 1/150. But of this 5 cc. only 1/50 cc. is sufficient to induce typical disease in a 30 kg. hog in two days after intravenous injection. So that there is now a dilution of $1/150 \times 1/50$, or 1/7,500 which is active. If then the active toxin content of the lymph is assumed to be like that of tetanus broth, or 1/500, then there would be $1/7{,}500 \times 1/500 = 1/3{,}750{,}000$ of toxin in 30,000 g. of hog or 2,000 g. of blood. The activity of the toxin would then be such that one part in 7,500,000,000 g. of hog blood would be active. Even this calculation of the amount of toxin present in the lymph seems to indicate an exceedingly high activity. It therefore seems more appropriate to conclude that the activity of the filtrate is not due to the presence in it of a soluble substance, but due to the presence of a causal agent capable of reproducing. This agent must then be obviously so small that the pores of a filter which will hold back the smallest bacterium will still allow it to pass. The smallest bacterium presently known is the influenza bacillus of Pfeiffer. It has a length from 0.5 to 1.0 microns. If the supposed causal agent of foot-and-mouth disease was only 1/10 or even 1/5 as large as this, which really does not seem impossible, then this agent would not be resolved in our microscope, even with the most modern immersion system, according to the calculations of Professor Abbé in Jena. This would explain very simply why it has been impossible to see the causal agent in the lymph under the microscope, even after the most extensive search. Through careful studies we may finally be able to decide definitely over the question of the presence of the causal agent in the bacterial free lymph filtrate. In order to continue this study, the commission asks for the granting of materials and supplies. This study, aside from its purely scientific value, seems to offer eminently practical possibilities. If it is confirmed by further studies of the commission that the action of the filtrate, as it appears, is actually due to the presence of such a minute living being, this brings up the thought that the causal agents of a large number of other infectious diseases, such as smallpox, cowpox, scarlet fever, measles, typhus, cattle plague, etc. which up to now have been sought in vain, may also belong to this smallest group of organisms. If a bacterial free cowpox lymph could be produced, this would ease the agitation against vaccination for smallpox.

The bacterial free filtrates of infectious material probably offer the most suitable material for the acquisition of important new conclusions on the nature of the diseases named above.

All of these speculations indicate that it would be highly desirable to continue the studies on the action of the filtrate in a larger number of animals as soon as possible.

Comment

This paper presents the first evidence for the occurrence of pathogens which we now call filterable viruses. The data in the paper are clear and easily understood. The authors have considered all of the possibilities known to them, and there can be little doubt that foot-and-mouth disease is caused by a parasite too small to be seen under the microscope or to be retained by bacterial filters. The parasite will also not grow on ordinary culture media, but this point is not stressed by the authors. We know now that all of the filterable viruses are apparently obligate parasites and have never been cultured away from living hosts.

The authors are well aware of the importance of their discovery. They were also correct in speculating that certain other infectious diseases, such as smallpox, cowpox, cattle plague, and measles, are caused by filterable viruses. Of the other diseases which they mention as possibly being caused by submicroscopic organisms, scarlet fever is caused by a streptococcus, while typhus is caused by a Rickettsia.

Again we must point out that the reason the present authors were able to obtain such clear-cut results was because of the disease they were studying. It is a disease of domestic animals, so that experimental infections can be easily performed. The symptoms are quite characteristic, and the virus is present in the lesions which develop on the skin. This makes it possible to obtain preparations relatively free of other material. These characteristics made it possible for the authors to study the agent so successfully.

A *Contagium vivum fluidum* * as the cause of the mosaic disease of tobacco leaves

1899 · M. W. Beijerinck

Beijerinck, M. W. 1899. Ueber ein Contagium vivum fluidum als Ursache der Fleckenkrankheit der Tabaksblätter. *Centralblatt für Bacteriologie und Parasitenkunde*, Part II, Vol. 5, pages 27–33.

THE LEAF SPOT DISEASE OF TOBACCO, also called the mosaic disease, is manifested first as a bleaching of the chlorophyll, occurring in spots over the leaf blade. This is followed later by the death of a part or all of the tissue of the spots. The discoloration first appears right next to the leaf veins and is manifested then by a strong increase in the amount of chlorophyll. Later the spaces between the spots become bleached usually to a yellow color, but in isolated cases they can become completely albino. The dark green patches grow at the beginning more rapidly than the other parts of the leaf, leading to wart-like growths which arise from the upper surface of the leaf. However this phenomenon is observed more often in artificial infections than in tobacco fields, where the diseased leaves usually remain completely flat.

The third phase of the disease consists of localized death of the hundred or thousand small spots which are distributed randomly over the leaf. These then assume a brown color and become very fragile, so that holes are

* [*Contagium vivum fluidum* could be most likely translated: "living germ that is soluble."]

formed easily during the harvest of the leaves. These spots are the fear of the Dutch tobacco farmer, because they make the leaf worthless for cigar wrappers. . . .

Herr Adolf Mayer showed in 1887 that this disease is contagious. He expressed the sap from sick plants, placed it in capillary tubes and stuck these in healthy plants. He found that after 2–3 weeks, the latter plants became diseased.

In 1887 I attempted to discover if there was not a parasite which could be demonstrated to be the cause of the disease. Since microscopic studies were completely negative, the only type of bacteria that could be considered were those which could not be observed directly. But culture procedures showed that aerobic bacteria were completely absent, either from healthy or diseased plants. I later showed that anaerobic bacteria were also absent.

It seemed certain, therefore, that we were dealing here with a disease which was caused by a *contagium* which was not a *contagium fixum* in the usual sense of the words. This encouraged me to carry out new experimental infections in 1897 and 1898, in order to understand the properties of the *contagium* better. I would like to present here briefly the main results which were obtained in these studies.

It was first shown that the juice expressed from sick plants did not lose its virulence even after being filtered through a porcelain filter that was so fine that it rendered the juice completely sterile. This filtrate was tested for the presence of both aerobes and anaerobes, so that the experiment was completely unobjectionable. This filtrate was kept three months and remained completely bacterial free during this time but was repeatedly shown to induce the identical mosaic disease when inoculated into plants. I do not know how long the virulence of this filtrate can be maintained.

The following experiments were designed to answer the question as to whether the virus should be considered particulate or soluble.

Pulverized tissue of diseased leaves was spread on thick agar plates and diffusion allowed to occur. A virus which was particulate would remain on the surface of the agar, since it could not diffuse into the molecular-sized pores of the agar plate. The deep layers of the agar would therefore not become virulent. But a water soluble virus ought to be able to penetrate to a certain depth in the agar plate. The experiment was discontinued after a diffusion time of about 10 days, which could be considered to be long enough, since I knew that diastase and trypsin would diffuse a considerable extent in this time. The upper surface of the plate was first washed with water and then with a strong solution of mercuric bichloride. After this, a sharp platinum needle was used to remove part of the agar, so that the inner layers could be reached, care being taken not to disturb the upper surface. Healthy plants were then infected with agar from these deep layers. The infection was just as extensive with this material as when the sterile filtrate was used. It can hardly be doubted, therefore, that the *contagium* must be considered to be fluid, or, more accurately, water soluble.

The experimental infections using plant juices were performed using the hypodermic needle of Pravaz. The most suitable place to infect is the youngest part of the stem which can be manipulated easily without causing extensive damage, since the closer the infection is to the meristem of the terminal bud, the earlier the results are seen. It has been shown that the virus moves slowly through the plant, and further, that only the portions of young leaves that are undergoing cell division are sensitive to the infection. Both the mature leaves as well as the young leaves in which the cells have already stopped dividing are completely insensitive to the virus, even though they are able to transport it towards the meristematic regions. If stem internodes that are enlarging are infected, after 10–12 days the first symptoms of the disease can be observed in the young leaves which are coming out of the apical meristem. However, if an infection is carefully made as close as possible to the apical meristem, even after 3–4 days yellow spots and crisp distorted areas can be observed in the youngest little leaves that are still within the bud.

The amount of virus which is sufficient to infect a large number of leaves is quite small. It is then possible to obtain material from these diseased leaves which can be used to infect unlimited numbers of new plants. It is therefore quite clear that the virus is reproducing within the plant. From the above it is clear that this reproduction is not in the mature plant cells but in those tissues where cell division is occurring. . . .

Although the virus can exist outside the tobacco plant, it cannot reproduce under these conditions. I conclude this from the following facts: If a sterile filtrate of the virus is mixed with fresh plant sap of young tissues of healthy tobacco plants, it can be determined by experimental infections that no reproduction of the virus is obtained. Instead the virus is diluted in the same way as if pure water had been used instead of plant sap.

It is not difficult to determine the accuracy of this statement, since the amount of virus used to infect plants has a great influence on the development of the symptoms of the mosaic disease. With a small amount of virus, the usual results are obtained as described above. With large amounts of virus, highly deformed leaves of characteristic shape are obtained. In order to obtain these deformed leaves, it is necessary to inject much more of a diluted virus than of one not diluted. In this way it is easy to tell whether the virus has reproduced or stayed the same in any type of fluid. As mentioned above, I have not observed reproduction under artificial conditions, so that I believe the only mode of reproduction of the virus is in cells of the plant that are dividing. . . .

The ability of the virus to reproduce only when combined with the living protoplasm of the host plant may be related to its soluble or fluid nature. It is not easy to understand why a *contagium fixum*, even if so small that it could not be seen by direct microscopic examination, could not still reproduce away from the host, like ordinary parasitic bacteria. In addition, it also would seem probable that a microscopically invisible *contagium*, if particulate, could develop into macroscopically visible colonies on gelatin plates.

A soluble and diffusible virus, such as the mosaic virus, should bring about some coloration or change in refractive index of a gelatin or agar medium, if the chemical nature of the medium were altered when used as a nutrient by a reproducing virus. Such changes could not be seen when the virus was seeded onto malt extract gelatin or onto plates containing 10 per cent gelatin dissolved in a plant decoction containing 2 per cent cane sugar—both excellent media in my hands for the growth of parasitic and saprophytic plant bacteria. It also seems to me that reproduction or growth of a soluble body is not inconceivable, although difficult to imagine. It would not seem wise to assume a division process of molecules which would lead to their reproduction, and the idea of molecules which feed themselves, which must be assumed to explain this, seems to me an unclear concept, if not actually contrary to nature.

A partial explanation would be the view that the *contagium* must be incorporated into the living protoplasm of the cell in order to reproduce, and its reproduction is so to speak passively brought about with the reproduction of the cell. But this would then leave us one mystery instead of two, since the incorporation of a virus into the living protoplasm, even if shown to be a fact, can in no way be viewed as an understandable process. . . .

If the soil in which a tobacco plant is growing is infected with the virus, after a time the disease is seen to appear in the apical bud. The length of time for its appearance is primarily dependent on the size of the plant. In young plants, I saw the first symptoms in two weeks, while in larger and older individuals, 4–6 weeks occurred before the first symptoms appeared in the newly formed leaves of the terminal meristem. Therefore, the roots and stem must be able to transmit the virus considerable distances. . . .

It is possible to infect the plants through the roots only when they are two or more decimeters high. It is uncertain whether wounds in the roots are necessary, or whether the uptake of the virus can occur through the intact surface of the root. Since the *contagium* can only attack the leaves that form after the infection begins, the number of healthy leaves below the infected ones can be used to approximate the time of infection of plants growing naturally which have taken up the virus through their roots.

The virus can be dried without any change in its virulence. It could therefore overwinter in soil, where it perhaps would be partially destroyed, like so many bacteria and yeasts.

An alcohol precipitate of virulent plant juice, dried at 40°C, retains its virulence.

The virulence is also maintained in dried leaves, so that two-year-old herbarium leaves are still suitable for experimental infections. Therefore the dried dust which forms easily during the harvest from the broken dead tissue of the leaf spots must undoubtedly be able to spread the disease.

As expected, in a moist environment the virus was inactivated by boiling water, as well as at 90°C. I have not determined the lowest temperature at which inactivation will occur but expect it would be between 70 and 80°C. . . .

It is possible that there are a whole series of plant diseases which are caused by a *contagium fluidum*, in a similar manner to the mosaic disease of tobacco plants. The diseases of peach trees described in America by Erwin Smith in 1894 under the names peach yellows and peach rosette seem, from his description, undoubtedly to belong here, although it is not yet certain if these diseases can be transmitted only through budding or grafting, as he describes, or, what is more likely, they can also be transmitted through the juice of the dead tissues.

Comment

The filterable nature of tobacco mosaic virus had been discovered in 1892 by Ivanowsky, but apparently Beijerinck was unaware of his work. Beijerinck went much further in his investigations than Ivanowsky; therefore his paper is presented here.

Although Beijerinck was wrong about

the diffusibility of the tobacco mosaic virus, the rest of his observations are quite valid. He describes an agent which can pass through the smallest filters, can apparently reproduce only in the living plant, and seems to be quite stable. In attempting to explain these observations, he finds himself in a dilemma, since the physiological and biochemical facts of cell function were not yet available for him to use in explaining his observations. It is interesting that he comes as close to hitting the nail on the head as he does. His postulate that the virus becomes incorporated into the living protoplasm of the host plant is one which is about as close to current thinking on virus multiplication as it would be possible to get in 1899. This hypothesis shows brilliant insight into the problem.

Beijerinck's descriptions of his experimental infections are detailed enough so that they could be readily reproduced. He showed that the amount of virus in a filtrate could be crudely quantitated. This is an important aspect of virology research, since it is necessary to have some idea of how much infectious material is present in a sample. Later methods of quantitation of tobacco mosaic virus were much more precise and ultimately made it possible to study the physics and chemistry of this virus, leading to its crystallization by Stanley (see page 160).

An invisible microbe that is antagonistic to the dysentery bacillus

The original French article appears in the Appendix, pages 273–274.

1917 • F. d'Herelle

d'Herelle, F. 1917. Sur un microbe invisible antagoniste des bacilles dysentériques. *Comptes rendus Acad. Sciences,* Vol. 165, pages 373–375.

From the feces of several patients convalescing from infection with the dysentery bacillus, as well as from the urine of another patient, I have isolated an invisible microbe endowed with an antagonistic property against the bacillus of Shiga. This discovery is particularly easy to make in cases of ordinary enteritis following upon dysentery infections. In convalescent cases in which enteritis does not occur, the antagonistic microbe disappears very soon after the disappearance of the pathogenic bacillus. In spite of a number of examinations, I have never found this antagonistic microbe in the feces of dysentery cases during the period of infection, nor in normal subjects.

It is very simple to isolate the anti-Shiga microbe. A tube of broth is seeded with four to five drops of feces, and the tube is placed in the incubator for 18 hours at 37°, and the contents then filtered through a Chamberland candle filter, L_2. When a small quantity of an active filtrate is added to a broth culture of the Shiga bacillus, or an emulsion of this bacillus in broth or in physiological saline, the culture is inhibited and the death of the bacillus through complete lysis occurs after a period of time which varies from several hours to several days, depending upon the amount of filtrate added and the size of the bacterial inoculum.

When the invisible microbe is cultivated in a culture of Shiga bacillus and transferred to a new culture of Shiga bacillus by a drop of the liquid, it reproduces the same phenomenon again with the same intensity. Up until today I have carried the first isolated strain through 50 successive transfers. Moreover, the following experiment offers visible proof that the antagonistic action is produced by a living germ. If one adds to a culture of the Shiga bacillus a dilution of a previously lysed culture such that the new culture contains only one-millionth of a part of the original lysate, and then, immediately, a small drop of this is spread on an agar slant, one obtains, after incubation, a layer of dysentery bacilli covering the surface, but in which there are a certain number of clear areas of about 1 mm. diameter, in which there are no bacteria. These clear areas can only represent colonies of the antagonistic microbe: a chemical substance would not be able to concentrate itself in such definite spots. By doing this experiment with measured quantities, I have been able to show that a lysed culture of Shiga bacillus contains around 5 to 6 billion filterable germs per cubic centimeter. One three-billionth of a cc. of a lysate of Shiga bacillus, containing therefore a single germ, fully inhibits a culture of Shiga bacillus similarly seeded. The same quantity added to a 10 cc. culture sterilizes and lyses it in 5 or 6 days.

Various strains of the antagonistic microbe which I have isolated were originally only active against the bacillus of Shiga. By culture in symbiosis with the dysentery bacilli of Hiss or Flexner, I have been able, after several passages, to render them active against these bacilli. I have never obtained an activity against other microbes: typhoid and paratyphoid bacilli, staphylococci, and so on. The appearance of antagonistic activity against the bacilli of Flexner and Hiss is accompanied by a reduction and then a loss of power against the Shiga bacillus. This power reappears with its original intensity, however, after several transfers in symbiosis with the latter. Its specificity of action is therefore not inherited in the nature of the invisible microbe, but is acquired in the organism which is attacked during the symbiotic culture with the pathogenic bacilli.*

The antagonistic microbe can never be cultivated in media in the absence of the dysentery bacillus. It does not attack heat-killed dysentery bacilli, but is cultivated perfectly in a suspension of washed cells in physiological saline. This indicates that the anti-dysentery microbe is an obligate bacteriophage.†

The anti-Shiga microbe exhibits no pathogenic action against experimental animals. Lysed cultures of the bacillus, which are in reality cultures of the antagonistic microbe, exhibit the property of immunizing rabbits against doses of the Shiga bacillus which kill the controls in five days.

I have attempted to discover evidence for an antagonistic microbe in convalescents from typhoid fever. In two cases, one in the urine, the other in the feces, I have isolated a filterable microbe able to lyse nicely the paratyphoid A bacillus but always less markedly than in the cases of the

* [This statement is not exactly true. Bacterial viruses continue to breed true.]

† [Apparently the first use of this word.]

Shiga bacillus. These latter properties were attenuated in succeeding cultures.

In summary, in certain convalescents from dysentery I have shown that the disappearance of the dysentery bacilli is coincident with the appearance of an invisible microbe endowed with antagonistic properties against the pathogen. This microbe, really a microbe of immunity, is an obligate bacteriophage. Its parasitism is strictly specific, but if it is restricted to a certain species, it is gradually able to acquire activity against various germs. It seems therefore that in the dysentery bacillus, as well as a homologous immunity coming directly from the person infected, there exists a heterologous immunity coming from an antagonistic microbe. It is probable that this phenomenon is not restricted to dysentery, but that it is of general significance, because I have been able to observe a similar situation, even though weaker, in two cases of paratyphoid fever.

Comment

This discovery of bacteriophages, or bacterial viruses, was actually first made by Twort in 1905. But his work is not too clear and was not followed up, so that I have included here the paper by d'Herelle, which was really responsible for the beginning of extensive scientific work on these interesting organisms.

Most species of bacteria that have been studied have been shown to be hosts for bacterial viruses. As d'Herelle suspected, the phenomenon is widespread. But his early hopes that bacterial viruses would be useful in treating bacterial infections or in conferring immunity have not been fulfilled. d'Herelle had little understanding of the word "immunity" as we use it, since the injection of a bacterial virus could not be expected to induce antibody formation against a bacterium. The bacterial viruses have remained essentially laboratory curiosities. They have been used as model systems for the study of viral reproduction and have made great impact on molecular genetics. They have also been used in diagnostic laboratories in the typing of certain widely occurring pathogens, such as *Salmonella typhosa* and *Staphylococcus aureus*. But all trials of their use as therapeutants have been unsuccessful.

Isolation of a crystalline protein possessing the properties of tobacco-mosaic virus

1935 • Wendell M. Stanley

Stanley, W. M. 1935. Isolation of a crystalline protein possessing the properties of tobacco-mosaic virus. *Science*, Vol. 81, No. 2113, pages 644–645.

A CRYSTALLINE MATERIAL, WHICH HAS the properties of tobacco-mosaic virus, has been isolated from the juice of Turkish tobacco plants infected with this virus. The crystalline material contains 20 per cent. nitrogen and 1 per cent. ash, and a solution containing 1 milligram per cubic centimeter gives a positive test with Millon's, biuret, xanthoproteic, glyoxylic acid and Folin's tyrosine reagents. The Molisch and Fehlings tests are negative, even with concentrated solutions. The material is precipitated by 0.4 saturated ammonium sulfate, by saturated magnesium sulfate, or by safranine, ethyl alcohol, acetone, trichloracetic acid, tannic acid, phosphotungstic acid and lead acetate. The crystalline protein is practically insoluble in water and is soluble in dilute acid, alkali or salt solutions. Solutions containing from 0.1 per cent. to 2 per cent. of the protein are opalescent. They are fairly clear between *p*H 6 and 11 and between *p*H 1 and 4, and take on a dense whitish appearance between *p*H 4 and 6.

The infectivity, chemical composition and optical rotation of the crystalline protein were unchanged after 10 successive crystallizations. In a fractional crystallization experiment the activity of the first small portion of crystals to come out of solution was the same as the activity of the mother liquor. When solutions are made more alkaline than about *p*H 11.8 the opalescence disappears and they become clear. Such solutions are devoid of activity and it was shown by solubility tests that the protein had been denatured. The material is also denatured and its activity lost when solutions are made more acid than about *p*H 1. It is completely coagulated and the activity lost on heating to 94°C. Preliminary experiments, in which the amorphous form of the protein was partially digested with pepsin, or partially coagulated by heat, indicate that the loss in activity is about proportional to the loss of native protein. The molecular weight of the protein, as determined by two preliminary experiments on osmotic pressure and diffusion, is of the order of a few millions. That the molecule is quite large is also indicated by the fact that the protein is held back by collodion filters through which proteins such as egg albumin readily pass. Collodion filters which fail to allow the protein to pass also fail to allow the active agent to pass. The material readily passes a Berkefeld "W" filter.

The crystals are over 100 times more active than the suspension made by grinding up diseased Turkish tobacco leaves, and about 1,000 times more active than the twice-frozen juice from diseased plants. One cubic centimeter of a 1 to 1,000,000,000 dilution of the crystals has usually proved infectious. The disease produced by this, as well as more concentrated solutions, has proved to be typical tobacco mosaic. Activity measurements were made by comparing the number of lesions produced on one half of the leaves of plants of Early Golden Cluster bean, *Nicotiana glutinosa* L., or *N. langsdorfii* Schrank after inoculation with dilutions of a solution of the crystals, with the number of lesions produced on the other halves of the same leaves after inoculation with dilutions of a virus preparation used for comparison.

The sera of animals injected with tobacco-mosaic virus give a precipitate when mixed with a solution of the crystals diluted as high as 1 part in 100,000. The sera of animals injected with juice from healthy tobacco plants give no precipitate when mixed with a solution of the crystals. Injection of solutions of the crystals into animals causes the production of a precipitin that is active for solutions of the crystals and juice of plants containing tobacco-mosaic virus but that is inactive for juice of normal plants.

The material herein described is quite different from the active crystalline material mentioned by Vinson and Petre and by Barton-Wright and McBain, which consisted, as Caldwell has demonstrated, largely of inorganic matter having no connection with the activity. These preparations were less active than ordinary juice from diseased plants, and the activity they possessed diminished on further crystallizations.

The crystalline protein described in this paper was prepared from the juice of Turkish tobacco plants infected with tobacco-mosaic virus. The juice was brought to 0.4 saturation with ammonium sulfate and the precipitated globulin fraction thus obtained was removed by filtration. The dark brown globulin portion was repeatedly fractionated with ammonium sulfate and then most of the remaining color was removed by precipitation with a small amount of lead subacetate at *p*H 8.7. An inactive protein fraction was removed from the light yellow colored filtrate by adjusting to *p*H 4.5 and adding 2 per cent. by weight of standard celite. The celite was removed, suspended in water at *p*H 8, and the suspension filtered. The active protein was found in the colorless filtrate. This procedure was repeated twice in order to remove completely the inactive protein. Crystallization was accomplished by adding slowly, with stirring, a solution containing 1 cubic centimeter of glacial acetic acid in 20 cubic centimeters of 0.5 saturated ammonium sulfate to a solution of the protein containing sufficient ammonium sulfate to cause a faint turbidity. Small needles about 0.03 millimeters long appeared immediately and crystallization was completed in an hour. Crystallization may also be caused by the addition of a little saturated ammonium or magnesium sulfate to a solution of the protein in 0.001 N acid. Several attempts to obtain crystals by dialyzing solutions of the protein gave only amorphous material. To date a little more than 10 grams of the active crystalline protein have been obtained.

Although it is difficult, if not impossible, to obtain conclusive positive proof of the purity of a protein, there is strong evidence that the crystalline protein herein described is either pure or is a solid solution of proteins. As yet no evidence for the existence of a mixture of active and inactive material in the crystals has been obtained. Tobacco-mosaic virus is regarded as an autocatalytic protein which, for the present, may be assumed to require the presence of living cells for multiplication.

Comment

Here we have the first demonstration that an agent which has some of the properties we associate with living organisms can be crystallized like a chemical. This work can be looked upon as directly descending from Beijerinck's original work on tobacco mosaic virus.

The procedures that Stanley used for this crystallization are not novel. They are procedures that had been used previously for the crystallization of a number of proteins. What is novel is that the crystals obtained retain their infectivity for tobacco plants, causing typical mosaic disease at extremely high dilutions.

Many plant viruses give very high yields of active virus when grown in susceptible hosts. Because of these high yields, it is possible to obtain plant juices that are very rich in virus particles. This has made it fairly easy to crystallize these viruses and explains why the first success was with tobacco mosaic. Again we see the importance of choosing the right experimental system. Animal viruses have proven to be much more difficult to crystallize.

In later work, Stanley and others showed that crystalline tobacco mosaic virus contained more than protein. Some ribose nucleic acid was also found. But it is important to remember that only these two substances, protein and nucleic acid, have been found in highly purified tobacco mosaic virus. Many other viruses also contain only these two substances, although many animal viruses are apparently more complex. All bacterial viruses so far studied have been shown to consist of protein and desoxyribose nucleic acid.

Stanley's work initiated a whole new field of study which is now probing at the very basis of life itself.

PART V
Chemotherapy

The chemical foundations of the study of disinfection and of the action of poisons

1897 • B. Krönig and Th. Paul

Krönig, B. and Paul, Th. 1897. Die chemischen Grundlagen der Lehre von der Giftwirkung und Desinfection. *Zeitschrift für Hygiene und Infectionskrankheiten,* Vol. 25, pages 1–112.

THE STUDY OF DISINFECTING AGENTS has only been carried out on a scientific basis since the time that the agents were allowed to act on definite species of bacteria under the simplest conditions. The leader in this field has been R. Koch, who, in cooperation with his students, has acquainted us with a series of chemical substances which possess significant disinfecting power.

With the help of improved methods, later workers have attempted to discover certain laws for the action of various substances, and from this to use the current chemical theories to offer leads to the production of new chemical disinfectants.

The study of physiological-chemical processes through the changing properties of the cell wall and protoplasm has presented exceptional difficulties, but even so the researches of a number of different workers has made it possible to give us a systematic account of the knowledge of disinfecting agents. This has been done by Behring in his book "Control of Infectious Diseases," which presents as well a large number of his own observations. He has presented there a general viewpoint and has given satisfactory explanations for a large number of observations.

Through the progress in physical chemistry, our understanding in recent years of the state of substances in solution has been turned to completely new paths. The works of van 't Hoff, Ostwald, Arrhenius, Nernst, and others have allowed us to discover in the area of inorganic and organic chemistry a series of general laws. Since the physiological activity of substances is in general dependent upon chemical properties, one can expect that the study of disinfection will also find new viewpoints through these new chemical theories.

We have set for ourselves the task of studying the behavior of bacteria with chemical agents from the basis of these viewpoints, in order, perhaps, to be able to use the results obtained for the development of practical disinfection procedures.

Although there is in the literature a large number of observations on the action of different disinfectants on bacteria in pure solutions, we have carried out anew a large series of experiments, since the previous results have partly disagreed with each other and are often not directly comparable, since the experiments had been run under a wide variety of conditions, so that the temperatures were not the same, and the test organisms differed, and so on.

Before we proceed to our experiments, we must examine briefly how one can determine the disinfecting properties of a solution and what conditions must be kept constant, in order to be able to compare the disinfectant properties of different solutions.

1. First, equimolar amounts of the various substances must be used in the comparative experiments.

2. The bacteria to be used as test organisms must all have the same resistance.

3. The number of bacteria used in each experiment must be the same.

4. The bacteria must be placed in the disinfecting solutions in such a way that none of the nutrient materials on which they were cultured are carried over with them.

5. The disinfecting solutions must always have the same temperature.

6. After the action of the disinfecting agent, the bacteria must be again completely freed from this agent.

7. After the bacteria have been removed from the disinfecting solution, they must be placed on equal amounts of the same suitable medium at the same temperature, which if possible should be the optimum for their growth.

8. The number of remaining bacteria capable of reproducing into colonies on solid medium must be counted after the same length of incubation time.

We have attempted to follow these rules in the performance of our experiments, so far as possible.

1. Comparison of equimolecular quantities.

In the literature, the experiments on disinfection almost always report the concentration of the solutions in per cent by weight.

As long as it is not necessary to compare the activity of two agents, the expression of concentration in per cent is not only the easiest, but for practical use it is the most suitable. A simple consideration will show, however, that for comparative studies only equimolecular amounts of the substances may be used. From the theory of van 't Hoff, a substance in solution is in a comparable condition to a gas; in relation to temperature, pressure, and volume both types show a perfectly analogous behavior, so that we can apply the gas laws directly to solutions.

A gram-molecular weight of every gas at the same temperature and pressure assumes the same volume. Further, from Avogadro's law, all gases at the same volume and at the same temperature and pressure contain the same number of molecules. Also, solutions of the same volume containing one gram-molecular weight of a substance contain the same number of molecules.

Naturally then, the activity of two substances can only be compared with each other, when in both cases the same number of molecules are present in the liquid. This will be the case when the substances are dissolved at concentrations related to their molecular weights. . . .

2. Use of bacteria of equal resistances.

Various species of bacteria vary considerably in their resistance to disinfection. But even in the same species there are large differences in resistance, depending on the growth phase in which the bacteria are used, and whether the vegetative or resting form is used.

Esmarch has further shown through careful studies that both the vegetative and resting forms of the same species may show varying resistances, depending upon the source of the culture and the medium on which it is grown. . . .

But in our studies we have shown further that even bacteria from the same source do not show equal resistances against disinfection. . . .

In general, the resistance of spores increases quickly in the early stages of sporulation, and after a certain time, which varies with the various spores, reaches a maximum. From then on, there is ordinarily a slow but continuous decrease in the resistance. . . .

Since we were not able to maintain bacteria at constant resistance, even for a short time, it was necessary, in every series of experiments, to determine the resistance of the bacteria that were used. . . .

We tested the resistance in most cases against a solution of mercuric bichloride which contained one gram-molecular weight in 16 liters (1.69%). It must be noted that this is naturally only a measure of the resistance against mercuric bichloride, which can not be carried over to other disinfectants before this is also determined for the others.

3. Number of spores used.

As Geppert has already shown, individual spores of the same species in the same preparation show varying resistances. This can be easily shown by allowing a disinfecting agent to act on a large number of spores from the same preparation. If the resistance of all of the individuals were the same, then the viability of all of the spores should disappear at exactly the same time. As we shall show later, this is in no way the case.

This means that in comparisons of two agents, always the same number of spores must be used with each. . . .

None of the methods currently employed permit us to use exactly the same number of spores without at the same time causing other difficulties. If the spores are allowed to dry on a silk thread, a procedure which R. Koch first introduced, the number of spores varies with the surface area of the thread and the degree of imbibition. If the spores are suspended in the disinfecting solution, as Geppert has done, then it is not possible to remove the disinfecting agent without either diluting the suspension of bacteria or without carrying over to the nutrient test medium a large amount of the agent used to precipitate the disinfectant. If the spores are dried on glass fragments or glass spheres, we have the difficulty that the surface of the glass is never uniformly wetted by the disinfection solution, as we have been able to show in a large number of experiments.

The best material that we have

found for this purpose is a crude Bohemian garnet. . . .

The properly prepared garnets are shaken in a shaking apparatus with a filtered aqueous suspension of bacteria. The excess liquid is removed and the garnets dried in a sterilized metal box under anhydrous calcium chloride. The drying was done at a temperature of 7°C. in the icebox and was usually completed in 12 hours. The garnets were stored in a ground-glass stoppered bottle at 7°C. in the dark. . . .

Because each garnet had approximately the same surface area and would be wet uniformly by the bacterial suspension, it would be expected that each garnet would have about the same number of bacteria on it.

The bacteria are fastened rather tightly to the garnets, so that when they are placed in the disinfectant solution and through the steps to neutralize the disinfectant, only a few bacteria are washed off the garnets. However, if they are shaken with water, a large proportion of the adhering bacteria can be removed in a short time. . . .

For comparative studies it is not necessary to know the number of spores adhering to the garnets. After the first shaking with water a very large proportion of the spores are released, so that we can obtain a good average sample of the spores which were adhering to the garnets. Further, since the number of spores released by identical shaking of identically prepared and handled garnets is always about the same, the bacteria released by the first shaking give us a directly comparable value of the number of bacteria on the garnet which are capable of reproducing.

4. In order to eliminate interfering substances, the carrying over of large amounts of culture medium with the bacteria must be avoided.

Because in the experiments many dilute solutions are used, it is possible that protein materials carried over with the bacteria may precipitate metal salts. Even when no precipitation occurs, one cannot be certain that the chemical constitution of the solution has remained unchanged. An aqueous solution of chlorine of a concentration of one gram-molecular weight in 128 liters, which is strongly disinfecting, contains in 10 cc. only 5.5 mg. chlorine. This small amount of chlorine can be made inactive by quite small amounts of organic substances. . . .

Further, the bacteria must be directly in contact with the disinfecting agent and not be imbedded in clumps of organic substances. . . . It is therefore necessary to filter the bacterial suspension. Since filter paper withholds many bacteria, the bacterial suspension must be very dense to begin with. . . .

5. Constant temperature control.

The exact studies of Heider, Behring, Pane, and others have shown that the activity of a disinfecting agent varies with the temperature. In general, the higher the temperature, the more active is the agent. . . . This behavior is in agreement with the law that states that chemical reactions are accelerated at higher temperatures.

All of our experiments have been carried out at 18°C. We have selected this temperature because disinfecting agents are ordinarily used under practical conditions at room temperature.

For the control of temperature, we placed our solutions in a water bath, which kept the temperature constant within 1/10 of a degree. . . .

6. Neutralization of the disinfectant.

Geppert has shown that it is necessary when testing a disinfecting agent to render it ineffective through some chemical procedure, and that a simple dilution in water is not sufficient to remove the disinfectant.

Even very small amounts of the disinfectant, when carried over to the nutrient medium used to indicate the viability of the bacteria, are able to inhibit the bacteria which have been weakened by the disinfection.

We have taken these precautions in our experiments and have, for example, precipitated the metal salts with 3 per cent ammonium sulfide solution, neutralized the bases with dilute acetic acid, the acids with dilute ammonia, inactivated iodine with sodium thiosulfate, and chlorine and bromine with dilute ammonia, and so on. . . .

7. *Utilization of identical culture media.*

After disinfection and the neutralization of the disinfectant, the bacteria must be placed on a suitable medium for growth. Behring and many other authors have described the advantages for this purpose of liquid media such as broth, blood serum, or liquid gelatin. These media have the disadvantage that one is not able to show how many bacteria have remained viable after the disinfecting. Instead, one is limited to comparing the lengths of time necessary to kill all of the spores. As we have seen above, some individual bacteria are more resistant than others. Therefore, the time for complete disinfection is extended considerably. In addition, a single individual may escape the action of the agent and alter the results considerably.

We have therefore abandoned liquid media and have employed solid media containing agar. Gelatin could not be used because we wanted to be able to use the temperature that was most favorable for the species of bacteria. The temperature optimum for the bacteria we have used the most, *Bacillus anthracis* and *Staphylococcus pyogenes aureus* is around 37°C., at which temperature gelatin is fluid.

Composition and reaction [*p*H] of the agar can influence the growth of the bacteria. Osborne has performed experiments with agar media to which varying amounts of meat extract were added. The growth of the anthrax bacillus was roughly proportional to the concentration of meat extract. For example, if he took vegetative forms and partially killed them by heating to 65–70°, the number of colonies developing from the viable cells was dependent on the quality of the nutrient medium.

Because of this we attempted to use for all experiments a medium which was as nearly reproducible as possible.

The meat infusion medium of Koch was not too suitable for the preparation of agar for our experiments, because the quality of the meat varies considerably. If we made a large amount of agar at one time, it would become altered through storage during the time our experiments were in progress.

One can insure a reasonable similarity in various preparations through the use of Liebig's meat extract, which can be purchased commercially in large boxes. There would be sufficient material in one box for a large number of experiments.

The variable qualities of commercial agar-agar make it necessary to take a large amount of agar and cut it up fine and mix it well.

A special point is to be placed on maintaining uniform reaction [*p*H] of the medium. This can best be done when one follows the suggestion of Schultz and adds to the broth, before the agar is added, sufficient sodium hydroxide so that the liquid shows an alkaline reaction with phenolphthalein. . . .

The preparation of our agar follows:

To 15 liters of distilled water are added 300 g. peptone, 30 g. grape sugar [glucose] and 75 g. Liebig's meat extract. The mixture is boiled in an autoclave at 105–110° for two hours, then made weakly alkaline with sodium hydroxide, so that a sample taken in a beaker 9 cm. high and 5 cm. in diameter shows a distinct rose color when one drop of phenolphthalein indicator is added. Then 260 g. of finely cut agar-agar is added and the mixture boiled at 105–110° eight hours a day for two days, in such a way that the evaporated water is always replaced through the addition of distilled water. The mixture is finally poured through a hot-water filter and stored in glass flasks in 3 liter amounts.

In this way we believe we are always able to have agar of the same quality at our disposal. . . .

Because of the alterability of the medium in the course of time, experiments are only strictly comparable when they are performed with agar which has been taken from the same container at the same time and sterilized in the same way, while in different series of experiments, small variations can occur. . . .

It should be stated here that in general we do not speak of disinfection in the sense of the killing of bacteria. By successful disinfection, we understand this to mean that an agent has acted on the bacteria in such a way that these are no longer able to germinate and reproduce on a particular medium at a particular temperature. In no way does this mean that the bacteria are really dead. It is indeed possible to consider that these weakened bacteria might be able to develop under favorable conditions, perhaps in the body. In this way Geppert has observed that anthrax bacillus spores which have been treated with mercuric bichloride for a certain length of time are able, when injected into mice, to induce them to die from anthrax, while on agar, no colonies developed from these treated spores. . . .

8. Enumeration of the colonies.

The first colony count is performed 24 hours after the beginning of incubation. On the 2nd and 3rd day, more colonies develop, so that a second and third enumeration is performed. After this length of time, our experience has shown that only isolated new colonies develop, which will not influence the evaluation of the germicide. . . .

Counts on all three days are performed because we would like to ascertain if there is a relationship between the type of disinfecting agent and the time of germination. Also, with anthrax bacillus, after the very extensive development of colonies on the first day, the colonies appearing later cannot be so accurately counted.

If the number of colonies does not exceed 800, then each single colony on the plate is counted with the aid of a magnifying glass. If over 800 colonies appear, we then use the counting apparatus of Wolfhügel.*

SELECTION OF TEST BACTERIA

We have used different species of bacteria in our experiments. We have selected the spores of the anthrax bacillus as our resting form, following the lead of Koch. Although the use of this species brings with it the danger of accidental infection of the workers, we have been able to find no better substitute for it. The anthrax bacillus forms characteristic

* [Modern procedure limits the maximum number of colonies per plate to 300, owing to statistical requirements.]

colonies on our artificial medium, which are easy to distinguish from accidental contaminants. Further, the colonies of this species spread very slowly over the surface of the agar, so that it is possible, using a Petri plate of 64 square cm., to count with the glass 6000 isolated colonies. . . .

A further advantage of the anthrax spores is that they are able to stand the extensive drying over calcium chloride, which is necessary for the operation of our experiments.

In order to obtain sufficient spores for our experiments, we inoculated 30 agar tubes to obtain a spore suspension of 100 cc. The tubes were placed three times for 24 hours at 24°C. Then the colonies on the surface were removed with a platinum loop and suspended in water. They were then filtered and handled as previously mentioned. . . .

For a vegetative test organism, we used *Staphylococcus pyogenes aureus*. The staphylococci will not endure a long period of drying, so that the garnets were usually removed after one to two days' drying. Even in this short time, the staphylococci have lost the resistance which they possessed when they were fresh, but since we were mainly interested in comparative results, these weakly resistant organisms could be used.

EXPERIMENTAL PROCEDURE

The experimental procedure was as follows:

The test solutions were placed in small dishes of 3.5 cm. height and 7 cm. diameter and were placed in the constant temperature chamber. After they had reached the temperature of 18°C., we placed in each container 30 garnets. These garnets were handled with platinum forceps and were placed on a small platinum screen raised up from the bottom, which was 2 cm. long and 2 cm. wide. This was done so that the garnets would not rest directly on the bottom and therefore reduce the uniformity of the wetting. After the proper time had passed, the garnets were removed from the solution and were rinsed once or twice in 15 cc. of water, and then placed in a dish which contained around 20 cc. of the neutralizing solution. The garnets were in this reagent 12 minutes and were then removed and placed in a Petri dish in water, in which they remained a further 12 minutes. Finally they were placed 5 each in 3 cc. of water in a graduated test tube. All test tubes were shaken identically in a wire basket for 3 minutes and then placed in a water bath at 37.5°. To each test tube was added exactly 10 cc. of fluid agar at 42°C., and the mixture well mixed with a platinum stirrer.

This mixture was then poured into prewarmed Petri plates and allowed to harden on a horizontal surface, and finally placed in an incubator at 37.5°. . . .

Since we had frequently noticed that moisture condensed on the underneath surface of the Petri plate top and dropped down onto the medium and spread the colonies over the surface, we placed a piece of sterilized filter paper between the plate and the lid. This procedure was more preferable for our purposes than that of Miller. Miller avoided the collection of condensation on the lid by incubating the plates in the oven upside down. Because of the 3 cc. of water added to our agar, we could not do this.

As an indication of the exactness and reproducibility of our method, we present the following tables on the action of mercuric bichloride on anthrax spores. . . . In the summary table, the average value for the count on the third day of the six replicate plates is given. . . .

TABLE 11, Mercuric bichloride, $HgCl_2$

Each solution has one gram-molecular weight in the volume indicated

Time of action in minutes	Colony counts				
	16 liters 1.69%	*32 liters 0.84%*	*64 liters 0.42%*	*128 liters 0.21%*	*256 liters 0.11%*
2	549				
3	323	678		3829	
4	236				
5	138		961		
6	82	310		2069	
7	42				
8	19				
9	—	168			
10	10		397	520	2027
12	1	38			
14	0				
15		10	178	302	749
16					
18		5			
20			41	231	612
21		3			
22					
24		2			
25			9	121	432
27		1			
30		0	7	46	306
33					
35			3	21	227
40			2	7	183
45			1		151
50			1	5	133
55			1		
60			1	1	79
70				1	16
80				0	10
90					5
100					3
110					3
120					2

IV. General relationships between concentration and the toxicity of mercuric bichloride solutions. (After a personal communication of K. Ikeda.)

We did not originally intend to go into this question in any detail, but since our last publication we have had a letter from Herr K. Ikeda in Tokyo (Japan), in which he has examined for us the data which we published previously (which is also published above) from the very interesting viewpoint of physical chemistry. This examination indicates the presence here of completely general laws, even though the experimental data available are still quite limited. Because of this, we would like to include here the most important points from this communication to us.

The data on which these considerations are based are in Tables 5 to 9, and have been summarized in Table 11. The data of Table 11 have been presented graphically in Plate I.

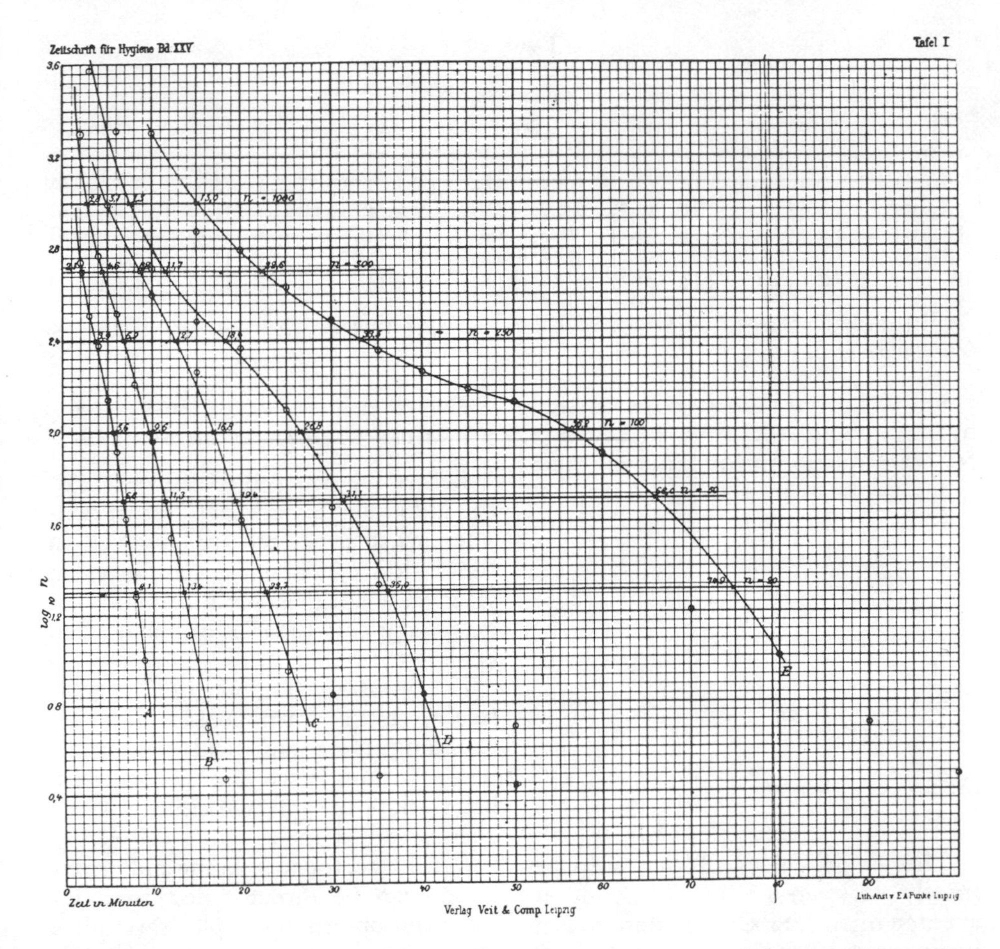

If one places on the ordinate of rectilinear coordinate paper the logarithm of the number of spores which are still viable after the action of the mercuric bichloride, and on the abscissa the time in minutes at which this value was taken, one obtains for the dilutions v = 16 liter, v = 32 liter, v = 64 liter, v = 128 liter, v = 256 liter, the five curves A, B, C, D, and E. These curves are steep for the strong concentrations and run in straight lines, while for the more dilute solutions they are flatter and more curved.

The time t, which is required to bring about the same disinfection effect for each dilution, can be obtained by taking a given number of viable spores, n, and reading from the parallel lines which cross the curves down to the abscissa. The distance from where this crosses the abscissa to the ordinate gives the time t in minutes. From this way, in Plate I the times for n = 1000, n = 500, n = 250, n = 100, n = 50, n = 20 can be found, and are brought together in Table 60.

If one compares the numbers in the vertical columns, which correspond to the identical disinfection effect at increasing dilutions of disinfectant, one finds that as the dilution doubles, the time to achieve this effect also doubles. When the time which is necessary for a given dilution v to reduce the viable

TABLE 60

v	n					
	1000	500	250	100	50	20
16 liter	—	2.1 (2.6)	3.9 (4.0)	5.6 (5.9)	6.8 (6.9)	8.1 (8.0)
32 liter	2.8 (2.9)	4.6 (4.4)	6.9 (6.9)	9.6 (9.9)	11.3 (11.7)	13.4 (13.6)
64 liter	5.1 (4.9)	8.8 (7.5)	12.7 (11.7)	16.8 (16.9)	19.4 (19.7)	22.7 (23.1)
128 liter	7.5 (8.3)	11.7 (12.7)	18.4 (19.8)	26.8 (28.7)	31.1 (33.7)	36.0 (39.3)
256 liter	15.0 (14.0)	22.6 (21.2)	33.5 (33.6)	56.2 (48.9)	66.0 (57.2)	74.9 (66.7)

[This table gives observed values, with the calculated values in parentheses of the time, *t*, in minutes.]

count to a given number n is expressed by the symbol $t_{n.v}$, then an equation for the relationships between these times can be expressed as follows:

$$\frac{t_{n.2v}}{t_{n.v}} = \frac{t_{n_1.2v}}{t_{n_1.v}}$$

For example:

$$\frac{t_{250.32}}{t_{250.16}} = \frac{6.9}{3.9} = 1.77$$

$$\frac{t_{50.64}}{t_{50.32}} = \frac{19.4}{11.3} = 1.72$$

The single values of this ratio lie mostly between 1.6 and 1.8, with an average of 1.70. Since the dilutions can be expressed by the formula $2^m \times 16$, it is possible to derive a general equation using the time value obtained from $t_{n.16}$ as a base. This equation is:

$$t_{n.16 \times 2^m} = 1.70^m \, t_{n.16}$$

The values obtained in this way are included in Table 60 in parentheses, along with the actual observed values. The agreement between the observed and the calculated values is quite good, with only a few exceptions. . . .

[Further discussion attempts to relate this equation to the dissociation constant of $HgCl_2$ and proceeds to other derivations.]

PROCESSES DURING THE ACTION OF SOLUBLE SUBSTANCES ON BACTERIA

Before we proceed to the results of the individual experiments, let us consider what processes are involved in the action of chemical substances in solution against bacteria, and to what extent the new theories can be used to explain them.

Every bacterium, whether vegetative or resting form, consists of a surrounding membrane and the protoplasm within. If we place a bacterium in solution, one of two situations can occur:

In the one, the membrane and protoplasm can be directly destroyed by the solution, for example, through a strongly corrosive mineral acid or alkali, or through a strong oxidizing agent, like a concentrated solution of permanganate.

In the other, the membrane can remain intact, and the solution will act on the protoplasm after it has penetrated the cell envelope.

Both situations can take place at the same time.

If the membrane remains intact, the solution can act on the protoplasm in various ways.

1. It can act merely through a concentration difference between the protoplasm and the solution, which results in the water of the protoplasm

being removed or added to.* In this way the viability of the bacteria will be more or less influenced.

2. The dissolved substance can penetrate the membrane and enter into chemical reaction with the protoplasm.

In the latter case, the speed of disinfection will depend upon the speed at which the dissolved material penetrates the membrane, and from the reaction of the agent with the protoplasm.

Since for disinfecting purposes ordinarily relatively dilute solutions are used, we must concern ourselves primarily with these last two factors.

Diffusion. The study of diffusion through organized membranes has been made exceedingly difficult because of the fact that various substances do not penetrate the membrane in the same manner. For example, sodium chloride diffuses unhindered through a fish bladder, while protein materials are withheld. However, by depositing certain precipitates in this membrane, it can be made more or less impenetrable for a large number of salts. In many experiments it has been shown that the living cell has similar semipermeable membranes.

In bacteria we must further distinguish between the membrane of a vegetative form and that of a spore. We may correctly assume that the latter is more poorly permeable than the former.

Above all, the large influence the nature of the cell wall has on the action of the disinfectant must be considered. We can only derive a clear picture of the action of a solution on a bacterial cell when we know both the rate at which the substance diffuses through the cell membrane, and the reaction rate which the substance undergoes with the cell body, i.e., the protoplasm.

* [Osmotic pressure effect.]

A certain evaluation of the action of different substances on the protoplasm of the bacterial cell can at the time be obtained, when we compare substances which because of their similar chemical properties can be expected to have similar diffusion rates through the membrane. In this way, the influence of diffusion can to a certain extent be ignored.

Reactions with the protoplasm. Although the chemical composition of protoplasm is at present almost unknown, its reaction with chemical agents can be assumed to follow the general laws which govern ordinary chemical reactions. Therefore the reaction rate will be dependent on the time, temperature, concentration, and the specific properties of the reacting substances. . . .

GROWTH INHIBITION

While the preceding has dealt exclusively with the lethal action of the various substances, we would like to consider in the following various observations which have been made on the growth inhibiting properties which metal salts may exhibit when dissolved in nutrient media.

The bactericidal action of a solution depends on the concentration of the active grouping and the time during which action is allowed.

In considering growth inhibiting properties, the time of action does not enter into the considerations, but it is only the concentration of the active substance which is determinative.

Therefore we can not use directly data on the bactericidal action of a substance to draw direct conclusions on the growth inhibiting properties. Because it is not possible to determine the growth inhibiting properties in pure solutions of the agent, it is necessary to add small amounts of the agent in question to a nutrient liquid of variable composition, which usually contains a large amount of organic substances. In this way, the disinfecting agent undergoes more or less severe changes and assumes quite different properties than it possesses in pure solution. . . .

It is probable that in growth inhibition, where the time does not enter in as a factor, the degree of dissociation of the compound plays a reduced role, and it is only the concentration of the metal dissolved in the nutrient fluid which is important.

Studies on the growth inhibiting properties of various metal salts in nutrient solutions have been reported many times in the literature.

Above all may we mention here the work of von Behring. . . . He tested the growth inhibiting properties of mercury salts in blood serum. For 14 different salts, he found an inhibiting concentration varying between one gram-molecular weight in 2200 liters (Hg = 1/11,000) and 6400 liters (Hg = 1/32,000). The agent giving the lowest inhibitory activity was a mixture of mercuric bichloride with 5 moles of tartaric acid. The most active compound was $K_2HgCyanide_4$.

Our experiments were performed in nutrient broth and gelatin. Although they are not completed, we have confirmed von Behring's results that $K_2HgCyanide_4$ inhibits better than mercuric bichloride. The activity of the former is one gram-molecule per 30,000 liters, while the latter is one in 20,000 liters, in nutrient gelatin. When these results are compared with those of von Behring, they indicate the great dependence on the nature of the medium. . . .

[There follow detailed data on the action of a variety of substances, and the results are summarized in the conclusions.]

CONCLUSIONS

We can summarize the results of the preceding studies briefly:

1. Comparative studies on the toxicity of various substances must be carried out with equimolecular concentrations.

2. The disinfectant action of solutions of metal salts depends not alone on the concentration of the dissolved metal but is also dependent on specific properties of the salt and the solvent.

3. Solutions of metal salt in which the metal portion is in a complex ion, and the concentration of the metal ion itself is very small, have very poor disinfecting powers.

4. The action of a metal salt depends not only on the specific action of the metal ion but also on the anion and the undissociated portion.

5. Halogen compounds of mercury (including thiocyanate and cyanate) act in proportion to their degree of dissociation.

6. The disinfecting power of aqueous solutions of mercuric bichloride are reduced by the addition of halogen compounds of other metals or hydrochloric acid.

7. The disinfecting power of aqueous solutions of mercuric nitrate, mercuric sulfate, and mercuric acetate is significantly increased by the addition of moderate amounts of sodium chloride.

8. Acids disinfect in general in relation to their degree of dissociation. Therefore, their action is a function of the hydrogen ion concentration in solution. Anions of the acids, and the undissociated molecules of hydrofluoric, nitric and trichloracetic acids exhibit a specific toxicity. This specific effect decreases with increasing

dilution as compared with the toxicity of the hydrogen ion.

9. The bases potassium, sodium, lithium, and ammonium hydroxide act in relation to their degree of dissociation, so that the action is due to the concentration of hydroxyl ions in solution.

Hydrogen ions are for anthrax spores, and to a greater degree for *Staphylococcus pyogenes aureus*, a stronger poison than hydroxyl ions.

10. The disinfecting action of the halogens chlorine, bromine, and iodine decreases more rapidly with increasing atomic weight than would be expected from their chemical properties.

11. The oxidizing agents nitric acid, dichromic acid, chloric acid, persulfuric acid, and permanganic acid disinfect in degrees corresponding to their positions in the electro-chemical series. Chlorine does not correspond to this relationship but exercises a strong specific activity.

12. The disinfecting action of various oxidizing agents is significantly increased by the addition of hydroacids of halogens (for example, potassium permanganate with hydrochloric acid).

13. We have confirmed the observation of Scheurlen that phenol solutions disinfect better after the addition of salts. A good explanation of this phenomenon could not be derived from the present experiments.

14. We have confirmed the well known fact that substances dissolved in ethyl alcohol, methyl alcohol, and ethyl ether have almost no effect on anthrax spores.

15. The disinfecting action of aqueous solutions of silver nitrate and mercuric bichloride is significantly increased by the addition of certain amounts of ethyl alcohol, methyl alcohol, and acetone.

16. The disinfecting action of aqueous solutions of phenol and formaldehyde decreases upon the addition of ethyl alcohol or methyl alcohol.

17. The disinfecting action of metal salts is in general weaker in broth, gelatin, body fluids, and the like, or in aqueous solutions containing these liquids, than it is in pure water. Probably this decrease in activity is due to a reduction in the concentration of the metal ion in solution.

18. No conclusions can be drawn about the growth inhibiting properties of a substance from the knowledge of its bactericidal activity.

19. It is probable that the degree of electrolytic dissociation of the metal salt has a less significant effect on the growth inhibiting properties of the substance than it does on the bactericidal properties. The former is more probably a function of the concentration of the metal in the nutrient solution.

20. General laws can be derived for the relationship between the concentration and the toxicity of mercuric bichloride solutions. It is probable that similar relationships will be discovered for the solutions of other substances.

We want to express our extreme thanks to Herr Geheimrath P. Zweifel and Herr Professor W. Ostwald, in whose laboratories the preceding work was carried out. They showed continual interest in our experiments and offered many helpful suggestions.

Comment

This paper tackles the problem of disinfection with typical German thoroughness. Although Koch and others had studied the action of chemicals on viability of bacterial cells and spores, there was a large amount of confusion about the basic processes involved. This paper clarified a number of concepts and was a basic paper for much of the later developments in antiseptics and disinfectants. The authors describe their methods in precise detail, apparently overlooking no point. This is a refreshing experience, since much of late nineteenth century microbiological work was published without adequate descriptions of methods. Although many of the details they describe have been long since improved upon, they describe for the first time a number of important precautions for this sort of work.

The influence of physical chemistry on this work is readily apparent. From the standpoint of disinfection theory, the most important point they present is that bacterial cells are not all killed instantaneously, but that populations of cells die at a logarithmic rate. In addition, the rate of kill is directly proportional to the concentration of disinfecting agent, so that one obtains a family of curves for the rate of kill at different disinfectant concentrations. This point is quite important. It shows first that not all bacterial cells are equally sensitive to the disinfectant but vary widely in their susceptibilities. The most susceptible die first, the others later. The second point is a practical one. Disinfection requires a finite length of time, and therefore one must be careful to leave the disinfectant in contact with the object being sterilized for a long enough period of time.

These observations and conclusions could not have been made without the precise quantitative procedures devised by the authors. It can be stated that Krönig and Paul's work opened up a new era of quantitative microbiology.

Modern chemotherapy

1908 • Paul Ehrlich

Ehrlich, Paul. 1908. Ueber moderne Chemotherapie. Vortrag gehalten in der X. Tagung der Deutschen Dermatologischen Gesellschaft. Frankfurt am Main, 8 Juni 1908. Published in "*Beiträge zur experimentellen Pathologie und Chemotherapie*. 1909. Leipzig, Akademische Verlagsgesellschaft m.b.H., pages 167–202.

IT GIVES ME GREAT PLEASURE TO BE invited to speak to you on the subject of modern trends in chemotherapy. As you know, this question has occupied me frequently for more than 25 years. In the last 5 years I have made a number of detailed studies of diseases caused by trypanosomes . . . which have revealed significant approaches to chemotherapy which differed from those of present-day pharmacology. Through studies on vital staining I was forced to the conclusion that the customary attempts to relate chemical structure and activity were too restricted, and it would be necessary to consider the way in which the chemical agent was distributed in the organism as an additional principle of chemotherapy.

It has been an almost self-evident conception, even since ancient times, that a particular agent can affect a particular organ, for example, the brain, . . . only when the agent is taken up by the cells of this organ. Indeed, the scientist of the middle-ages stated that the drug must have certain spiculae with which it could attach to the organ. However, this axiom has played no practical role in modern pharmacology.

My opponents will assert that what I have in mind is something self-evident and does not require a long dissertation to bring out. But, my dear sirs, in science we do not deal with words but with deeds. When King Heinrich IV made the statement that every citizen in his land should have a chicken in the pot on Sunday, this was indeed a noble and beautiful thought. How much more important it would have been if he had done something practical to make this wish a reality. An axiom is primarily important when it can be converted through labor into something useful, and one should not hide it in a shrine. Such, however, was the case in pharmacology, which has only in the most recent years been forced to recognize this principle and begun to modernize its approaches. The principle that every agent must be taken up by the cells to be active has been important in immunology for many years.

May it be recognized that I was the first to realize the necessity of carrying out therapeutic measures with reference to how the drug was distributed. This consideration also became the basis of the "side-chain" theory of immunology, which my opponents must also agree has exercised a very great influence on the progress of modern research in immunology. Immunology has showed us that the agents active in immunity–antitoxins, bacteriotropic substances–are fixed to the bacteria or bacterial secretion products with the help of specific groups. In this way it is very easy to understand the highly specific nature of the therapy which we obtain with antibodies. If we realize cures in a series of bacterial infections by the use of antiserums, this is very easy to understand. The protective materials which are present in the antiserums, whether they be of the amboceptor or opsonin type, find in infected organisms their point of attack only and exclusively in the bacteria, not in the tissues. These antibodies are exclusively "parasitotrophic," and not "organotrophic," and so it is not surprising that they seek out their targets like magic bullets. In this way I can explain the miraculous cures that are sometimes obtained.

It is therefore self-evident that the serum method, other things being equal, must be better than any other method of treatment because of the purely parasitotrophic nature of the curative agent.

But, gentlemen, this method of treatment by antibody action is effective only in a portion of the infectious diseases. In a large number of infections, and especially in those in which higher organisms are the parasites, the realization of a strong and long-lasting immunity is very difficult to obtain. This is true for malaria and for the trypanosome diseases. In these diseases it is necessary to develop chemical agents which can kill the parasites. That such a task is in principle possible can be seen from the cure of malaria with quinine and syphilis with mercury.

But, gentlemen, it should be made

clear that in general this task is much more complicated than that using serum therapy. These chemical agents, in contrast to the antibodies, may be harmful to the body. When such an agent is given to a sick organism, a difference must exist between the toxicity of this agent to the parasite and its toxicity to the host. "Bullseyes" such as the bacterial antibodies permit are no longer possible, so that we must always be aware of the fact that these agents are able to act on other parts of the body as well as on the parasite.

Therefore, if we wish to extend this problem further into modern pharmacology, then we have no other choice than to learn to shoot better. This we can do by making chemical modifications in our agents. Perhaps this will seem rather difficult to you. Let me give you an example to show you what I have in mind.

Because of the great interest which atoxyl therapy has excited in recent years, let me illustrate my views with this substance. As you know, decades ago Bechamp made, in the process of preparing fuchsin, a product by the fusion of aniline with arsenic acid. This product was later made available to medicine by the United Chemical Works under the name "Atoxyl." Bechamp and the United Chemical Works both considered this product to be an "arsenilic acid anilide." However, such anilides are very difficult to prepare by chemical means, and therefore not easy to use as starting materials for a program of chemical modifications. But a large field was opened when Bertheim and I showed that the structure of this substance was quite different, and that the arsenic acid moiety was not loosely attached to the nitrogen atom, but relatively strongly attached to the benzene ring. From our work it seemed that the compound was an amino derivative of phenyl arsenic acid.* The synthesis of Bechamp's substance followed the same process as that which occurs when sulfuric acid is fused with aniline. The sulfuric acid group attaches in the para position to the amino group and para-amino sulfonic acid is produced. This latter is called sulfanilic acid. We therefore call Bechamp's substance arsanilic acid, and this name will be used in the following instead of the fanciful name "atoxyl."

After the structure had been determined, it was now possible to produce a large series of related compounds, all of which were organic compounds of arsenic acid. Various groups could be introduced on the amino group, or it could be combined with various acidic groups, or it could be coupled with aldehydes. Further, with the help of nitrite, it could be diazotized and then a large number of azo dyes could be made containing arsenic acid. Then from the diazo compounds one could use well-known chemical procedures to produce chlorine, iodine, or cyano derivatives of the phenylarsenic acid. Also a series of homologs of atoxyl could be produced in which the aniline could be replaced with methyl and dimethylanilines. Over a period of time at the Georg Speyer Haus we produced a large series of such compounds with the help of Kahn, Bertheim, and

* [The chemical formula of atoxyl or arsanilic acid (sodium salt) is:

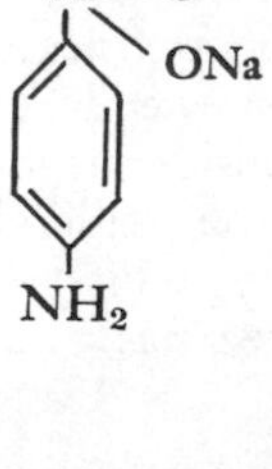

OH
As=O
ONa
NH_2

Schmitz. These have been tested biologically by Browning, Röhl and Fräulein Gulbransen.

It has been shown that by altering the composition or position in the arsanilic acid, it is possible at will to make the compound more toxic or less toxic. If I indicate the toxicity of sodium arsanilate as 1, certain products may be 20 times less toxic, while others may be 60–70 times more toxic. . . . We have thus a scale of toxicity which can vary over a 1500 fold range. Unfortunately it has not been possible in such a small animal as the mouse to analyse the nature of the toxicity, but we have made a number of interesting observations. A number of compounds, like acetarsanil, convert the mice into dancing mice, and this condition lasts for months or years in otherwise healthy mice. A series of compounds raise the icterus index; the mice become intensely yellow and this lasts for weeks. They excrete a urine which gives a test for bile and show, at autopsy, changes in the liver. Other compounds cause a profuse diarrhea, and in others there are changes in the kidney. It would not be wrong to assume that these various symptoms indicate that the compounds in question cause their main toxicity against different organs of the body.

After their toxicity is tested, all of these compounds are tested for their action against trypanosomes. The trypanosomes we have used were able to kill the experimental animals in three days. Twenty-four hours after infection, a small number were already present in the blood; after 48 hours they were very plentiful. Treatment was ordinarily begun on the first day, and 1 cc. of solution was injected for a 20 g. mouse. The maximum tolerated dose was used, as well as fractions of this.

We discovered in these experiments that certain alterations of arsanilic acid, like the introduction of a sulfuric acid group, remarkably detoxified the compound, so that it was less toxic than ordinary salt. But these substances did not show the slightest influence on the disappearance of the trypanosomes, even when used at almost toxic doses. It is apparent that with these substitutions we have completely missed our goal, since the trypanosomes are completely unaffected by them. In this way the relative nontoxicity of the preparations can be explained, since the preparations are only weakly fixed by the organs of the body, and for the most part appear in the excretions and secretions. These were, therefore, "shots into water."

Much more favorable were the results with the compound of acetyl arsanilic acid, which can be made easily by the action of acetic anhydride on the sodium salt of arsanilic acid. This compound is considerably less toxic for most, but not all, animal species.

Therapeutic results with the mouse have been exceptionally favorable. It has been possible with the help of this substance to cure mice which were on their second day of infection and would have died in a few hours. This is a result which a priori would not have been considered possible. We have found a series of these compounds which are able to quickly and in high proportion effect a cure which I have called "*Therapia sterilisans magna.*" That is to say, *a complete sterilization of a highly infected organism with one dose.*

I now come to an extremely important point concerning the cause of the action of arnsanil which has until now been clothed in darkness. As you all know, arsanil—as I call atoxyl—brings about a wonderfully rapid and surprising disappearance of the parasite in experimental as well as in natural infections with trypanosomes, especially in the case of sleeping sickness.

It seemed very probable, from the considerations above, that arsanil brought about directly the death of the parasites. However, while a number of other trypanosome active agents—fuchsin, triphenylmethane salts, salts of arsenious acid, tartar emetic—kill completely the trypanosomes in the test tube, such is not the case with arsanil. For example, Uhlenhuth has shown that the trypanosomes can remain for hours in a solution containing 8% arsanil without showing any damage. Levadities has made the same observation with the syphilis spirochete, and both of these excellent workers have arrived at the conclusion that arsanil does not act directly on the parasite, but indirectly. It may be either that the body modifies the compound in some way, or that it induces in the body the production of some other substance which is trypanosomicidal.

About one-and-a-half to two years ago I was able to shed a little light on the mode of action of arsanil, and the observations which first showed me the way were really quite ordinary. As some of you know, in my first studies I stated that arsanil displayed no reproducible effect in mice. While in general the tolerated dose for our mice was a 1/300 solution, there were mice that were able to tolerate double this concentration without any symptoms, while there were others which became sick from quite small doses, for example, 1/400, and could even die.

It seemed to me probable that this was not due to uncontrolled variables, but was a constitutional characteristic of the individual mice, since I was able to show that mice which had once survived a dose of 1/150, could tolerate a second injection of this same concentration, and so forth. On the other hand, I noticed that the therapeutic effect of arsanilic acid was closely related to the constitutional resistance of the mouse to arsanil. A mouse which had tolerated well an injection of a 1/150 solution, did not exhibit through this double dose any better therapeutic effect than the average mouse which received only half this dose, that is, 1/300. On the other hand, I noticed that the mice which became sick and died with a low dose (1/400 solution) showed, during the course of the trypanosome infection, an excellent trypanosomicidal effect. Indeed, the mice died of toxicity of the drug, but their blood remained free of parasites for days.

From these observations it seemed probable that in the organism the arsanil was being converted into another substance which was toxic, but which at the same time exerted a strong effect on the parasites. For anyone who had worked earlier on the reducing power of animal tissue, as I had done, it was an easy step to conclude that this phenomenon rested on a reducing process in the animal. You are aware that the oxygen-containing acids of arsenic occur in two types. In one, the arsenic has a valence of 5, while in the other it has a valence of 3. In arsanil the acidic complex of arsenic attached to the benzene ring has a valence of 5. Therefore arsanil is an aromatic arsenic acid. It therefore seemed necessary to convert this substance into the reduction product in which arsenic had a valence of 3.

In association with Bertheim, two reduction products of arsanil were obtained. The first was a white compound, soluble in acid and alkali, and has the chemical name of para-aminophenyl arsenic oxide, with the formula:

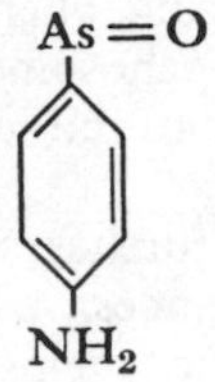

$As{=}O$

NH_2

The second compound comes from the first by a further reduction, in which the oxygen is removed and two arsenic groups come together to form a dimer:

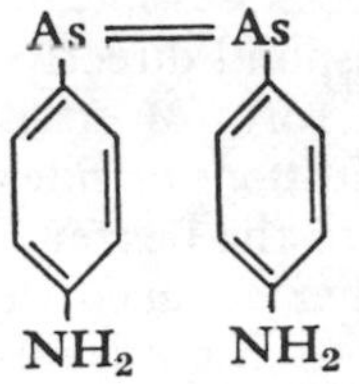

$As{=}As$

NH_2 NH_2

Both these compounds, and also others related to them, are completely altered in their biological properties through these small changes. From the relatively nontoxic substance a highly toxic product has been produced. The arsenic oxide product kills mice in a 1/12,000 solution, while the dimer kills them at 1/6,000. But the most important point is that through these changes a striking trypanosomicidal effect has been realized. The arsenic oxide compound kills trypanosomes in the test tube immediately at 1/100,000, and in one-half hour at a dilution of 1/1,000,000. When one considers that arsanil itself has not the slightest influence on the parasites in a 5% solution, it is easy to see what a colossal conversion and intensification of the activity is obtained merely by the removal of a single oxygen atom from the arsanil. We thus must assume that in the animal arsanil itself exerts no effect, but a very small portion of it undergoes reduction in the living organism to the actual toxic agent.

In this way it is possible to explain the results with arsanil in the mouse which were mentioned above. Clearly, different mice vary in their power to reduce arsanil. Those mice which have a strong reducing power reduce a relatively large amount of the compound and convert it into the toxic and trypanocidal substance. Thus they become sick, although free of parasites. On the other hand, the resistant mice possess a weak reducing power and form on the average a small amount of the reduction product. They remain unpoisoned and show an insufficient influence on the parasites.

If these considerations are correct, then it is obvious that we should seek to produce for therapeutic purposes the actual therapeutic agent and save the animals the trouble of making it themselves, especially since individual mice vary in this ability.

For experimental pharmacology our first duty is to clarify not only the questions "What?" and "How?" but also "Why?" I have therefore striven to discover the cause of the action of arsenic. To be sure, this has not been directly possible as yet, but an alternate path has been discovered which offers a considerable aid for the progress of our knowledge. This is through the study of drug-resistant strains of trypanosomes.

As you know, our Institute in Frankfurt was the first to show that unsuccessful treatments of trypanosome infections during long-term therapy with the various chemotherapeutic agents were due to the gradual development of trypanosome varieties which were resistant to the agent in question. In this way I have been able to obtain in collaboration with Browning and Röhl a strain of trypanosome which was resistant to fuchsin and a series of other basic triphenylmethane dyes. Also a strain which was resistant to trypan red and the related substances trypan blue and trypan violet, and a strain resistant to arsanil were obtained.

I would like to add here that this resistance is specific, in the sense that the fuchsin resistant strain is only resistant to substances related to it, and not to arsenic or trypan dyes. Further, it is an established fact that this resistance property, once acquired, remains unaltered for a long time. My arsanil resistant strain has been passed 300 times through normal animals over a period of 700 days, and in spite of this it is still completely resistant. The characteristic is therefore inherited.

Similar results on the inheritance of acquired resistance have been obtained by others. Effront has succeeded in obtaining a yeast which is tolerant to hydrogen fluoride and at the same time shows other biological changes from normal yeasts. He has also showed that this acquired resistance is maintained through a large number of generations, just as in the trypanosome strains.

It was then necessary to consider the question of the way in which this arsanil resistance came about.

Because of the fact that the reduction product of arsanil is able to kill the microorganisms at high dilution, while arsanil itself is not, one is inclined to assume that the trivalent arsenic group attached to the benzene ring is able to attach to the trypanosomes. Through therapeutic studies as well as through studies in the test tube, I have convinced myself that not only the reduction products of arsanil, but a large number of substances of other chemical types, such as those in which the amino group has been substituted in various ways, or in which the other groups, like the hydroxyl, are substituted, or in which other substitutions have been made on the benzene ring, behave similarly against the trypanosomes, when they are converted into the reduction product.

It has been further shown that all of these substances, when examined in the test tube against normal and arsanil-resistant strains, act stronger on the normal than on the resistant strains. From this, it must be assumed, that there is no mere physical distribution of the drug between the parasite and the body fluids, but that the trypanosome body possesses specific receptors, which correspond directly to those of the reduced form of arsanil, whether or not substitutions or side chains have been added to the benzene ring.

I had earlier assumed that "drugs," in contrast to the "toxins," are not bound to specific receptors. But the study of trypanosomes has caused me to discard this assumption and to postulate that *chemical groups must exist in the protoplasm which are necessary for the uptake of specific drugs*. I find myself in harmony with the views which Langley has put forth. I designate groups of this sort as *chemoreceptors* and would like to assume that in general, in relation to the simpler functions which they exert, they are of less complicated structure than those responsible for the binding of bacterial toxins. The drugs are further different in that they are bound tighter to the cells and not so easily removed. Through this sessile nature of the chemoreceptors it can be easily explained that, in contrast to the bacterial products, the crystalline drugs in general will induce no antibody formation.

If you accept this view, then you will ask the question: What changes take place in the chemoreceptor apparatus of the trypanosomes which make them become arsanil resistant?

The simplest would be that the receptors for the reduced arsanil would be completely eliminated from the trypanosomes. This cannot be considered, because in the test tube, when the trypanosome-containing blood is mixed with an aqueous solution of the drug in question, the resistant varieties are killed! It is therefore necessary to search further! To decide this question I have set up the following experiment: I took the arsanil resistant strain and divided it into two parts. One part was passed through mice treated with arsanil, the other part was passed through normal mice. After 160 passages, I tested both strains comparatively against arsanil. A priori, one would have expected that the strain would have become more strongly resistant through passage in the treated mice and would have become less strongly resistant through passage in normal mice. A quantitative determination of resistance showed to my great surprise that both strains had remained absolutely the same: the one strain had gained nothing, while the other had lost nothing.

The explanation of this result is not difficult, however. The arsanil resistant strain was not in the slightest inhibited by the high doses of arsanil and acetarsanil, because the avidity of the receptors had been highly reduced. The trypanosomes had remained free of the drug, in spite of the treatment, and therefore there was no difference between those passed through the normal mouse. Therefore it remains that both strains behaved exactly alike, in spite of the long and diverse passages.

We can interpret this phenomenon best by assuming that the cause of the arsanil resistance is due to the reduced avidity of the chemoreceptors, so that they are no longer able to take up arsanil. This alteration in avidity occurs naturally only in the species with which the experiments were done. Nature exhibits a wonderful fineness in only doing what is strictly necessary for the case at hand. . . .

Now, gentlemen, if we must seriously consider the possibility that through long treatment in men and animals arsanil resistant strains can arise, the important question arises, what therapeutic measures can be taken in such a case?

Resistance to arsanil is due to a reduction in the avidity of the chemoreceptors. If we wish to treat such cases therapeutically, we must find chemical substances which possess an increased avidity through changes in their constitution . . . and therefore be able to attach to the trypanosomes. This is naturally a question for chemical attack. At the Speyer-Haus I have been able, with the help of the chemical section, to discover an effective substance which belongs to the family of arsenic compounds, which I have called substance No. 418. This compound is light yellow in color, soluble in water, and slightly toxic for most animals. . . . This substance is effective against ordinary trypanosomes as well as against arsanil resistant trypanosomes. . . .

Mice which were in the second day of infection and would have died in a few hours were cured through a single injection of substance 418 at a dose which was only a fraction of the lethal dose. When one thinks what it means to sterilize an animal organism which contains millions, billions of parasites, and do this with one dose, this result must be considered very hopeful. . . .

I am well aware that animal experiments allow initially no conclusion for the therapy of man, since here the individual sensitivity and the degenerative character by individuals, and especially in such which suffer from latent damage to certain vital organs make the difficulties of therapy exceedingly greater. But these favorable curative results are possible in various animals with all trypanosome infections, and it seems certain that this should also make this therapy of benefit to man. If this substance does not prove to be suitable in human pathology, this does not mean that we should throw in the sponge and give up all hope. Rome was not built in a day! Therefore, we must continue to stride forward on the path which has now been clearly revealed before us.

Comment

It is difficult to select one paper of Ehrlich's that is representative of his work on chemotherapy. He published so many papers and ranged so widely over so many fields, that choosing one paper will do him an injustice. The paper chosen was presented at a time when Ehrlich's long work in chemotherapy was coming to fruition with the discovery of Salvarsan (compound 606).

Ehrlich's early work had been on the staining of cells and microorganisms with dyes. He acquired here an appreciation of the selective combination of chemicals with different types of cells, since microorganisms would often be stained by dyes that would have little effect on animal cells. He then went to work in Robert Koch's Institute in Berlin, where von Behring (see page 141) had just discovered diphtheria antitoxin. He worked with von Behring and was instrumental in working out a practical process for making diphtheria antitoxin in commercial amounts. He did much additional research in immunology and was impressed by the highly specific action of antibodies on toxins or on bacterial cells. He formulated a theory to explain this specific action based on the presence on the bacterial cell or toxin of specific groupings, which he called receptors, to which the antibodies would specifically attach by haptophores. He viewed the antibodies as magic bullets which would seek out the parasite or toxin and destroy it without affecting the host cell. This theory has been essentially discarded today but was very important in stimulating research in immunology for many years.

But Ehrlich was aware that there were many infectious diseases in which acquired immunity could not be induced. Thinking back to his earlier work on dyes, he began to think about the possibility of discovering chemical magic bullets which could be useful in these diseases. The most important idea that he had in regard to this work was that one should not be content to study just a few chemical compounds, but should embark on an extensive synthetic program, in which all possible chemical modifications would be made and carefully tested in experimental animals. As he outlines in the present paper, the modifications were usually made with some logic behind them. Ehrlich was the first to formulate this concept, in which he was greatly aided by the extensive developments of the German chemical industry.

As he mentions, compound 418 was synthesized. This compound, which is arsenophenylglycine, was the best compound which had been discovered for trypanosome infections. A number of other compounds were synthesized, including 606, which has the following structure:

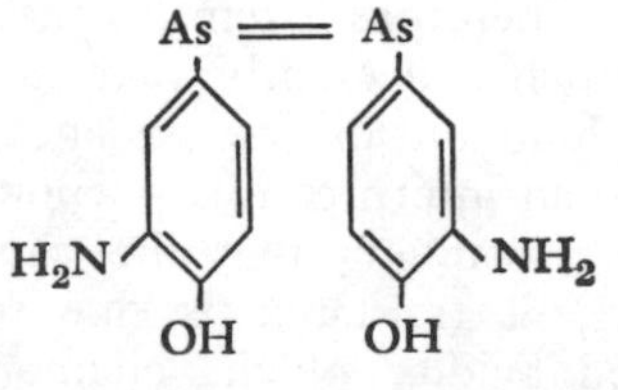

As = As

H_2N NH_2

OH OH

Its similarity to the dimer compound mentioned in the present paper is evident. For some reason, the activity of 606 against trypanosomes was missed. In

1909, a year after the present paper was given, a Japanese worker, Hata, who had developed techniques for infecting experimental animals with the syphilis spirochete, came to work in Ehrlich's laboratory. Ehrlich insisted that he try all of the arsenical compounds for activity against syphilis. During this search, it was discovered that 606 was amazingly active. This discovery was Ehrlich's most brilliant success and seemed to indicate that his original experimental conception was correct. The impact of this work on the field of chemotherapy was tremendous, and it can safely be stated that our present successes with sulfa drugs (see page 195) and antibiotics (see page 185) would not have been made without the pioneering work of Ehrlich.

In the same year the present paper was published Ehrlich was awarded the Nobel Prize in Medicine jointly with Metschnikoff (see page 132).

On the antibacterial action of cultures of a Penicillium, with special referencc to their use in the isolation of *B. influenzae*

1929 · Alexander Fleming

Fleming, Alexander. 1929. On the Antibacterial Action of Cultures of a Penicillium, with Special Reference to Their Use in the Isolation of *B. influenzae. British Journal of Experimental Pathology*, Vol. 10, pages 226–236.

WHILE WORKING WITH STAPHYLOCOCCUS variants a number of culture-plates were set aside on the laboratory bench and examined from time to time. In the examinations these plates were necessarily exposed to the air and they became contaminated with various micro-organisms. It was noticed that around a large colony of a contaminating mould the staphylococcus colonies became transparent and were obviously undergoing lysis.

Subcultures of this mould were made and experiments conducted with a view to ascertaining something of the properties of the bacteriolytic substance which had evidently been formed in the mould culture and which had diffused into the surrounding medium. It was found that broth in which the mould had been grown at room temperature for one or two weeks had acquired marked inhibitory, bactericidal and bacteriolytic properties to many of the more common pathogenic bacteria.

CHARACTERS OF THE MOULD

The colony appears as a white fluffy mass which rapidly increases in size and after a few days sporulates, the centre becoming dark green and later in old cultures darkens to almost black. In four or five days a bright yellow colour is produced which diffuses into the medium. In certain conditions a reddish colour can be observed in the growth.

In broth the mould grows on the surface as a white fluffy growth changing in a few days to a dark green felted mass. The broth becomes bright yellow and this yellow pigment is not extracted by $CHCl_3$. The reaction of the broth becomes markedly alkaline, the *p*H varying from 8.5 to 9. Acid is produced in 3 or 4 days in glucose and saccharose broth. There is no acid production in 7 days in lactose, mannite or dulcite broth.

Growth is slow at 37°C. and is most rapid about 20°C. No growth is observed under anaerobic conditions.

In its morphology this organism is a penicillium and in all its characters it most closely resembles *P. rubrum*. Biourge (1923) states that he has never found *P. rubrum* in nature and that it is an "animal de laboratoire." This penicillium is not uncommon in the air of the laboratory.

IS THE ANTIBACTERIAL BODY ELABORATED IN CULTURE BY ALL MOULDS?

A number of other moulds were grown in broth at room temperature and the culture fluids were tested for antibacterial substances at various intervals up to one month. The species examined were: *Eidamia viridiscens*, *Botrytis cineria*, *Aspergillus fumigatus*, *Sporotrichum*, *Cladosporium*, *Penicillium*, 8 strains. Of these it was found that only one strain of penicillium produced any inhibitory substance, and that one had exactly the same cultural characters as the original one from the contaminated plate.

It is clear, therefore, that the production of this antibacterial substance is not common to all moulds or to all types of penicillium.

In the rest of this article allusion will constantly be made to experiments with filtrates of a broth culture of this mould, so for convenience and to avoid the repetition of the rather cumbersome phrase "mould broth filtrate," the name "penicillin" will be used. This will denote the filtrate of a broth culture of the particular penicillium with which we are concerned.

METHODS OF EXAMINING CULTURES FOR ANTIBACTERIAL SUBSTANCES

The simplest method of examining for inhibitory power is to cut a furrow in an agar plate (or a plate of other suitable culture material), and fill this in with a mixture of equal parts of agar and the broth in which the mould has grown. When this has solidified, cultures of various microbes can be streaked at right angles from the furrow to the edge of the plate. The inhibitory substance diffuses very rapidly in the agar, so that in the few hours before the microbes show visible growth it has spread out for a centimetre or more in sufficient concentration to inhibit growth of a sensitive microbe. On further incubation it will be seen that the proximal portion of the culture for perhaps one centimetre becomes transparent, and on examination of this portion of the culture it is found that practically all the microbes are dissolved, indicating that the anti-bacterial substance has continued to diffuse into the agar in sufficient concentration to induce dissolution of the bacteria. This simple method therefore suffices to demonstrate the bacterio-inhibitory and bacteriolytic properties of the mould culture, and also by the extent of the area of inhibition gives some measure of the sensitiveness of the particular microbe tested. Figure 2 shows the degree of inhibition obtained with various microbes tested in this way.

The inhibitory power can be accurately titrated by making serial dilutions of penicillin in fresh nutrient broth, and then implanting all the tubes with the same volume of a bacterial suspension and incubating them. The inhibition can then readily be seen by noting the opacity of the broth.

For the estimation of the antibacterial power of a mould culture it is unnecessary to filter as the mould grows only slowly at 37°C., and in 24 hours, when the results are read, no growth of mould is perceptible. Staphylococcus is a very suitable microbe on which to test the broth as it is hardy, lives well in culture, grows rapidly, and is very sensitive to penicillin.

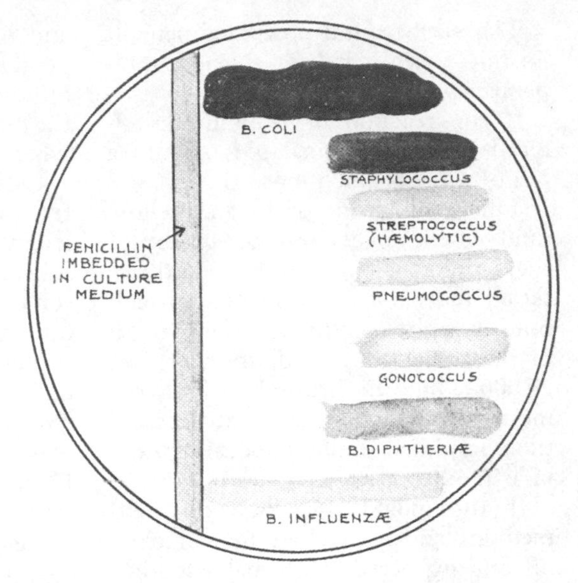

Fig. 2.

The bactericidal power can be tested in the same way except that at intervals measured quantities are explanted so that the number of surviving microbes can be estimated.

PROPERTIES OF THE ANTIBACTERIAL SUBSTANCE

Effect of heat. Heating for 1 hour at 56° or 80°C. has no effect on the antibacterial power of penicillin. Boiling for a few minutes hardly affects it. Boiling for 1 hour reduces it to less than one quarter its previous strength if the fluid is alkaline, but if it is neutral or very slightly acid then the reduction is much less. Autoclaving for 20 minutes at 115°C. practically destroys it.

Effect of filtration. Passage through a Seitz filter does not diminish the antibacterial power. This is the best method of obtaining sterile active mould broth.

Solubility. It is freely soluble in water and weak saline solutions. My colleague, Mr. Ridley, has found that if penicillin is evaporated at a low temperature to a sticky mass the active principle can be completely extracted by absolute alcohol. It is insoluble in ether or chloroform.

Rate of development of inhibitory substance in culture. A 500 cc. Erlenmeyer flask containing 200 cc. of broth was planted with mould spores and incubated at room temperature (10° to 20°C.). The inhibitory power of the broth to staphylococcus was tested at intervals.

After	Complete inhibition in
5 days	1 in 20 dilution.
6 "	1 in 40 "
7 "	1 in 200 "
8 "	1 in 500 "

Grown at 20°C. the development of the active principle is more rapid and a good sample will completely inhibit staphylococci in a 1 in 500 or 1 in 800 dilution in 6 or 7 days. As the culture ages the antibacterial power falls and may in 14 days at 20°C. have almost disappeared.

The antibacterial power of penicillin falls when it is kept at room temperature.

If the reaction of penicillin is altered from its original *p*H of 9 to a *p*H of 6.8 it is much more stable.

The small drops of bright yellow fluid which collect on the surface of the mould may have a high antibacterial titre. One specimen of such fluid completely inhibited the growth of staphylococci in a dilution of 1 in 20,000 while the broth in which the mould was growing, tested at the same time, inhibited staphylococcal growth in 1 in 800.

If the mould is grown on solid medium and the felted mass picked off and extracted in normal salt solution for 24 hours it is found that the extract has bacteriolytic properties.

If this extract is mixed with a thick suspension of staphylococcus suspension and incubated for 2 hours at 45°C. it will be found that the opacity of the suspension has markedly diminished and after 24 hours the previously opaque suspension will have become almost clear.

Influence of the medium on the antibacterial titre of the mould culture. So far as has been ascertained nutrient broth is the most suitable medium for the production of penicillin. The addition of glucose or saccharose, which are fermented by the mould with the production of acid, delays or prevents the appearance of the antibacterial substance. Dilution of the broth with water delays the formation of the antibacterial substance and diminishes the concentration which is ultimately reached.

INHIBITORY POWER OF PENICILLIN ON THE GROWTH OF BACTERIA

Tables II and III show the extent to which various microbes, pathogenic and non-pathogenic, are inhibited by penicillin. The first table shows the inhibition by the agar plate method and the second shows the inhibitory power when diluted in nutrient broth.

Certain interesting facts emerge from these Tables. It is clear that penicillin contains a bacterio-inhibitory substance which is very active towards some microbes, while not affecting others. The members of the colityphoid group are unaffected as are other intestinal bacilli such as *B. pyocyaneus* [Pseudomonas], *B. proteus* [Proteus] and *V. cholerae* [Vibrio]. Other bacteria which are insensitive to penicillin are the enterococcus, some of the Gram-negative cocci of the mouth, Friedlander's pneumobacillus [Klebsiella], and *B. influenzae* (Pfeiffer) [Hemophilus], while the action on B. dysenteriae (Flexner) [Shigella], and *B. pseudotuberculosis rodentium* is almost negligible. The anthrax bacillus is completely inhibited in a 1 in 10 dilution but in this case the inhibitory influence is trifling when compared with the effect on the pyogenic cocci.

It is on the pyogenic cocci and on bacilli of the diphtheria group that the action is most manifest.

Staphylococci are very sensitive, and the inhibitory effect is practically the same on all strains, whatever the colour or type of the staphylococcus.

Streptococcus pyogenes is also very sensitive. There were small differences in the titre with different strains, but it may be said generally that it is slightly more sensitive than staphylococcus.

Pneumococci are equally sensitive with *Streptococcus pyogenes*.

The green streptococci vary very considerably, a few strains being almost unaffected while others are as sensitive as *S. pyogenes*. Gonococci, meningococci, and some of the Gram-negative cocci found in nasal catar-

TABLE II.—*Inhibitory Power of Penicillin on Various Microbes (Agar Plate Method)*

Type of microbe	Extent of inhibition in mm. from penicillin embedded in agar, serum agar, or blood agar plates
Experiment 1:	
Staphylococcus pyogenes	23
Streptococcus pyogenes	17
Streptococcus viridans (mouth)	17
Diphtheroid bacillus	27
Sarcina	10
Micrococcus lysodiekticus	6
Micrococcus from air (1)	20
Micrococcus from air (2)	4
B. anthracis [Bacillus]	0
B. typhosus [Salmonella]	0
Enterococcus [Streptococcus?]	0
Experiment 2:	
Staphylococcus pyogenes	24
Streptococcus pyogenes	30
Streptococcus viridans (mouth)	25
Pneumococcus	30
Diphtheroid bacillus	35
B. pyocyaneus [Pseudomonas]	0
B. pneumoniae (Friedlander) [Klebsiella]	0
B. coli [Escherichia]	0
B. paratyphosus A [Salmonella]	0
Experiment 3:	
Staphylococcus pyogenes	16
Gonococcus [Neisseria]	16
Meningococcus [Neisseria]	17
Experiment 4:	
Staphylococcus pyogenes	17
Staphylococcus epidermidis	18
Streptococcus pyogenes	15
Streptococcus viridans (faeces)	5
B. diphtheriae (2 strains) [Corynebacterium]	14
Diphtheroid bacillus	10
Gram-negative coccus from the mouth (1)	12
Gram-negative coccus from the mouth (2)	0
B. coli [Escherichia]	0
B. influenzae (Pfeiffer) 6 strains [Hemophilus]	0

Table III.—*Inhibitory Power of Penicillin on Different Bacteria.**

	Dilution of penicillin in broth											
	1/5.	1/10.	1/20.	1/40.	1/80.	1/100.	1/200.	1/400.	1/800.	1/1600.	1/3200.	Control.
Staphylococcus aureus	0	0	0	0	0	0	0	0	±	++	++	++
” *epidermidis*	0	0	0	0	0	0	0	0	±	++	++	++
Pneumococcus	0	0	0	0	0	0	0	0	0	++	++	++
Streptococcus (haemolytic)	0	0	0	0	0	0	0	0	0	±	++	++
” *viridans* (mouth)	0	0	0	0	0	0	±	++	++	++	++	++
” *faecalis*	++	++	++	++	++	++	++	++	++	++	++	++
B. anthracis [Bacillus]	0	0	+	+	++	++	++	++	++	++	++	++
B. pseudo-tuberculosis rodentium	+	+	++	++	++	++	++	++	++	++	++	++
B. pullorum [Salmonella]	+	+	++	++	++	++	++	++	++	++	++	++
B. dysenteriae [Shigella]	+	++	++	++	++	++	++	++	++	++	++	++
B. coli [Escherichia]	++	++	++									++
B. typhosus [Salmonella]	++	++	++									++
B. pyocyaneus [Pseudomonas]	++	++	++									++
B. proteus [Proteus]	++	++	++									++
V. cholerae [Vibrio]	++	++	++									++

	1/60.	1/120.	1/300.	1/600.	Control.
B. diphtheriae (3 strains) [Corynebacterium]	0	±	++	++	++
Streptococcus pyogenes (13 strains)	0	0	0	++	++
” ” (1 strain)	0	0	±	++	++
” *faecalis* (11 strains)	++	++	++	++	++
” *viridans* at random from faeces (1 strain)	0	0	0	++	++
” ” ” ” (2 strains)	0	0	±	++	++
” ” ” ” (1 strain)	0	±	++	++	++
” ” ” ” (1 strain)	+	++	++	++	++
” ” ” ” (1 strain)	++	++	++	++	++
” ” at random from mouth (1 strain)	0	±	++	++	++
” ” ” ” (2 strains)	0	0	++	++	++
” ” ” ” (1 strain)	0	0	0	++	++

* Legend: 0 = no growth; ± = trace of growth; + = poor growth; ++ = normal growth.

rhal conditions are about as sensitive as are staphylococci. Many of the Gram-negative cocci found in the mouth and throat are, however, quite insensitive.

B. diphtheriae [Corynebacterium] is less affected than staphylococcus but is yet completely inhibited by a 1% dilution of a fair sample of penicillin.

It may be noted here that penicillin, which is strongly inhibitory to many bacteria, does not inhibit the growth of the original penicillium which was used in its preparation.

THE RATE OF KILLING OF STAPHYLOCOCCI BY PENICILLIN

Some bactericidal agents like the hypochlorites are extremely rapid in their action, others like flavine or novarsenobillon are slow. Experiments were made to find into which category penicillin fell.

To 1 cc. volumes of dilutions in broth of penicillin were added 10 c.mm. volumes of a 1 in 1000 dilution of a staphylococcus broth culture. The tubes were then incubated at 37°C. and at intervals 10 c.mm. volumes were removed and plated with the following result:

	Number of colonies developing after sojourn in penicillin in concentrations as under:				
	Control	*1/80*	*1/40*	*1/20*	*1/10*
Before	27	27	27	27	27
After 2 hours	116	73	51	48	23
After 4½ hours	∞	13	1	2	5
After 8 hours	∞	0	0	0	0
After 12 hours	∞	0	0	0	0

It appears, therefore, that penicillin belongs to the group of slow acting antiseptics, and the staphylococci are only completely killed after an interval of over 4½ hours even in a concentration 30 or 40 times stronger than is necessary to inhibit completely the culture in broth. In the weaker concentrations it will be seen that at first there is growth of the staphylococci and only after some hours are the cocci killed off. The same thing can be seen if a series of dilutions of penicillin in broth are heavily infected with staphylococcus and incubated. If the cultures are examined after 4 hours it may be seen that growth has taken place apparently equally in all the tubes but when examined after being incubated overnight, the tubes containing penicillin in concentrations greater than 1 in 300 or 1 in 400 are perfectly clear while the control tube shows a heavy growth. This is a clear illustration of the bacteriolytic action of penicillin.

TOXICITY OF PENICILLIN

The toxicity to animals of powerfully antibacterial mould broth filtrates appears to be very low. Twenty cc. injected intravenously into a rabbit were not more toxic than the same quantity of broth. Half a cc. injected intraperitoneally into a mouse weighing about 20 gm. induced no toxic symptoms. Constant irrigation of large infected surfaces in man was not accompanied by any toxic symptoms, while irrigation of the human conjunctiva every hour for a day had no irritant effect.

In vitro pencillin which completely inhibits the growth of staphylococci in a dilution of 1 in 600 does not interfere with leucocytic function to a greater extent than does ordinary broth.

USE OF PENICILLIN IN THE ISOLATION OF *B. INFLUENZAE* (PFEIFFER) AND OTHER ORGANISMS

It sometimes happens that in the human body a pathogenic microbe may be difficult to isolate because it occurs in association with others which grow more profusely and which mask it. If in such a case the first microbe is insensitive to penicillin and the obscuring microbes are sensitive, then by the use of this substance these latter can be inhibited while the former are allowed to develop normally. Such an example occurs in the body, certainly with *B. influenzae* (Pfeiffer) and probably with Bordet's whooping-cough bacillus and other organisms. Pfeiffer's bacillus, occurring as it does in the respiratory tract, is usually associated with streptococci, pneumococci, staphylococci and Gram-negative cocci. All of these, with the exception of some of the Gram-negative cocci, are highly sensitive to penicillin and by the addition of some of this to the medium they can be completely inhibited while *B. influenzae* is unaffected. A definite quantity of the penicillin may be incorporated with the molten culture medium before the plates are made, but an easier and very satisfactory method is to spread the infected material, sputum, nasal mucus, and so forth, on the plate in the usual way and then over one half of the plate spread 2 to 6 drops (according to potency) of the penicillin. This small amount of fluid soaks into the agar and after cultivation for 24 hours it will be found that the half of the plate without the penicillin will show the normal growth while on the penicillin treated half there will be nothing but *B. influenzae* with Gram-negative cocci and occasionally some other microbe. This makes it infinitely easier to isolate these penicillin-insensitive organisms, and repeatedly *B. influenzae* has been isolated in this way when they have not been seen in films of sputum and when it has not been possible to detect them in plates not treated with penicillin. Of course if this method is adopted then a medium favourable for the growth of *B. influenzae* must be used, *e.g.* boiled blood agar, as by the repression of the pneumococci and the staphylococci the symbiotic effect of these, so familiar in cultures of sputum on blood agar, is lost and if blood agar alone is used the colonies of *B. influenzae* may be so minute as to be easily missed.

From a number of observations which have been made on sputum, postnasal and throat swabs it seems likely that by the use of penicillin, organisms of the *B. influenzae* group will be isolated from a great variety of pathological conditions as well as from individuals who are apparently healthy.

DISCUSSION

It has been demonstrated that a species of penicillium produces in culture a very powerful antibacterial substance which affects different bacteria in different degrees. Speaking generally it may be said that the least sensitive bacteria are the Gram-negative bacilli, and the most susceptible are the pyogenic cocci. Inhibitory substances have been described in old cultures of many organisms; generally the inhibition is more or less specific to the microbe which has been used for the culture, and the inhibitory substances are seldom strong enough to withstand even slight dilution with fresh nutrient material. Penicillin is not inhibitory to the original penicillium used in its preparation.

Emmerich and other workers have shown that old cultures of *B. pyocyaneus* acquire a marked bacteriolytic power. The bacteriolytic agent, pyocyanase, possesses properties similar to penicillin in that its heat resistance is the same and it exists in the filtrate of a fluid culture. It resembles penicillin also in that it acts only on certain microbes. It differs however in being relatively extremely weak in its action and in acting on quite different types of bacteria. The bacilli of anthrax, diphtheria, cholera and typhoid are those most sensitive to pyocyanase, while the pyogenic cocci are unaffected, but the percentages of pyocyaneus filtrate necessary for the inhibition of these organisms was 40, 33, 40 and 60 respectively (Bocchia, 1909). This degree of inhibition is hardly comparable with 0.2% or less of penicillin which is necessary to completely inhibit the pyogenic cocci or the 1% necessary for *B. diphtheriae*.

Penicillin, in regard to infections with sensitive microbes, appears to have some advantages over the well-known chemical antiseptics. A good sample will completely inhibit staphylococci, Streptococcus pyogenes and pneumococcus in a dilution of 1 in 800. It is therefore a more powerful inhibitory agent than is carbolic acid and it can be applied to an infected surface undiluted as it is non-irritant and non-toxic. If applied, therefore, on a dressing, it will still be effective even when diluted 800 times which is more than can be said of the chemical antiseptics in use. Experiments in connection with its value in the treatment of pyogenic infections are in progress.

In addition to its possible use in the treatment of bacterial infections penicillin is certainly useful to the bacteriologist for its power of inhibiting unwanted microbes in bacterial cultures so that penicillin insensitive bacteria can readily be isolated. A notable instance of this is the very easy isolation of Pfeiffer's bacillus of influenza when penicillin is used.

In conclusion my thanks are due to my colleagues, Mr. Ridley and Mr. Craddock, for their help in carrying out some of the experiments described in this paper, and to our mycologist, Mr. la Touche, for his suggestions as to the identity of the penicillium.

SUMMARY

1. A certain type of penicillium produces in culture a powerful antibacterial substance. The antibacterial power of the culture reaches its maximum in about 7 days at 20°C. and after 10 days diminishes until it has almost disappeared in 4 weeks.

2. The best medium found for the production of the antibacterial substance has been ordinary nutrient broth.

3. The active agent is readily filterable and the name "penicillin" has been given to filtrates of broth cultures of the mould.

4. Penicillin loses most of its power after 10 to 14 days at room temperature but can be preserved longer by neutralization.

5. The active agent is not destroyed by boiling for a few minutes but in alkaline solution boiling for 1 hour markedly reduces the power. Autoclaving for 20 minutes at 115°C. practically destroys it. It is soluble in alcohol but insoluble in ether or chloroform.

6. The action is very marked on the pyogenic cocci and the diphtheria group of bacilli. Many bacteria are quite insensitive, *e.g.* the coli-typhoid group, the influenza-bacillus group, and the enterococcus.

7. Penicillin is non-toxic to animals in enormous doses and is non-irritant. It doses (sic) not interfere with leucocytic function to a greater degree than does ordinary broth.

8. It is suggested that it may be an efficient antiseptic for application to, or injection into, areas infected with penicillin-sensitive microbes.

9. The use of penicillin on culture

plates renders obvious many bacterial inhibitions which are not very evident in ordinary cultures.

10. Its value as an aid to the isolation of *B. influenzae* has been demonstrated.

Comment

The history of the development of penicillin is now quite well known. This paper presents the first scientific observations of this substance.

Many earlier workers had noted the effects of various microorganisms on pathogenic bacteria. Fleming was the first to actually develop a clear conception of what this action was due to. It is obvious that he considers penicillin to be a substance, and although he does not say this in so many words, he considered it to be a chemical compound with a definite structure. He presented sufficient basic information so that later workers could develop penicillin into a practical substance.

He noted that this substance was produced only by a specific species of mold. This observation is important because it showed that the action of the mold broth filtrate was not due to some nonspecific effect, such as a change of *p*H during growth.

He devised a crude assay for determining the potency of the substance. With this assay he was able to study the stability and certain chemical properties of penicillin. He was able to follow its production in culture. By current standards of penicillin production, Fleming's material was of very low potency. His broths probably had around 10 μg./ml. of penicillin or less, while current commercial production gives yields of at least 5000 μg./ml. But because of the high activity of penicillin against certain bacteria, he was able to study the effects of these crude broths. He showed that not all bacteria were affected equally by penicillin, indicating a high degree of specificity in its action. He further described the bactericidal and bacteriolytic action of the antibiotic and showed that a period of growth was needed before either action was manifest. We know now that this is because penicillin specifically inhibits the synthesis of the cell walls of sensitive bacteria. This means that as they continue to grow without synthesizing cell walls, they are killed or lysed because they become osmotically fragile.

Fleming perceived only dimly the practical applications of penicillin. He obviously was aware that such a substance would be highly useful if it would kill pathogenic bacteria in humans, but such a hope seemed much in the future in view of the low broth potencies he was obtaining. It took the development of the sulfa drugs (see page 195) and the impetus of World War II before the tremendous technological and medical problems could be adequately solved which led to the development of penicillin as a practical chemotherapeutant.

A contribution to the chemotherapy of bacterial infections

1935 · Gerhard Domagk

Domagk, Gerhard. 1935. Ein Beitrag zur Chemotherapie der bakteriellen Infektionen. *Deutsche medizinische Wochenschrift*, Vol. 61, pages 250–253.

IT IS CURRENTLY THE GENERAL OPINION that only protozoal infections can be attacked by chemotherapeutic means. For protozoal infections, a number of effective drugs are available; for example, germanin for trypanosome infections, neostibodan for kala azar, plasmochin and atebrin for malaria, and salvarsan and its derivatives for spirochetes, especially syphilis.

For coccal infections, there have been no reasonably effective chemotherapeutants known. Protozoa and spirochetes represent relatively advanced groups of living organisms, and the more highly developed an infectious agent is, the more loci it offers for attack by chemotherapeutants. An advance in the chemotherapy of pneumococcus infections has been made by Morgenroth, but optochin is used mostly for direct application on the infection focus, as is vuzin—another derivative of hydroquinine—in streptococcal infections. In systemic infections we have found no clear-cut effect of these preparations in our experimental animals. Also, the silver compounds recommended for therapy of septic infections have proved to be inadequate in practice. Indeed, critical observation often indicates that they cause a detrimental effect on the course of the illness.

A prerequisite for the systematic search for chemotherapeutically effective substances is always a suitable model system. With streptococci it is possible to produce reproducibly in mice a fatal infection. We have used for our studies a hemolytic strain of streptococcus which came from a fatal human infection.

The first chemical compounds which we found to be effective in streptococcus infections were a series of compounds of gold. These compounds produced a significant effect in animals and also showed an unquestionably favorable influence on streptococcus infections in humans. These gold compounds however had an important disadvantage. They could not be used in doses high enough to produce a certain chemotherapeutic effect and could not be used over a long period of time. For in long-term treatment, there was the danger of gold toxicity developing. Skin rashes and kidney damage appeared, which disappeared when the drug was stopped, but returned when the therapy was begun again.

Success with gold compounds in

the treatment of streptococcus infections as well as syphilis has also been reported by Feldt, who recommends as the most effective agent a gold thioglucose preparation.

Because of the disadvantages mentioned, we turned our attention to other chemical classes which were pure organic compounds without any metallic groups and which showed in mouse experiments an indication of activity. We were aware of a series of azo and acridine compounds which had shown a relatively good effect during *in vitro* disinfection experiments against streptococci. This occasional excellent *in vitro* activity almost completely disappeared when these substances were injected into the animal body.

Azo compounds have often excited therapeutic interest. Among the acidic azo compounds, trypan blue has been found effective against trypanosomes as well as against leprosy. Of the neutral azo compounds, diacetylaminoazotoluene has been used as an agent for the promotion of healing after accidents. The oldest of the bactericidal basic azo compounds is 2, 4-diaminoazobenzene, whose hydrochloride has been known for a long time under the name chrysoidin and has been used as a bacterial stain. Eisenberg showed in 1913 that chrysoidin inhibited the growth of Gram-negative bacteria in a dilution of 1/1000 and Gram-positive in a dilution of 1/10,000. Eisenberg himself had entertained the thought that this occasionally highly active dye might have a chemotherapeutic effect in infections. He emphasized at the same time what a long way it is from the most excellent *in vitro* results to a decisive *in vivo* result. This demonstrable *in vitro* bactericidal activity of such basic azo compounds was mentioned again by Lockemann and Ulrich in 1934. They tested preparations which contained as their important ingredient 2,4-diaminoazobenzene. A further development in the basic azo compounds over chrysoidin was the use of the pyridine component in phenylazo-2,6-diaminopyridine and 2′-butoxypyridyl-5,5′-azo-2,6-diaminopyridine. These and other related compounds were used in practice as urinary antiseptics and were also recommended by the manufacturer for gonorrhoea. We could not demonstrate any therapeutic effect of these compounds in animals infected with streptococci and staphylococci.

In the course of our studies, we hit upon a group of very nontoxic azo compounds which indeed had no significant disinfection action *in vitro* against streptococci but showed a clear effect in mouse experiments. In this group was Prontosil, which Mietzsch and Klarer had synthesized in 1932. With Prontosil we can show the best chemotherapeutic effect in streptococcus infections in animals that we have ever seen. Prontosil is the hydrochloride salt of 4′-sulfonamid-2,4-diaminoazobenzene.* This is a red crystalline powder with a melting point of 247–251°, which is soluble to 0.25% in cold water but is more readily soluble in warm water.

The nontoxicity of this preparation can be seen from the toxicological data. The animals tolerated orally the following doses: mice, 500 mg./kg.; rabbits, 500 mg./kg.; cats, 200 mg./kg. Higher doses were vomited up.

Subcutaneously, a 20g. mouse can be given 1–2 cc. of a 0.25% solution. As a suspension, a 4% solution of the dye can be injected under the skin of mice in 1–2 cc. without any skin necrosis developing and without any generalized toxic symptoms occurring. . . .

Weese and Hecht have shown that Prontosil is pharmacologically an ex-

* [See page 199 for this structure.]

tremely indifferent compound. Even with rapid injections of 10 mg./kg. intravenously there was no change in blood pressure or heart function in cats and rabbits. The smooth muscle of the uterus, large intestine and small intestine were not affected either *in situ* or isolated. The physiological function of these organs was not influenced by Prontosil. In subcutaneous injections up to 1 g./kg., no toxicity in animals was observed, while intravenous injections showed no production of thrombosis as the result of damage of the blood vessel walls.

Prontosil shows in mice a selective chemotherapeutic effect against streptococci.

In our studies we have used a highly pathogenic *Streptococcus hemolyticus* from a human case of streptococcus sepsis. We cultured the streptococci in egg broth. Dilutions of 1/1000, 1/5000, 1/10,000 and 1/100,000 of a 24-hour culture were used. In general 0.3 cc. of a 1/10,000 dilution was sufficient to kill mice within 24 hours when injected intraperitoneally. For the experiments we usually infected the animals with 10 times the lethal dose, so that all of the infected, untreated controls died with certainty within 24 hours, or at latest 48 hours, from the sepsis which developed in them. In the sick animals hemolytic streptococci could be demonstrated in the heart blood and in almost all organs, even 1–2 hours after infection.

For subcutaneous and oral treatment of the infected mice we used the dye dissolved in water up to 0.5% or suspended in water up to 1–4%. For a treatment with a single dose, 1/10 to 1/50 of the highest tolerated dose was sufficient to show a clear effect. If the infected mice were treated with this dose for 3–5 days, then they generally showed a complete cure from the otherwise fatal infection. From time to time we could show a clear effect at a dose 1/100–1/500 of the tolerated dose, especially when the infection was not quite so acute and the controls did not die until later than 24–48 hours.

The table on page 198 shows an experiment in which the animals were infected with a 1/1000 dilution of a 24 hour culture, although the infection was somewhat slower than usual. . . .

The results of our animal experiments have been confirmed in the clinic. Prontosil will be tested in the clinic under the name "Streptozon."

Whether Prontosil acts directly or indirectly against the pathogen in the body cannot be decided as yet.* It is remarkable that *in vitro* it shows no noticeable effect against streptococci or staphylococci. It exerts a true chemotherapeutic effect only in the living animal. In pneumococcus and other infections Prontosil shows no noteworthy effect, so that it seems to be specific for streptococci, while acting somewhat on staphylococci. These observations argue against a nonspecific, general, and indirect effect of Prontosil on the organism, such as a nonspecific activation of the reticuloendothelial system.

Therefore, in the future the physician will have to determine also with bacterial infections immediately the specific character of the infection. From the chemotherapy of the protozoal infections it has been known for a long time that our most effective chemotherapeutic preparations affect only certain protozoa, while others, which cannot be distinguished morphologically from those which are affected, show no sensitivity to the agent.

In order to determine the specific character of a bacterial infection as soon as possible, a close cooperation

* [See comment, page 199.]

Streptococcus experiment of Dec. 20, 1932.

Infected with 1/1000 dilution of egg broth culture, 0.3 cc. intraperitoneally.

Treated 1½ hour after the infection, orally.

Animal No.	*Weight*	*Dose*	*Date*						
			Dec. 21	*Dec. 22*	*Dec. 23*	*Dec. 24*	*Dec. 25*	*Dec. 26*	*Dec. 27*
201	14 g.	0	+	(+)	(+)	—			
202	14 g.	0	+	—					
203	14 g.	0	+	—					
204	17 g.	0	+	—					
205	19 g.	0	+	—					
206	14 g.	0	+	+	—				
303	18 g.	0.2 cc. 0.01%	+	+	+	+	+	+	+
304	19 g.	0.2 cc. 0.01%	+	+	+	+	+	+	+
305	18 g.	1.0 cc. 0.01%	+	+	+	+	+	+	+
306	14 g.	1.0 cc. 0.01%	+	+	+	+	+	+	+
307	16 g.	0.2 cc. 0.1%	+	+	+	+	+	+	+
308	15 g.	0.2 cc. 0.1%	+	+	+	+	+	+	+
309	17 g.	1.0 cc. 0.1%	+	+	+	+	+	+	+
310	17 g.	1.0 cc. 0.1%	+	+	+	+	+	+	+
311	14 g.	0.2 cc. 1.0%	+	+	+	+	+	+	+
312	17 g.	0.2 cc. 1.0%	+	+	+	+	+	+	+
313	18 g.	1.0 cc. 1.0%	+	+	+	+	+	+	+
314	14 g.	1.0 cc. 1.0%	+	+	+	+	+	+	+

Key: + alive and well, (+) sick, — dead

between the clinician or physician on one side and the bacteriologist on the other side will be necessary. But even this ideal cooperation will not always achieve the desired result, since many times it is not possible to demonstrate the pathogen in the blood, especially in the early stages. . . .

Since this problem will not be quick to be solved, it should be given our attention immediately.

Summary. Prontosil shows an effect on animals infected with streptococci such as has not been observed before. It is possible to protect infected mice which would die within 24 hours by giving them subcutaneous or oral doses of Prontosil. Also chronic infections with streptococci in rabbits show a favorable result. Prontosil seems to show the best effect on streptococcal sepsis in the mouse but has some effect in staphyloccoccus infections. In pneumoccoccus infections as well as in other infections, it has not been possible to show any effect in experimental animals.

Comment

Domagk's discovery of Prontosil is a direct fulfillment of the dreams of Ehrlich when he first formulated his concepts of chemotherapy (see page 176). The methods used were based directly on Ehrlich's, and the success of the work results directly from the use of these methods. The basic idea was to test the action of chemical compounds directly in the infected animal, without regard to whether they had any effect in the test tube. In addition, a large number of compounds should be tested, representing as many of the possible modifications of certain basic structures as could be used.

Prontosil is not active against bacteria

in vitro, as Domagk noted. It is quite active in infected animals. This difference is due, as we know now, to the fact that Prontosil is broken down in the animal into the active agent, sulfanilamide, as shown in the formulas below.
With the discovery of this fact, it was possible to embark on a program of synthesis around the sulfanilamide structure. This program has led to the large number of sulfa drugs known today.

The sulfa drugs were the first chemical substances that had any real effectiveness against bacterial infections. In particular, their action against streptococcal infections is noteworthy, since streptococcal infections are among the most frequent and most acute that occur in man. When it was possible to conquer these infections with chemical compounds, a major breakthrough in medicine had occurred. In addition, this success was probably instrumental in encouraging a group at Oxford University to attempt to make practical use of Fleming's discovery of penicillin (see page 185). Therefore, although the sulfa drugs have been replaced in many medical uses by antibiotics, their historical importance is quite great. In addition, they still have wide usage for certain infectious diseases, especially urinary tract infections, where they are often the drugs of choice.

The mode of action of sulfanilamide and the other sulfa drugs against bacteria has been understood since the work of Woods, whose paper is presented next.

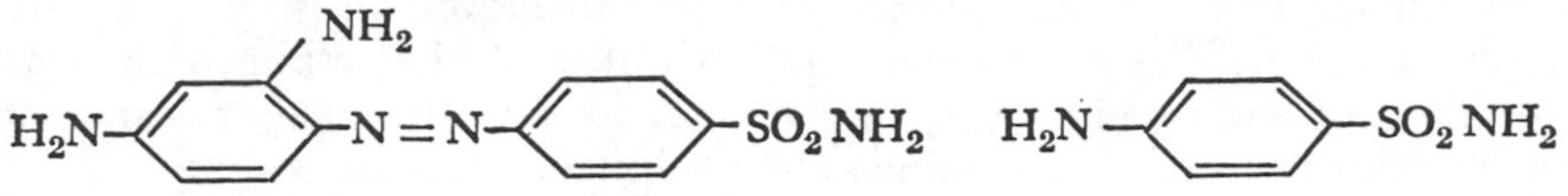

NH_2
$H_2N-C_6H_3-N=N-C_6H_4-SO_2NH_2$ $H_2N-C_6H_4-SO_2NH_2$

Prontosil *Sulfanilamide*

The relation of para-aminobenzoic acid to the mechanism of the action of sulphanilamide

1940 · Donald D. Woods

Woods, D. D. 1940. The Relation of *p*-Aminobenzoic Acid to the Mechanism of the Action of Sulphanilamide. *British Journal of Experimental Pathology*, Vol. 21, pages 74–90.

FILDES (1940) DISCUSSED A GENERAL hypothesis that disinfectants and other substances preventing the growth of bacteria act by interfering in some way with substances essential to the growth of the organism. In the present work the mode of action of sulphanilamide is investigated from the point of view of this general hypothesis.

A large number of substances

known to be associated with bacterial metabolism were tested to determine whether they had an antagonistic relation to sulphanilamide akin to that of –SH and Hg * but without conclusive results. While this work was in progress, Stamp (1939), working on a similar hypothesis, found that extracts of streptococci were able to antagonize the action of sulphanilamide and later, while the present paper was actually in preparation, Green (1940) obtained a preparation from *Brucella abortus*. Following Stamp's procedure it was found that yeast extracts contained a substance which, like that of Stamp and Green, reversed the inhibitory action of sulphanilamide. The chemical properties of this substance and its behaviour in growth tests indicated that it might be chemically related to sulphanilamide itself. As a result of this suggestion *p*-aminobenzoic acid was tested and found to have high anti-sulphanilamide activity. A preliminary report of this work has already been given (Woods and Fildes, 1940). The bearing of these results on the possible mode of action of the drug is discussed.

BACTERIOLOGICAL TECHNIQUE

The organism used was *Streptococcus haemolyticus* (Richards). Stock cultures were maintained on meat infusion peptone agar and were passaged through mice every 14 days.

Medium.

The test medium was buffered Bacto-peptone (McIlwain, Fildes, Gladstone and Knight, 1939), tubed in 7 ml. lots and autoclaved, with the addition of the following materials to each tube at the time of testing:

	ml.
Phosphate buffer $M/18$, *p*H 7.6	1.175
List A (in $N/5$ HCl)	0.5
N NaOH	0.1
$M/50$ glutamine	0.125
Cocci (in peptone water)	0.1
Sulphanilamide $M/300$, or water	1.0
Test solution, or water	1.0

List A contained glucose, KH_2PO_4, aneurin,† nicotinamide, β-alanine, pimelic acid and riboflavin in quantities given by McIlwain *et al.* (1939) and also 0.24 g. cystine/100 ml. The mixture was made up, stored and sterilized (by filtration) in $N/5$ HCl. The total volume of the test medium was thus 11 ml. and the final concentration of sulphanilamide $M/3300$.

At the time this work was begun the above medium was one of the simplest known on which the streptococcus will grow satisfactorily (McIlwain *et al.*, 1939). It was felt desirable to use as simple a medium as possible, as it is known that the inhibitory action of sulphanilamide is less complete in complex media containing, for example, serum or meat broth.

Inoculum.

For satisfactory testing it was necessary that sulphanilamide inhibition should be as complete as possible, and a standard of no growth for at least five days with $M/3300$ sulphanilamide was aimed at. To attain this it was found that both the size and age of the inoculum were of importance (Table I). Young cultures and large inocula both reduced the inhibitory power of the drug. As a routine therefore the inoculum chosen was *ca.* 1000 cells from a 2- or 3-day stock culture. It was also found that if the meat

* [Fildes had shown that mercury (Hg) inhibited bacterial growth by combining with sulfhydryl (SH) groups on bacterial proteins, and that this combination could be reversed by any other free sulfhydryl compound added to the medium.]

† [Vitamin B_1 or thiamine.]

TABLE I

Effect of Size and Age of Inoculum on Sulphanilamide Inhibition

Cultures on meat infusion peptone agar used as source of inocula.

All tubes contained 3.03×10^{-4} *M* sulphanilamide.

Age of agar cultures (hrs.)	*Number of cells inoculated*	*Growth after days* 1.	2.	3.	5.
24	10^6	+	+++	+++	+++
	10^5	0	+++	+++	+++
	10^4	0	0	+++	+++
	10^3	0	0	0	0
48	10^6	+	+++	+++	+++
	10^5	0	++	+++	+++
	10^4	0	0	0	0
	10^3	0	0	0	0

Here and elsewhere + signs are roughly proportional to the mass of growth. Above experiment repeated in absence of sulphanilamide gave +++ in all cases in one day.

infusion broth used for the stock culture medium was freshly prepared inhibition was again unsatisfactory; freshly-made medium was therefore avoided. With these precautions inhibition was complete, with few exceptions, for the duration of the test (5 to 6 days).

Tests.

Activity in antagonizing the inhibitory effect of sulphanilamide was followed by determining the minimum amount of material necessary to promote full growth in the presence of a standard concentration (*M*/3300) of the drug. In each test the following series of tubes were set up: (1) medium alone (full growth normally attained in one day or less), (2) medium + sulphanilamide (in duplicate), (3) medium + top concentrations of test materials used (to test for any inhibitory or growth-stimulating effect), (4) medium + sulphanilamide + test material in falling concentration from a 1 in 5 serial dilution. Tubes were incubated in air + 5 per cent CO_2 at 37° C. and readings taken daily for 5 days. It was observed throughout that if growth occurred at all, it always became maximal within 24 hours of the appearance of the first trace.

ANTI-SULPHANILAMIDE ACTIVITY OF CELL EXTRACTS, ET CETERA

The failure to obtain any significant activity with known essential substances led to a consideration of the possibility that the substance postulated might be one whose importance in cell metabolism had not yet been recognized and a survey was made of cell extracts, et cetera. At this point Stamp (1939) demonstrated that extracts of streptococcal cells contained a substance which powerfully antagonized the inhibitory action of sulphanilamide. . . . As Stamp points out, this substance, which appears normally to be present in the cell, may be a metabolite with which the action of sulphanilamide is concerned. By applying Stamp's method of extraction (incubation in *N*/25 NH_3 at 37°C.) to baker's yeast we were able to obtain more active preparations than by other methods. Such extracts were also more active than the other cell extracts tried (Table II), and seemed

TABLE II

Anti-Sulphanilamide Activity of Cell Extracts, etc.

Activity expressed as minimum volume required to give growth in 3.03×10^{-4} *M* sulphanilamide.

Material	*Method of extraction*	*Concentration (g./ml. extract)*	*Activity*
Baker's yeast	Stamp	3.5	0.0016
Yeast wash	..	≡3.5	0.0016
Brewer's yeast	Stamp	3.5	0.04
Brewer's yeast	H_2O, 100°C.	3.5	0.04
Brewer's yeast	Na_2SO_4*	3.5	1.0
Ox muscle	H_2O, 100°C.	1.0	0.2
Ox muscle	Tryptic digest	1.7	0.2
Ox liver	H_2O, 100°C.	1.0	0.2
Turnip	"	1.7	0.2
Urine	..	..	0.04
Serum (horse)	..	..	0.04
Laked blood, 50 per cent. (ox)	..	..	0.2
Albumin (egg)	HCl hydrolysed	0.1	<1.0

* Method of Deutch, Eggleton, and Eggleton (1938).

to be a suitable starting-point for larger scale work. Less active extracts were obtained if the yeast was washed prior to extraction, and it seemed possible that a greater amount of factor might be obtained from the medium on which the yeast had been grown. A sample of yeast wash (I am indebted to British Fermentation Products, Ltd., for the gift of this material.) was tested (after precipitation of inactive material with alcohol), but was found to have no higher activity (on basis of weight of yeast obtained from it) than the yeast extracts. It also contained a greater bulk of inactive material. . . . [Chemical fractionation studies were reported and are omitted here.]

PROPERTIES OF THE FACTOR IN GROWTH TESTS

Concurrently with the chemical fractionation some experiments on the possible mode of action of the anti-sulphanilamide factor were carried out with partly purified material.

Table VI shows the results of a typical titration for determination of activity. The following points may be noted:

TABLE VI

Titration of Anti-Sulphanilamide Activity of Yeast Extract A.

	ml. extract A	*Growth after hours and days* 18.	42.	3.	5.
Medium alone		+	+++	+++	+++
Medium + sulphan. (duplicate)		0	0	0	0
" no sulphan. + Extract A	1.0	+++	+++	+++	+++
" no sulphan. + "	0.2	++	+++	+++	+++
" + sulphan. + "	1.0	+++	+++	+++	+++
" + " + "	0.2	+++	+++	+++	+++
" + " + "	0.04	tr.	+++	+++	+++
" + " + "	0.008	0	tr.	+++	+++
" + " + "	0.0016	0	0	+++	+++
" + " + "	0.0003	0	0	0	0

(1) The extract has some growth-promoting activity at the higher concentrations. This type of activity became reduced with increasing fractionation, although anti-sulphanilamide activity remained unchanged.

(2) The bacteriostatic nature of sulphanilamide inhibition is confirmed. With the smaller amounts of factor the organism grew up after 2–3 days' incubation with sulphanilamide during which time no visible growth occurred.

(3) In the titration each tube contains ⅕ of the amount of factor in the preceding tube and this is the limit of accuracy of the method. If finer differences are used the results are less consistent.

It was soon evident from the sharpness of the chemical fractionation obtained with many of the precipitants and solvents used that the factor was probably a single substance and not a mixture of components each accelerating growth. This is important, as it is known that vigorous growth, particularly during the early stages, diminishes the inhibitory effect of sulphanilamide in a non-specific manner.

It was first considered whether the factor might be working by reacting chemically with the sulphanilamide, and thus removing it from the medium. In this case it would be expected that the molar concentration of the factor would be of the same order as that of the sulphanilamide used. Comparison of the dry weight and activity of even crude factor preparations showed that this cannot be the case. 0.003 mg. of an acid ether extract of Extract B was sufficient to reverse the inhibitory action of 1 ml. 0.0033 *M* sulphanilamide. If the factor is equimolar with sulphanilamide, then 0.003 mg./ml. must be equivalent to 0.0033 *M* factor, and from this the mol. wt. of the factor works out at *ca.* 1. As even this result was obtained by assuming a crude preparation to be pure factor, it is certain that any molecular reaction can be excluded. It remained possible that the factor might act catalytically in promoting the removal of sulphanilamide.

In other experiments it was found that if the concentration of sulphanilamide is increased, it is necessary to raise the concentration of factor in the same proportion in order to reverse the inhibition. This constant quantitative relationship has been demonstrated with a number of different factor preparations at various stages of purification and also using *Bact. coli* * in place of the streptococcus as the test organism. A selection of such experiments is given in Table VII. The fact that the concentration ratios are *exactly* constant is due to the fact that sulphanilamide and factor were each used in ⅕ falling concentration. As the titration cannot be used accurately with finer differences in amount of factor, it was considered better to carry out a number of separate titrations than to attempt to work with smaller differences. Sulphanilamide is insufficiently soluble for work with higher concentrations than those recorded and smaller amounts tend to give incomplete inhibition.

As there was no possibility of a direct molecular reaction between sulphanilamide and the factor, this constant quantitative relationship between inhibitor and active substance was reminiscent of the competitive inhibition of enzyme reactions by substances chemically related to the substrate or product. From this point of view the factor would be considered to be the substrate (or product) of the enzyme reaction in question and sulphanilamide the substance of related chemical structure inhibiting the reaction. It will be recalled that consideration of

* [*Escherichia coli.*]

TABLE VII

Relation between Concentration of Sulphanilamide and Concentration of Factor Required to Reverse the Inhibition.

Quantities of Extracts A and B are expressed as ml. "standard strength."

Organism:	Streptococcus		*Bact. coli**		Streptococcus	
Factor used:	Extract A		Extract B		*p*-aminobenzoic acid	
Conc. sulphan. ($M \times 10^{-3}$) (a)	ml. required ($\times 10^{-1}$) (b)	Ratio: † (b)/(a)	ml. required ($\times 10^{-1}$) (c)	Ratio: † (c)/(a)	Conc. required ($M \times 10^{-7}$) (d)	Ratio: (d)/(a)
0.303	0.016	5.3	0.08	26.5	0.58	1.92
1.515	0.08	5.3	0.4	26.5	2.91	1.92
7.575	0.4	5.3	2.0	26.5	14.54	1.92

* The strain of *Bact. coli* and the medium employed (lactate + inorganic salts) were those used by Fildes (1940).

† These arbitrary ratios are an expression of the ratio: conc. factor required/conc. sulphanilamide used.

the chemical properties of the factor also indicated the possibility that the factor might be chemically related to sulphanilamide. Examples of known cases of competitive inhibition of enzyme reactions by substances related to the substrate or products are: (a) succinic dehydrogenase by malonic acid, (b) lipase (hydrolysis of ethyl butyrate) by acetophenone and other non-polar compounds containing a carbonyl group, and (c) invertase by α and β-galactose and β-l-arabinose.

As it was possible to interpret these two distinct lines of evidence (chemical properties and behaviour in growth tests) in the same way, it seemed worth while at this stage to test for anti-sulphanilamide activity some compounds which are structurally related to sulphanilamide, and whose properties are reasonably in accord with those of the yeast factor.

ANTI-SULPHANILAMIDE ACTIVITY OF *p*-AMINOBENZOIC ACID

In view of the probability that the acid group of the factor was carboxylic the first substance tested was *p*-aminobenzoic acid. In this acid the *p*-NH_2 of sulphanilamide is unchanged but $-SO_2NH_2$ is replaced by $-COOH$. *p*-aminobenzoic acid proved to have very high activity, and a final concentration of $1.2–5.8 \times 10^{-8}$ M was sufficient to reverse the inhibition caused by 3.03×10^{-4} M sulphanilamide (i.e., 0.02–0.1 μg. in 11 ml. medium compared with 570 μg. sulphanilamide). The results of a number of quantitative determinations are given in Table VIII. The agreement between separate determinations (maximum variation by a factor of 5) is satisfactory in view of the high dilutions used and possible differences in the state of the inoculum. Duplicates in the same determination did not show this variation.

Purity of p-aminobenzoic acid.

Table VIII also shows that the activity of *p*-aminobenzoic acid is unaffected by recrystallizing five times. . . . From the method of preparation the most likely impurities are the *o*-

TABLE VIII

Anti-Sulphanilamide Activity of p-*Aminobenzoic Acid*

Sulphanilamide conc. = 3.03×10^{-4} *M*

Sample of p-*aminobenzoic acid*		*Commercial*	*5X recrystallized*
Number of determinations:			
Total		7	5
	Molarity		
Titrating to	2.91×10^{-7}	0	1
"	5.82×10^{-8}	4	2
"	1.16×10^{-8}	3	2

and *m*-isomers and *p*-nitrobenzoic acid; these were tested (Table IX) and found to be inactive, or very feeble active compared with *p*-aminobenzoic acid. (It is equally likely that the *m*-isomer is contaminated with traces of the active *p*-compound.) There is little doubt, therefore, that the active substance is *p*-aminobenzoic acid and not an impurity in the specimen used.

Quantitative relationship with sulphanilamide.

As with the yeast factor, there was a constant relationship between the concentration of sulphanilamide used and the concentration of *p*-aminobenzoic acid required to reverse the inhibition (Table VII). In order to avoid the variation in titration found in different experiments (see above), it is necessary that the titrations be carried

TABLE IX

Anti-Sulphanilamide Activity of Substances Related to p-*Aminobenzoic Acid*

Conc. of sulphanilamide = 3.03×10^{-4} *M*

Substance	*Active at* M *conc.*
p-aminobenzoic acid	$1.2 - 5.8 \times 10^{-8}$
o-aminobenzoic acid	—
m-aminobenzoic acid	0.9×10^{-3}
p-nitrobenzoic acid	1.8×10^{-4}
p-acetaminobenzoic acid	1.8×10^{-4}
ethyl *p*-aminobenzoate (Benzocaine)	3.6×10^{-5}
Novocaine	5.8×10^{-8}
p-hydroxybenzoic acid	—
p-toluic acid	—
benzoic acid	—
benzamide	—
p-aminobenzamide	1.4×10^{-6}
2-(*p*-aminobenzylamino)pyridine	0.9×10^{-3}
p-hydroxylaminobenzoic acid *	5.8×10^{-8}
p-aminophenol	— †
p-aminophenylarsonic acid (arsanilic acid)	—
sulphanilic acid	— ‡

— Indicates substance inactive at 10^{-3} *M*.

* It is difficult to be sure that this compound is free from traces of *p*-aminobenzoic acid.

† Inhibits growth down to 3.6×10^{-5} M.

‡ Inhibits growth at 10^{-3} *M*.

out in the same experiment and with the same source of inoculum and serial dilution factor. In a number of separate experiments with the same concentration of sulphanilamide (Table VIII) the ratio molar conc. sulphanilamide/molar conc. *p*-aminobenzoic acid required ranged from 5000–25,000, reflecting the usual factor of 5 variation.

Experiments with Bact. coli.

The action of *p*-aminobenzoic acid was also demonstrated with *Bact. coli* as test organism and an entirely synthetic basal medium of lactate and inorganic salts (Fildes, 1940). On this medium the usual amount of sulfanilamide (3.03×10^{-4} *M*) inhibited completely for five days, and the inhibition was overcome by the same amount of *p*-aminobenzoic acid (5.8×10^{-8} *M*) as was effective with the streptococcus. This was also the case when the activity of a particular yeast extract was measured with the two organisms. . . .

Growth-promoting activity.

p-aminobenzoic acid had no significant effect on the rate or mass of growth under the test conditions. There was slight acceleration at high concentrations (0.2×10^{-3} *M*), but above this there was slight inhibition. No growth stimulation was found with *Bact. coli* on the ammonium lactate medium which is certainly free from preformed *p*-aminobenzoic acid. The crude yeast extracts had considerable growth-accelerating action, but this was greatly diminished as the factor was purified. The anti-sulphanilamide factor ("P") obtained from *Brucella abortus* by Green (1940) was reported to have high growth-promoting activity for a number of organisms, but the material had not at that time been much purified.

ANTI-SULPHANILAMIDE ACTIVITY OF SUBSTANCES RELATED TO *p*-AMINOBENZOIC ACID

. . . In general both carboxylic and amino groups in *p*-positions to one another appear to be necessary for anti-sulphanilamide activity (Table IX). Acetylation of the amino group of *p*-aminobenzoic acid leads to a tenthousandfold decrease in activity, and ethyl esterification of the carboxylic group to a decrease of a thousandfold. These facts were useful in comparing the properties of *p*-aminobenzoic acid with those of the yeast factor. . . .

A POSSIBLE MECHANISM OF SULFANILAMIDE INHIBITION

The present investigation was based on a general working hypothesis that anti-bacterial substances act by interfering with some substance essential to the bacterial cell ("essential metabolite," Fildes, 1940). The experiments are also in accord with the suggestion, based on indirect evidence, that the interference is connected with inactivation of bacterial enzymes, and provide strong evidence that the inactivation is due to competition for an enzyme between the essential metabolite and the inhibitor. A clearer hypothesis of the possible mode of action of sulphanilamide may now be built up and may prove useful as a basis for further work. Throughout the following argument *p*-aminobenzoic acid should be taken to include the probably related, if not identical, naturally occurring materials, such as yeast factor or Stamp factor.

In the first place it is suggested that *p*-aminobenzoic acid is essential for the growth of the organism. It is, however, normally synthesized in sufficient quantity by the present strain of streptococcus (and by *coli*), since it is not necessary to add it to a medium containing only known substances or preparations known to be free from anti-sulphanilamide activity (McIlwain, unpublished). It can also be extracted from the streptococcal cell (Stamp, 1939). On the basis of the experimental work it is next suggested that the enzyme reaction involved in the further utilization of *p*-aminobenzoic acid is subject to competitive inhibition by sulphanilamide, and that this inhibition is due to a structural relationship between sulphanilamide and *p*-aminobenzoic acid (which is the substrate of the enzyme reaction in question). Examples of similar competitive inhibition have already been quoted. It was found that the concentration of *p*-aminobenzoic acid required to overcome this inhibition is 1/5000–1/25,000 of the concentration of sulphanilamide used. The further course of events in a culture may now be considered as follows:

(1) p-*aminobenzoic acid is present preformed in the medium.* Growth occurs. As the anti-sulphanilamide factor appears to be widely distributed in small amount (Table II), this may account in part for the difficulty in getting complete inhibition on more complex media.

(2) p-*aminobenzoic acid is absent from the medium or present in insufficient concentration.* This would be the position under the test conditions used here. There are now two possibilities:

(2a) *The organism is unable to synthesize enough* p-*aminobenzoic acid.* Growth is therefore inhibited. This would normally be the case with the streptococcus under the conditions used in the present experiments.

(2b) *The organism is able to make sufficient* p-*aminobenzoic acid.* In this case the competitive inhibition is overcome and growth occurs. This is presumed to be the case with organisms that are insensitive to sulphanilamide.

The conditions determining whether (2a) or (2b) shall take place are delicately balanced, and this may explain why different organisms, and even the same organism under differing growth conditions, exhibit many degrees of sensitivity to sulphanilamide. It is here suggested that such differences in sensitivity are correlated with quantitative differences in ability to synthesize *p*-aminobenzoic acid. On meat infusion broth (in which its growth rate approaches optimum) *Bact. coli* is almost indifferent to sulphanilamide, whilst on the ammonium lactate medium (where the growth rate is suboptimal) inhibition is well marked. Similarly, complete inhibition of streptococcal growth is not obtained (a) with rich media on which growth is very rapid, or (b) with poorer media if the inoculum is large or consists of young actively dividing cells (Table I). To account for such variability it is suggested that the original inoculum contains sufficient *p*-aminobenzoic acid to reverse the inhibition and permit some (non-visible) growth to take place; such growth is known to occur in the early stages of inhibition. When this supply of *p*-aminobenzoic acid becomes exhausted by the further enzyme reaction under discussion, subsequent growth will depend on the rate at which more can be synthesized, and this in turn on the number of organisms present and thus on the initial growth rate. Presence of precursors of *p*-aminobenzoic acid in the medium may also influence the rate of synthesis.

Another possible interpretation of the experimental results is that sulphanilamide inhibits the enzyme reaction involved in the *synthesis* (and not, as above, the further utilization) of *p*-aminobenzoic acid, and that it does so this time by virtue of its chemical similarity to the product of the reaction. The balance of evidence is against this view, as in this case it would be expected, on any simple interpretation, that the addition of just sufficient *p*-aminobenzoic acid for the needs of the organism should cause growth; the amount of *p*-aminobenzoic acid needed should thus be independent of sulphanilamide concentration.*

SUMMARY

(1) Yeast extracts contain a substance which reverses the inhibitory action of sulphanilamide on the growth of haemolytic streptococci.

(2) Examination of the chemical properties of this substance and its behaviour in growth tests suggested that it might be chemically related to sulphanilamide.

(3) *p*-aminobenzoic acid has high activity in antagonizing sulphanilamide inhibition.

(4) There is strong circumstantial evidence that the yeast factor may be *p*-aminobenzoic acid.

(5) On the basis of these results a suggestion is put forward regarding the possible mode of action of sulphanilamide.

The work was carried out at the suggestion of Dr. P. Fildes, to whom I am deeply indebted for advice and criticism throughout. I am grateful to all members of the Department and to Dr. G. M. Richardson for suggestions and to Mr. D. E. Hughes for valuable assistance with some of the chemical work.

* [In other words, the inhibition in this case would be noncompetitive.]

Comment

Read Domagk's paper (page 195) for the discovery of sulfanilamide.

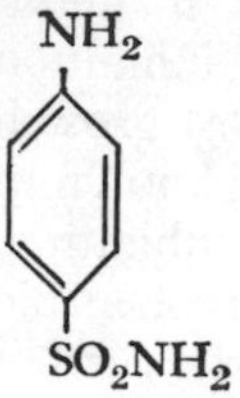

NH_2 / SO_2NH_2 — *Sulfanilamide*

NH_2 / $COOH$ — *Para-amino-benzoic acid*

H_2N, HN, $C{=}O$, $C{-}CH_2{-}NH$, CH, $CO{-}NH{-}CH{-}COOH$, CH_2, $CH_2{-}COOH$ — *p*-amino-benzoic acid

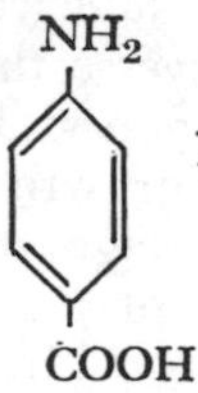

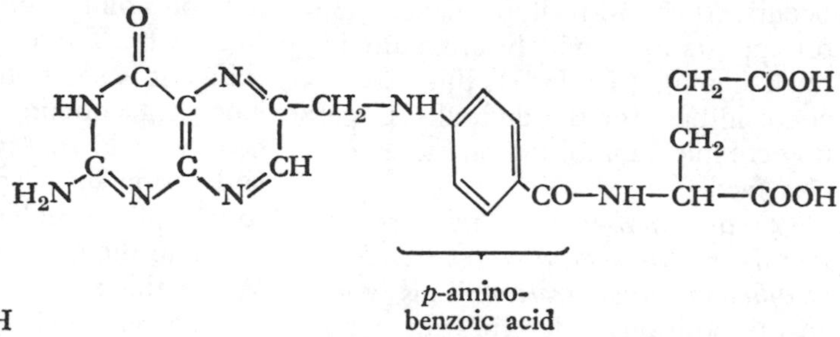

Folic acid (Pteroyl L-glutamic acid)

It is now known that para-aminobenzoic acid (PABA) acts to reverse sulfanilamide action exactly as Woods hypothesized. PABA is a structural part of the essential vitamin, folic acid, and sulfanilamide prevents the incorporation of PABA into folic acid. Thus the essentials of Woods' hypothesis have been proven.

An important concept that is considered in this paper is that of competitive inhibition. It had been known some years earlier that malonic acid inhibited the action of the enzyme succinic dehydrogenase on succinic acid, and that these two acids are structurally related. A rela-

```
COOH          COOH
 |             |
CH2           CH2
 |             |
COOH          CH2
               |
              COOH
```

Malonic acid *Succinic acid*

tionship exisits between the concentration of succinic acid and that of malonic acid, so that as greater amounts of succinic are added, greater amounts of malonic must be added to achieve inhibition. The ratio of malonic to succinic for any degree of inhibition remains constant. It is thought that malonic acid, because of its structural similarity to succinic acid, combines with the active site of the enzyme and prevents the succinic acid from being acted upon.

Woods is proposing essentially the same hypothesis to explain the action of sulfanilamide. Here the enzyme in question would be one involved in the synthesis of folic acid (unknown to Woods). It is easy to see that if this explanation is correct, chemotherapeutic agents could be designed on the basis of other essential cell metabolites, by synthesizing molecules that are structurally related to the metabolites. This has been found to be true, and at present thousands of such compounds (antimetabolites) have been synthesized. Many have proved quite useful in studying fundamental processes in the cell, while a few have been useful in therapy both in microbial and in other diseases.

Woods' paper has been influential in a wide area of microbiology and medicine. It is interesting to read Ehrlich's work (page 176) and see how far chemotherapy has advanced since he first devised the concept.

PART VI
General Microbiology

Studies on bacteria

1875 • Ferdinand Cohn

Cohn, Ferdinand. 1875. Untersuchungen über Bacterien. *Beiträge zur Biologie der Pflanzen*, Vol. 1, pages 127–222.

FOR A NUMBER OF YEARS I HAVE endeavored, in association with my friend Herr Oberstabsarzt Dr. Schroeter, to study bacteria with the aid of the more perfect optical systems now available to us. Since Schroeter left this plant physiological institute in the summer of 1870, I have carried on these studies alone. I have first attempted to discover the biological relationships of the bacteria as well as to arrive at a decision concerning the differentiation of species. In addition, I have considered general questions, above all the fermentative activity of bacteria. Several preliminary communications of my results have already been presented to the Verhandlungen der Schlesischen Gesellschaft. Although these studies are still uncompleted, I believe it would be useful to present here a detailed discussion of them.

1. SYSTEMATICS (CLASSIFICATION)

What organisms belong to the group known as bacteria? What genera, what species, can be distinguished in this group? My first studies have been directed to this question.

Whoever is acquainted with the literature in recent years knows that there is a great confusion in the nomenclature of bacteria. Almost every observer has given new names to the forms that he sees, without consideration for his predecessors. The rule of priority, which is universally used as the basis of nomenclature, has been completely disregarded here.

It is true that there are tremendous difficulties in differentiating and naming these organisms. Only Ehrenberg and Dujardin have attempted to classify the whole group of bacteria and related organisms and divide them into genus and species, and their works must serve as a point of departure. But aside from the fact that the principles which these workers used for separating out the groups leave much to be desired, they were further handicapped by the magnifications with which they could study the organisms. It is not surprising, therefore, that Ehrenberg has indicated structural relationships which we can no longer observe.

Even with the strong immersion systems available on microscopes today, we must admit that most bacteria are still at the limits of resolution, so that we cannot observe clearly their forms, the organization of their interiors, and the details of their reproduction. Even the very existence of some of the smallest forms would be in doubt if it were not for the fact that they occur in very large numbers.

An important difficulty lies in the small number of characteristics which are available for the classification of bacteria. In other organisms, the separation of genera is based on differences in reproduction, while the bacteria have not revealed as yet any true reproduction (egg or spore formation). So far as we can differentiate, their bodies show no diversity in arrangement and no characteristics of membrane or interior. Only the size, and within certain limits, the form of the members, as well as their combination into groups, offer certain characteristics which might be used, although we cannot always know how much these differences are due to species differences and how much they are due to the effect of external conditions, or indeed whether they are different stages in the development of the same organism. It is the most easy to differentiate bacteria by their size. But since they are usually composed of two or more members in a chain, we have the question of whether we should measure the size of the whole chain or of the single members. The former shows very marked variations in number of members, while the latter is difficult to measure due to the small size. It is impossible to isolate single bacteria and observe them for a long time under different conditions. But in mass culture there is no certainty that the inoculum was composed of a single type, or whether several types were inoculated at the same time. Therefore we possess no methods as yet for distinguishing age and developmental states, varieties, and species.

All of these difficulties arise when we attempt to separate bacteria into natural genera. The genera of bacteria do not have the same significance as do the genera of higher plants and animals, since bacteria only reproduce by vegetative reproduction, not sexually. It is therefore necessary to use in many cases a technique which has been used for a long time in mycology when it has not been possible to arrive at culture methods which will reveal the entire life history. This technique has also found applications today in the field of paleontology. It consists of calling every form which shows wide differences a genus. Then every small deviation from this is called a different species. In this way the possibility is not eliminated that various of such species may have arisen from one and the same parent form, and even that different genera may be only stages in the life history of one and the same individual. In this way we differentiate species of *Uredo*, *Puccinia* and *Aecidium*, without knowing whether all three genera might only be single stages in one life history.* We speak of *Oidium* and *Aspergillus*, of *Achorion* and *Microsporon*, of *Stigmaria*, *Sigillaria*, and *Sigillariostachys*, without any certainty over the separateness of these "form genera." In the bacteria as well, we cannot avoid differentiating into "form genera and form species," except for a certain number of natural types. These form species must be accepted for every form showing deviation from the type, when this deviation under certain conditions is the exclusive or predominant form. The task of further research will then be to discover which of these form genera and species are perhaps merely stages in one life history.

Although Leeuwenhoek had observed bacteria in the seventeenth century, and O. F. Müller had recognized and described the most important forms in the eighteenth century, the first separation of forms on a scientific basis began with Ehrenberg. . . . In the basic work on the animal infusoria in 1838, he separated the family Vibrionia into four genera in the following manner:

Straight, rigid filaments: *Bacterium*

Straight filaments, twisted, nonrigid: *Vibrio*

Spiral filaments, nonrigid: *Spirochaeta*

Spiral filaments, rigid: *Spirillum*

He described 3 species of *Bacterium*, 9 of *Vibrio*, 1 of *Spirochaeta*, and 3 of *Spirillum*. . . .

Dujardin accepted Ehrenberg's family Vibrionia as the lowest in the series of infusoria in his work on the natural history of the zoophytes in 1841. . . .

* [It was later shown that these three genera were merely different stages in the life history of one species.]

However clear the characteristics for the differentiation of the genera *Bacterium*, *Vibrio*, and *Spirillum* might have seemed through Ehrenberg's description, in practice their use is quite difficult. . . .

All of those who have studied bacteria in the last 30 years have either accepted the genera of Ehrenberg or Dujardin without question, or they have designated the forms they have observed with indeterminate and occasionally completely obvious names (*Microphytes*, *Microzoaires*, etc.). This has been especially true of Pasteur, who sometimes speaks of *végétaux cryptogames microscopiques*, sometimes of *animalcules*, of *Champignons* or *Infusoires*, or of "*Torulacées*, *Bacteries*, *Vibrioniens*, *Monades*" without any sharp distinctions. . . .

If we turn then from the genera to the species, we find that even O. F. Müller, in spite of the low magnifications available to him, has named and illustrated the most surprising forms. But we should mention Ehrenberg, who continued the work of Müller, and who brought light and order into this confused area with his remarkable insight, and who not only used for his species precise and reliable characteristics, but also gave us a series of illustrations that have not been surpassed, which enable us to recognize these forms when we see them again. . . .

One can therefore ask the question if, in the bacteria, species generally occur in the usual sense that we find them in higher organisms. Even those who do not agree with the doctrine of some mycologists that everything comes from everything and develops into everything will despair when they look at a mass of bacteria, to perceive a separation into natural species of these countless little bodies.

It seems, therefore, that all of these forms are only stages in the life history of one and the same organism, and intermediate stages can be found between the different forms varying in shape and size. Actually most of the recent workers on bacteria have come to accept this opinion as more or less proven. (Perty, Hoffman, Karsten.)

However, I have become convinced that the bacteria can be separated into just as good and distinct species as other lower plants and animals, and that it is only their extraordinary smallness and the variability of the species which makes it impossible for us with our present day methods to differentiate the various species which are living together in mixed array. I base this opinion upon the fact that in the larger bacterial species, always, even under different conditions, the same forms can be found in countless numbers and without intermediate forms. This is especially true for the spirilla which remain different from true rod-shaped bacteria. Also the individual species of spirilla are constant, in the same way as a "good" species of algae or infusoria. If in the smaller bacteria we cannot always delineate natural species but must be limited to the construction of form species, I consider this to be due to the inadequacy of our experimental methods. In general it will be difficult to determine the species of individual bacteria with certainty. However, when one and the same form is present in huge numbers without other organisms being present, the constancy of this type will usually be quite easy to determine.

A special difficulty arises from the fact that there are forms which cannot be distinguished at all on morphological grounds but nevertheless differ in important ways and show constant physiological differences, whether this may be in the environments in which they live, or in the products which they produce, or in the characteristics of their motility. . . .

Pasteur, who has already remarked that one cannot with certainty distinguish the nature of the organized ferments through microscopic structure, but only through physiological function, has cited the extraordinary similarity between the lactic acid and the acetic acid fermentations, as well as between the ammonia fermentation of urine and the slimy alcoholic fermentation (vin filant). The bacteria which produce red, yellow, orange, blue and other pigments can hardly be distinguished from each other under the microscope but when inoculated each always produces the same pigment. The bacteria present in various contagions agree in their forms at times with those of the urine or butyric acid fermentation, at other times with those that produce pigments. Should one consider each form which always occurs in a special environment, or which brings about a characteristic fermentation, to be a particular species, even when it cannot be distinguished from others microscopically? If we say yes, we will be erecting purely physiological species, which are characterized on exclusively physiological grounds, and not like the "good" morphological species.

I believe that it is not yet time to attempt an absolute answer to this question. . . . It is perhaps to be expected that amongst many apparently similar organisms which differ physiologically, we will find by more precise microscopical examination that morphological differences will be evident which can be used for primary differentiation. On the other hand, I consider it possible that we will find bacteria which cannot be differentiated by morphological characteristics but show chemical physiological differences, and these will be varieties or races which originally arose from the same germ, but through constant, natural, or artificial culture under the same conditions and on the same medium always produce the same product. Since all bacteria reproduce only by asexual methods, such as budding or fission, such a fixation of a race characteristic is easier to accept. In the various types of yeast the production of races through artificial culture has been shown by Rees. In the same way that summer rye cannot be used as winter rye, although both races are of the same origin and could be returned to the same race through continued culture over a long period of time, top yeast cannot be used for the preparation of Bavarian beer (bottom yeast), and almost every wine or beer producer has his own yeast, so that it seems probable that many alcohol-producing yeasts are only a large number of cultural races of the same species. I assume that the bacteria which cause different chemical and pathological processes consist of a small number of individual species which have developed into a large number of natural and cultural races, which, since they only reproduce by asexual means, are able to maintain their physiological characteristics with more firmness.

2. ORGANIZATION AND DEVELOPMENT OF BACTERIA

The general characteristics of all of the organisms which I have included together in the bacteria appear to me as follows:

The bacteria are cells without chlorophyll, spherical, oblong, or cylindrical, containing also twisted or curved forms, which reproduce exclusively by transverse fission, and are either single or in families of cells. . . .

I divide the bacteria into four groups (Tribes) in each of which are one or more genera. In the nomenclature of the genera, I have retained throughout the older names, so that the nomenclature will not be overburdened, but I have used more clear cut characteristics and at times used other bases.

Tribe I. Sphaerobacteria (Sphere bacteria)
- Genus 1. *Micrococcus char. emend.*

Tribe II. Microbacteria (Rod bacteria)
- Genus 2. *Bacterium char. emend.*

Tribe III. Desmobacteria (Filament bacteria)
- Genus 3. *Bacillus n. g.*
- Genus 4. *Vibrio char. emend.*

Tribe IV. Spirobacteria (Corkscrew bacteria)
- Genus 5. *Spirillum Ehrenberg*
- Genus 6. *Spirochaete Ehr.* . . .

9. RELATIONSHIPS OF THE BACTERIA

Are the bacteria animals or plants? A review of the literature shows that the bacteria were earlier considered to be animals, but now most of the researchers consider them to be plants. . . .

In relationship to this question I can indicate the conclusion which I have already published in 1853:

"The bacteria (Vibrionien) seem to belong to the plant kingdom, because they are in direct and close relationship with undisputed algae."

On the other hand, the bacteria have no relationship with clear-cut animals. . . .

Most writers who include the bacteria amongst the plants consider them to be fungi. This is correct if one includes amongst the fungi all cellular plants or thallophytes which do not have chlorophyll or an equivalent pigment and do not assimilate carbon dioxide. But bacteria have no relationships with a typical fungus which develops a filamentous mycelium and reproduces either through basidiospores or ascospores.

Comment

This paper illustrates some clear thinking regarding the problem of bacterial taxonomy. Considering the limited knowledge of the times and the absence of pure culture methods, it is amazing that Cohn was able to analyze the problem as accurately as he did.

Throughout the nineteenth century a controversy raged regarding the variability of bacteria. Some workers thought bacteria were highly variable (pleomorphic) and that all of the different forms that could be seen under the microscope were different stages of one species. Using modern genetic concepts, this would mean that all bacterial cells contained exactly the same genes, and the different appearances which they sometimes revealed were due to environmental influences. Other workers felt that different forms of bacteria were actually separate species with different genetic backgrounds. Cohn belonged to this latter group and presented his case here.

The controversy could not be ended until the pure culture methods of Koch became available (see page 101). Only then could it be shown that different bacterial types bred true and could be considered separate species. Cohn's attempt here to delineate several bacterial tribes and genera was premature but set the stage for later discoveries.

The problems of bacterial taxonomy are not yet solved today. Our current bacterial classification, as presented in *Bergey's Manual of Determinative Bacteriology*, was devised on the assumption that genetic recombination between bacteria did not occur, making classification strictly artificial. We know now that genetic recombination can occur. Future taxonomic studies will have to attempt to include this concept. Thus some day Cohn's objection that: "The genera of bacteria do not have the same significance as do the genera of higher plants and animals, since bacteria only reproduce by vegetative reproduction, not sexually," will no longer be valid.

The differential staining of Schizomycetes in tissue sections and in dried preparations

1884 · Christian Gram

Gram, C. 1884. Ueber die isolirte Färbung der Schizomyceten in Schnitt-und Trockenpräparaten. *Fortschritte der Medicin*, Vol. 2, pages 185–189.

(I WISH TO THANK HERR DR. RIESS, Director of the General Hospital of the city of Berlin for the facilities and equipment to perform the following studies.)

The differential staining method of

Koch and Ehrlich for tubercle bacilli gives very excellent results either with or without counter-staining, since the bacilli stand out very clearly due to the contrast effect.

It would be very desirable if a similar method for the differential staining of other Schizomycetes were available which could be used routinely by the microscopist.

My studies—as associate of Herr Dr. Friedländer in the morgue of the city hospital in Berlin—have attempted to demonstrate cocci in tissue sections of lungs of those who have died of pneumonia as well as in experimental animals. As Friedländer has already mentioned briefly in his paper on the micrococci of pneumonia, I have discovered by experimentation a procedure for the differential staining of pneumococci. In my procedure the nucleus and other tissue elements remain unstained, while the cocci are strongly stained. This makes them much easier to locate than previously, since in ordinary preparations from pneumonia patients, where such a large amount of exudate occurs, they are impossible to see.

Further studies on the usefulness of this method for other Schizomycetes has gradually shown that this method is an almost general method for all tissue sections and dried preparations. . . .

For staining the ordinary aniline-gentian violet solution of Ehrlich is used. The appropriate sections must be carried up to absolute alcohol and taken from this directly into the dye solution. They should remain in the dye for 1–3 minutes (except tubercle bacilli, which should remain as usual 12–24 hours). Then they are placed in a solution of iodine-potassium iodide in water (iodine 1.0 part, potassium iodide 2.0 parts, water 300.0 parts) with or without a light rinse with alcohol and allowed to remain there for 1–3 minutes. During this time, a precipitate forms, and the previously dark blue-violet sections now become dark purple-red. (Footnote: This purple-red color is not soluble in water but dissolves very easily in alcohol. The chemical studies will be continued at a later time.) They are then placed in absolute alcohol until they are completely decolorized. It is well to change the alcohol once or twice during this step. Then the sections are cleared as usual in clove oil, in which the rest of the dye is dissolved. The nucleus and fundamental tissue is stained only a light yellow (from the iodine) while the Schizomycetes, if any are present in the preparation, are an intense blue (often almost black). The intensity of the staining has not been equaled by any of the current staining methods. This presents another great advantage of our method. It is possible after the decolorization in alcohol to place the sections for a moment in a weak solution of bismarck brown or vesuvine, and then dehydrate again with alcohol, in order to achieve a counterstain. The nucleus will appear brown, while the Schizomycetes will remain blue.

In this way it is possible to prepare doubly-stained preparations that are just as excellent as those of the tubercle bacillus made after the method of Koch and Ehrlich. Permanent preparations in Canada balsam-xylene or gelatine-glycerol remain unchanged after 4 months.

This method is very quick and easy. The whole procedure takes only a quarter-hour, and the preparations can remain many days in clove oil without the Schizomycete cells becoming decolorized.

It is also useful for dried preparations. It is performed exactly as for tissue sections, except that one treats the cover slip just like a section.

I have tried many times different aniline dyes (rubine-aniline, fuchsin-aniline and simple gentian violet solution), but without success. In addition, tincture of iodine or potassium iodide solution, as opposed to iodine-potassium iodide solution, are also ineffective, since the Schizomycetes then are also decolorized. When the sections are treated with water or dilute alcohol, the results are variable. . . .

I. After iodine treatment, the following forms of Schizomycetes are not decolorized by alcohol.

(a) The coccus of bronchial pneumonia (19 cases). . . .

(b) Pyogenic Schizomycetes (9 cases). . . .

(c) Cocci of a liver abscess . . . (1 case). . . .

(d) Cocci and small bacilli in circumscriptive infiltration of the lungs . . . (1 case). . . .

(e) Cocci of osteomyelitis (2 cases). . . .

(f) Cocci of suppurative arthritis following scarlet fever (1 case). . . .

(g) Cocci of suppurative nephritis following cystitis (3 cases). . . .

(h) Cocci of multiple brain abscesses following empyema (2 cases). . . .

(i) Cocci of erysipelas (1 case). . . .

(k) Tubercle bacilli (5 cases). . . .

(l) Anthrax bacilli (3 cases) (in mice). . . .

(m) Putrefactive Schizomycetes (bacilli and cocci). . . .

II. The following forms of Schizomycetes are decolorized in alcohol after iodine treatment.

(a) 1 case of bronchial pneumonia with cocci that formed capsules. . . .

(b) 1 case of bronchial pneumonia with cocci that did not form capsules. . . .

(c) Typhoid bacilli (5 cases) are decolorized either with or without iodine treatment very easily by alcohol. I have attempted to leave the sections in the dye for as long as 24 hours but without any better results.

I would like to make one closing remark. The behavior of the cell nucleus and the Schizomycetes to aniline dyes in other methods are almost identical, whereas with the present staining method a very distinct difference is visible.

Studies on Schizomycetes have been significantly improved by the use of this method. It is because of this that I publish my results, although I am well aware that they are brief and with many gaps. It is to be hoped that this method will also be useful in the hands of other workers.

Editor's note. I would like to testify that I have found the Gram method to be one of the best and for many cases the best method which I have ever used for staining Schizomycetes.

Comment

Presented here is the first report of the bacteriological staining method most widely used today. As first devised by Gram, the method was useful in staining bacteria in tissue sections. In his time this was an important discovery, because studies of the pathogenesis of different species of bacteria was just in its infancy. The first of Koch's postulates (see page 116) was that the suspected causal organism should always be found in association with the disease. However, this presupposed a method for staining the minute bacteria in lesions so that they could be adequately visualized. Because of the fact that many bacteria exhibit the peculiar staining reaction which Gram describes here, it was possible to detect them much easier with his method.

For many years the main use of the Gram stain has been to differentiate species of bacteria. In the present paper, Gram describes several organisms that were not stained by his technique. We would call these Gram-negative, and the number of Gram-negative bacteria is probably larger than the number of Gram-positive bacteria. The Gram stain is one of the first procedures learned by beginning bacteriology students and is one of the first procedures carried out in any laboratory where bacteria are being identified. Its importance to bacterial taxonomy is therefore obvious.

The mechanism of the Gram stain is still a partial mystery. As Gram himself noted, the iodine-potassium iodide solution is essential in the reaction. We know that this solution must follow, and not precede, the gentian violet. We know that the iodine and the gentian violet form a complex inside the cell (Gram also noted this complex formation) which is insoluble in water but is soluble in alcohol. Apparently Gram-positive bacteria are those which are able in some way to keep the alcohol from reaching this insoluble complex. We know that the Gram stain is not an all-or-nothing phenomenon, but that quantitative variations in Gram-positivity exist between different species, and within the same species during different parts of the growth cycle or under different environmental conditions. We know that only intact cells are Gram-positive, so that cells which are even gently broken become Gram-negative. We know that bacterial protoplasts, devoid of cell wall, are still Gram-positive, indicating that it is probably the semipermeable membrane which is somehow involved in the reaction. Finally, we know that Gram-positivity is restricted almost exclusively to the bacteria, with only a few other groups, such as the yeasts, exhibiting this reaction. We can truly say that the implications of Gram's discovery have been widespread.

A minor modification of the plating technique of Koch

1887 • R. J. Petri

Petri, R. J. 1887. Eine kleine Modification des Koch'schen Plattenverfahrens. *Centralblatt für Bacteriologie und Parasitenkunde,* Vol. 1, pages 279–280.

IN ORDER TO PERFORM THE GELATIN plate technique of Koch, it is necessary to have a special horizontal pouring apparatus. The poured plates are then placed over one another in layers on small glass shelves in a large bell jar. In many cases it would be desirable to carry out the procedure with less equipment, especially without the pouring apparatus. Since the first of the year I have been using flat double dishes of 10–11 cm. in diameter and 1–1.5 cm. high. The upper dish serves as a lid as usual and has a somewhat larger diameter. These dishes are sterilized by dry heat as usual and after cooling the nutrient gelatin containing the inoculum is poured in. The upper lid is lifted only slightly and used as a shield while the tube containing the gelatin, its edge previously flamed and cooled in the usual manner, is emptied into the bottom of the dish. Under these conditions contamination from airborne germs rarely occurs. The poured layer of gelatin soon hardens into a layer several millimeters thick which can be kept and observed for a long time because of the protecting upper lid. In studies of soil samples, sand, earth, and similar substances, it is advantageous to place the material in the dish and then pour the liquid gelatin over it. The material is well mixed with the gelatin by rotating the dish with short, intermittent movements. With the dimensions given, every spot on the gelatin surface is accessible with the low power microscope. Only when high power lenses are used is the area at the edge of the dish no longer accessible. The gelatin dries in these dishes quite slowly. They can be kept moist longer if 5–6 dishes are placed on top of one another on a disc of moist filter paper in a flat dish over which a bell jar is inverted. These dishes can be especially recommended for agar-agar plates, since agar-agar sticks poorly to simple glass plates unless special means are used. In addition, it is quite simple to count the colonies that have grown on the plates. The upper lid is replaced by a glass plate that has etched on it squares of known area. The colonies are then counted against a black background using a magnifier. The total area of the plate can be calculated from the diameter.

Comment

We have here the first description of the Petri dish, a simple yet effective device for culturing microorganisms on solid media. The original idea of Petri has not been improved upon to this day, and in bacteriology laboratories all over the world dishes are used of almost the identical features as those first described by Petri. The Petri dish is such a simple idea that if Petri had not thought of it, someone else probably would have conceived of it later. But because of its great usefulness and universality, its first description warrants inclusion in the present collection.

We also have an interesting sidelight in this article on the procedures used for viable counting. The methods did not differ much from those we use today, except that in the time of Koch and Petri, many more viable cells were put in a plate, resulting in a larger number of colonies and overcrowding. Because of statistical considerations, we do not use such large numbers of cells, but dilute our samples until there are between 30 and 300 viable colonies developing. In this way competition for nutrients is eliminated, and the apparent colony count is higher, being closer to the true number of viable cells.

The root-nodule bacteria

1888 • Martinus W. Beijerinck

Beijerinck, M. W. 1888. Die Bacterien der Papilionaceenknöllchen. *Botanische Zeitung*, Vol. 46, pages 725–804.

SINCE ROOT NODULES HAVE BEEN discussed widely in recent years, I assume that their structure and morphological properties are quite well known. However, since I am going to present a new viewpoint on these bodies, a short discussion of their relationships would seem useful.

Following Brunchorst, I shall call the characteristic bodies within the root nodules "bacteroids." However, to avoid any ambiguity, let me say that I consider these bacteroids to develop from a species of bacteria which enters the roots from the outside, and not an autonomous formation of the protoplasm of the plant, as Brunchorst believed. I shall call this bacterial species *Bacillus radicicola.* The bacteroids are derived from bacteria by a metamorphic process, have lost their ability to reproduce, and function in the plant as proteinaceous bodies. They are derived from normal *Bacillus radicicola* by a stepwise loss in their power of reproduction. . . .

Bacteria that are still capable of growth on gelatin plates can be isolated in large numbers from the very young root nodules, as well as from the actively growing regions of older root nodules. The bacteria in the nodules eventually lose their ability to multiply on gelatin plates.

At the end of the period of growth, the root nodules can behave in one of two ways. They can lose completely their protein content through dissolution, in which case the whole nodule including the bacteroids that have been formed from the bacteria becomes emptied, and the protein material is used by the plant for growth. Alternatively, the nodules can be weakened through extensive growth of bacteria, in which case large numbers of bacteria remain alive within the cells and make use of the nodule as a habitat for growth and nutrition. As soon as the dissolution begins, it is ordinarily impossible to culture bacteria from these nodules, . . . while in the second case it is always very easy to culture bacteria. The two processes are not always sharply differentiated. . . .

In sterilized soil, the nodules do not develop. For this experiment, I used a closed sheet-iron container, in which the soil could be heated extensively in a double-boiler on a stove. Small leguminous plants, such as *Vicia hirsuta* and *Lathyrus aphaca*, could germinate and form roots quite easily in this container, and when the atmosphere was kept moist with boiled canal water, these plants could mature and form flowers and seeds without the slightest trace of nodules. This result has been previously reported also by Frank in 1879. . . . Even a short heating is sufficient to prevent nodule formation, showing that spore-forming bacteria in the soil are not responsible for the process.

From these results one can conclude that the nodules develop because of an infection by an organism that is killed by boiling water temperature. *Bacillus radicicola*, which is always present in soil, and which I have always been able to culture from nodules, does not form spores and is killed by 100°C., and is therefore the infective agent. . . .

Before discussing the bacteriological properties of the nodules in detail, I would like to describe briefly the culture conditions which have been used in this study.

Since *Bacillus radicicola* is an aerobe, I could use the plate technique of Koch in these studies. For my culture plate, I used the bottom of a glass dish which had a ground glass lid.* The gelatin was poured directly in the container. The gelatin was either inoculated before pouring by mixing into the gelatin pulverized nodules that had been carefully surface-sterilized before grinding, or the gelatin was inoculated after solidifying by spreading on the surface water that contained the test material, or finally, it was inoculated by streaking test material on the surface of the gelatin. Since *Bacillus radicicola* grows very poorly when imbedded in the gelatin, culture on the surface is preferable. . . . If one is not acquainted with *Bacillus radicicola*, it is necessary to begin a culture by washing a nodule in alcohol, burning off the alcohol, disrupting the nodule in water, and pouring the aqueous suspension on plates. Amongst the large number of *Bacillus radicicola* colonies, the occasional contaminants are readily identified and can be avoided when a pure culture is sought.

Meat extract-peptone-gelatin is too rich a medium for the initial culturing of nodule bacteria, since if growth occurs at all it is too slow. . . . More rapid growth of active forms takes place only on a poorer nutrient medium, such as an extract of the leaves of leguminous plants containing 7% gelatin. To this extract should be added 0.25% asparagine and 0.5% sucrose. . . .

The nutrient medium should not be too acid, since even 2 or 3 ml. of 1*N* acid per 100 ml.† prevents growth completely. Even Cohn's nutrient solution is too acid for *B. radicicola*. Alkaline and neutral reactions are also harmful, and I have found that for *Bac. rad.* from *Trifolium repens*, that the optimum growth was obtained when 0.6 ml. of 1*N* malic acid per 100 ml. was used. Growth is best at room temperature, while no growth occurs above 47°C., but growth is still fairly good between 0° and 10°C.

To prevent moisture from accumulating on the surface of the glass lid and making observations difficult, I have found it useful to incubate the dishes upside down. In this way, cultures can be incubated for weeks with no accumulation of water droplets and no drying out. . . .

From culturing nodule bacteria from a wide range of different leguminous plants, I have convinced myself that only one species of bacteria is found, which I have characterized by the name *Bacillus radicicola*. . . .

However, I must state that the isolates from different species of plants are always quite similar, although not necessarily always identical. This is

* [From Beijerinck's illustration, his culture dish seems to be quite similar to a Petri dish.]

† [The concept of *p*H to express hydrogen ion concentration was not devised until many years later.]

especially noticeable in the primary culture on gelatin, deriving directly from the nodule. It is probable that this variability is due to the influence of the nodule material present, which has exerted an influence on the bacteria. However, the inheritability of these observed differences makes it necessary to indicate these different cultures by different variety names. . . .

It is hardly necessary for me to add that the surface of the nodules harbors large numbers of soil bacteria of numerous types, which will appear in the cultures unless the nodules are carefully surface-sterilized. If the nodules are cleaned first with water, then with alcohol, and the alcohol quickly burned off in a flame, one almost always obtains pure cultures of nodule bacteria when this material is streaked onto plates. . . .

On a nutrient medium made from stems of *Vicia faba*, *Bacillus radicicola* forms colonies that are not spreading, white, hyaline, or turbid, hemispherical, and variable in size. Isolated colonies are usually very small, about 0.25 mm. in diameter but often smaller, and first can be seen only with a magnifying glass. The largest colonies are watery and faintly turbid, while the smaller ones are more firm and opalescent, while the smallest are usually spheres, which can be lifted in one unit from the gelatin. . . .

The large watery colonies consist in all cases of a mixture of rods and swarmers. In such colonies almost all the cells are motile, and the very smallest rods and double rods seek oxygen at the edge of the preparation. Except for occasional very long forms, the thick rods usually measure 4 microns long by 1 micron thick. . . .

The swarmers which occur in all preparations are exceedingly small, belonging to the smallest living beings which have been described. Exact measurements for these swarmers . . . indicate dimensions of 0.9 microns long and 0.18 microns thick. . . .

Because of the size of these swarmers and the known dimensions of the holes in the pericambial cells of the plant root, it seems possible that these bacteria could pass through the pericambial cells into the root without first inducing or requiring a lesion for their entry. . . .

The rapid movement of these forms is dependent on oxygen. In an Engelmann gas chamber with carbon dioxide or hydrogen, movement ceases, and begins again when oxygen is readmitted. Movement is not prevented in distilled water.

Although these swarmers are very small, they seem to move in the same way as rods, by the use of flagella. Although I could not see these flagella directly, I saw often the type of movement well known to microscopists, which consists of a backing up or rotating on invisible threads, and I also saw the characteristic snake-like movement of inert particles, as they were moved back and forth by the flagella of the swarmers. . . .

I conclude these general observations on *Bacillus radicicola* with the observation that this organism does not have any fermentative powers, and no special oxidizing or reducing ability, either of potassium nitrate, indigo blue, or induline. However, it attacks hydrogen peroxide rapidly with the liberation of oxygen. Spore formation has never been observed, and freezing and drying are not lethal. In liquid nutrient media, the cultures always die between 60 and 70°C., and often much lower.

It seems to me premature to consider the question of whether the bacteria have any influence on the protoplasm of the plant, and whether the formation of nodules is due to the excretion of some special product by the bacteria, or is due only to the nutritional conditions of the bacteria. . . .

Although the nodules cannot be considered as plant organs in the usual sense, such as roots, stems and leaves, these nodules seem to show such an extensive analogy to organs that the existence of some sort of physiological function in the plant seems quite likely. This possibility seems further likely from the observation that the protein content of the bacteroids of these nodules is apparently utilized by the plant under normal conditions. . . .

The accumulation of protein in the nodules and its later utilization seems to vary amongst the various species of legumes. It seems to be the most important in the annual herbaceous legumes, where the development of the nodules and the later utilization of their protein seems to be a regular event. . . . However in the woody forms the nodules usually appear late and at irregular times, and often are missing completely. . . .

Although it may seem improbable a priori that an organism with such weakly developed chemical processes as *Bacillus radicicola* would be able to oxidize ammonium salts to nitrate, or convert free atmospheric nitrogen into protoplasmic protein, it seems to me that such a conclusion should be verified by experimentation. . . . However, the results up until now have been completely negative.

The formation of nitrate was determined in nitrate-free liquid medium, as well as in agar medium, in which ammonium sulfate or asparagine were present as nitrogen sources. *Bacillus radicicola* . . . develops quite luxuriously in these media at 25°C., but at the end of the incubation period neither nitric acid nor nitrates could be detected. . . . I must conclude therefore that this species is different from the nitrifying bacteria of other authors. Further, neither nitrous oxide nor free nitrogen gas were formed in significant quantities by action on ammonium salts or asparagine, since no gas bubbles developed within nutrient gelatin.

Further, it was not possible to demonstrate the fixation of free nitrogen gas. When the nitrogen content was determined by Kjeldahl's method on an agar medium that contained no added nitrogen compounds except for the nitrogen of the agar itself, after 14 days no detectable increase in nitrogen could be observed. The growth of the organism in this agar culture was fairly extensive, but eventually stopped, apparently when the nitrogen compounds of the agar were used up. When a nutrient medium was used containing some asparagine, in which very good growth occurred, there was no indication of any increase in bound nitrogen. . . .

The possibility remains that there is only a very slow fixation of nitrogen, which could be detected only after several months, . . . or else it may be, as Frank has postulated, that the fixation of nitrogen is connected in some way with growth of the bacteria in the green plant. . . .

Because of the simple nutritional requirements of the organism, the symbiotic relationship with the leguminous plants seems all the more surprising, since except for sugar, its nutrient requirements are just as simple as those of the plant itself, and we know of many nonsymbiotic organisms, such as *Bacillus* [*Pseudomonas*] *fluorescens*, which have just as simple requirements. . . .

If we examine all of the data, mostly negative, which is available to us regarding the chemical requirements of these bacteria, this seems to indicate well the extremely complicated situation we have in the symbiotic relationship which exists in the nodule. When the living plant cell must live with another organism which is actually a part of its protoplasm, it is then necessary that a subtle balance must exist between the growth of the plant and the growth of the bacterium. It seems to me hardly likely that a bacterium which has strong chemical abilities, such as the ability to fix nitrogen, or to convert ammonium salts into nitrate, would be suitable in the development of such a delicate equilibrium. Therefore only an organism like *Bacillus radicicola*, which is similar in its chemical properties to that of the protoplasm of the plant cell, would be suitable for such a symbiotic relationship. . . .

Comment

Beijerinck became interested in root nodules of plants through his earlier work on leaf galls. These leaf galls are caused by larvae of certain insects which have hatched on the leaf, and apparently secret substances which stimulate the growth of the leaf and make tumor-like galls in which the larvae live. Because of his knowledge of the cause of these galls, it was probably easy for Beijerinck to postulate a similar cause for the development of root nodules in leguminous plants. He then proceeded through careful bacteriological studies to isolate the bacterium in these nodules. This group of bacteria is classified today in the genus *Rhizobium.*

Beijerinck found it difficult to determine the function of these organisms in the root nodules. The fixation of free nitrogen seemed an obvious possibility, and he made a number of atttempts to demonstrate that his pure cultures would carry out this process. However, the root-nodule bacteria are not able to fix nitrogen when they are free-living, but only when in symbiotic relationship with the plant. The reasons for this are still not understood today and would make a challenging research area. The free-living nitrogen fixing bacteria, such as *Azotobacter* sp., were later discovered by Beijerinck and Winogradsky (see page 237), and their ability to fix nitrogen is much easier to demonstrate, since the process can be carried out in pure culture.

Beijerinck's work on the root-nodule bacteria was a pioneer effort in a difficult field of endeavour, and his work opened the way for later studies that showed clearly that nitrogen fixation does take place when the bacteria are living symbiotically within the plant. The importance of this process in agriculture is very great, and most of the fixation of nitrogen that occurs under natural conditions is carried out by the symbiotic organisms, rather than by the free-living bacteria.

Auxanography, a method useful in microbiological research, involving diffusion in gelatin

1889 · Martinus W. Beijerinck

Beijerinck, M. W. 1889. L'auxanographie, ou la méthode de l'hydrodiffusion dans la gélatine appliquée aux recherches microbiologiques. *Archives Néerlandaises des Sciences Exactes et Naturelles,* Haarlem, Vol. 23, pages 367–372.

FOLLOWING CURRENT USAGE, I INCLUDE in the group "microorganisms" those forms which can be cultivated on nutrient gelatin. In the study of biological phenomena in microorganisms, one of the most important aspects is the discovery of those substances which are required for development and multiplication. In most cases, a new form is discovered by isolation on a particular mixture of nutrients, so that it is quite easy to use a similar mixture of nutrients to establish a permanent culture of the new form. However, in systematic studies one wishes to obtain a more extensive picture of the substances which are assimilable by a new form. In this instance, many obstacles are encountered. The usual method of approaching this problem is to incorporate the substances to be studied in a liquid or gelatin, inoculating this, and subsequently determining the weight or the number of cells in a liquid medium, or on gelatin, estimating the extent of development of colonies or lines of inoculation, and in this way determining the amount of newly formed cell substance. A study of the literature shows that only rarely have reproducible results been obtained in this manner. One is amazed at the small number of publications that exist on the nutrition of yeast! Aside from the classical work of Pasteur, and that of Cohn on the inorganic elements and nitrogen sources utilizable by various species of yeast and bacteria, the amount of work published by contemporary authors in this field has been extremely unsatisfactory. Often the methods used have been so complicated that the experiment is difficult to repeat, or the purity of the cultures used is not certain, or it is not clear if the proper concentrations of the substances examined have been used. Finally, the conditions used in comparing different materials are not standard from one worker to another. The use of the procedure which I shall describe below makes it possible to eliminate a number of these difficulties. This method is based on two simple properties of a solid culture surface. These two properties are: (1) Gelatin and agar are not nutritive materials for the majority of microorganisms. (2) In a layer of solidified gelatin or agar, diffusion of soluble materials proceeds in the same manner as it does in water. . . .

All bacteriologists know that pure gelatin or agar, or their solutions in distilled water, provide very poor substrata for the culture of bacteria, yeasts, and molds. This is because gelatin itself is not assimilable or digestible by the majority of species, and is absolutely devoid of assimilable nitrogenous compounds, such as amides and peptones. Also it is more or less completely devoid of necessary inorganic elements such as potassium phosphate and is also devoid of non-nitrogenous foods such as lactates and carbohydrates. Therefore when these solidifying agents are dissolved in distilled water, they will not support the growth of the majority of microorganisms. Further, if each of the three groups of nutrient substances—nitrogenous compounds, mineral salts, or carbon compounds—are added singly or in pairs to gelatin or agar, microorganisms that have been inoculated into the medium will not grow. In order to obtain growth, all three groups of nutrients must be added simultaneously.

Therefore, if one inoculates a species of yeast or bacterium in a gelatin medium lacking in inorganic elements but containing all other necessary nutrients, no growth occurs. However, if on the surface of such a gelatin layer a small amount of salts known to be required are added, in a short period of time each germ that is in the area in which the salts diffuse will develop into a colony, and there is formed, against the transparent background of the gelatin, a circular opaque zone, quite nicely circumscribed, which I would like to call an *auxanogram.* I have chosen the word *auxanography* to indicate briefly the method. . . .

I would also like to point out that the principle outlined above can be applied to other problems. For instance, if an organism carries out a certain function that is visible directly or can be made visible by chemical means, such as excreting an acid or an enzyme, forming a pigment or emitting light, actions which are not necessarily linked to growth, but which are carried out solely under the influence of definite substances, such processes may be studied by the auxanographic procedure. In these cases large numbers of organisms can be suspended in a gelatin substratum under conditions in which the process in question cannot be carried out. Then, by the application of materials to the surface, it would be possible to decide if these materials can elicit the phenomenon. . . .

It is apparent that the auxanographic method provides a simple means for studying the effect of certain substances on the vital activity of microbes without knowing the concentration of material that is the most favorable for this process, while in liquid culture it is almost always necessary to know precisely the concentration needed. In my procedure, diffusion in the gelatin serves to continually diminish the concentration, and leads to a gradient of concentrations, from high to low, so that there will be a certain area of the diffusion field in which the concentration is optimal. One obtains at the center an area of dead or arrested cells, and then an area with actively developing colonies. Further, the diffusion method eliminates some of the doubts regarding the purity of the culture. If one uses large gelatin plates, many different materials can be placed simultaneously on the same plate, so that their effects can be compared under identical conditions, eliminating variations due to temperature, light, availability of oxygen, or loss of carbon dioxide or water.

Comment

The auxanographic method has found wide applications in microbiology, just as Beijerinck visualized. It has been used extensively for determining the availability of sugars for various yeasts, and in this way has been used in the classification of these organisms. However, its most widespread use in recent years has been in the study of antibiotics and vitamins. Fleming's method for the study and assay of penicillin (see page 185) is based on the auxanographic method, and most later antibiotic assays involve diffusion of the antibiotic through an agar medium which has been inoculated with a test bacterium. Usually the antibiotic is applied to the agar on a filter paper disc, instead of directly as crystals, but the principle is the same. In vitamin assays, the test organism requiring a particular vitamin is inoculated into an agar medium containing everything for growth except the vitamin in question. When solutions containing the vitamin are applied in filter paper discs, the size of the zone of growth is a function of the concentration of vitamin.

Although the development of this method by Beijerinck did not involve any profound efforts, its simplicity and utility make it a useful contribution to microbiology. Note also that he made another use of agar diffusion in his later study of tobacco mosaic virus (see page 153).

Physiological studies on the sulfur bacteria

1889 • S. Winogradsky

Winogradsky, S. 1889. Recherches physiologiques sur les sulfobactéries. *Annales de l'Institut Pasteur*, Vol. 3, pages 49–60.

In 1887 I published the results of my research on the physiology of the organisms in sulfur-containing waters. In another extensive work I have shown further that these unusual organisms, so numerous that I can distinguish 15 genera and more than 25 species, form a distinct physiological group which is characterized by the role that sulfur plays in their economy. In order to indicate the physiological relationship between these organisms, I have called them the sulfur bacteria.

The study of these organisms is quite difficult because the usual methods used for the culture of lower organisms have not been applicable. I have not succeeded in culturing *Beggiatoa*, *Thiothrix*, *Chromatium*, and so on, in pure culture in flasks of nutrient liquid, so that it has not been possible to learn much about their nutrient requirements. In addition, these organisms die rapidly on a solid culture medium, which makes it difficult to isolate them easily.

I have shown that the reason for this is that these organisms have a nutrition which is different from all other non-chlorophyll containing organisms. They only develop in a solution containing a moderate amount of hydrogen sulfide. Since these organisms are aerobes, access to air is necessary. The amount of organic matter in the medium should be low but also constant. All of these conditions are only realized if one continually replenishes the culture medium, including the sulfur source. These observations explain the very extensive development of the sulfur bacteria in sulfur media, and at the same time the difficulties of cultivation.

These difficulties have led me to develop a simple method for physiological investigations in miniature, using floccules of filaments about the size of the head of a pin, placed in a drop of water. . . . It is always easy to isolate a floccule of *Beggiatoa* that is almost pure or which can be purified by mechanical means. This state of demi-purity would be poor in an ordinary culture but is sufficient for a culture under the microscope, since it is possible to observe the development from day to day, almost from hour to hour, and see easily the presence of contaminants. It is therefore possible to vary the composition of the nutrient liquid and by comparing the elongation or multiplication of the filaments of *Beggiatoa*, it is possible to determine almost as precisely as in a pure culture the nutrient requirements of the species studied.

By imitating conditions in nature where sulfur bacteria occur, I have been able to obtain growth in drops of water for weeks or months of *Beggiatoa*, *Thiothrix*, and other species. . . .

All those who have studied the subject have observed that if a sample of natural sulfur water is placed in a bottle and stoppered, hydrogen sulfide is invariably formed. It has been natural to attribute the production of this gas to the vital activity of organisms in the sulfur water. Although the observation is true, the interpretation is not. If one examines under the microscope the contents of this bottle at intervals, one finds that under these conditions the filaments that are present die rapidly. After 3–5 days, at which time hydrogen sulfide production begins, numerous dead filaments are already found, and several days later, when the liquid is saturated with hydrogen sulfide, no living organisms are found. The swollen filaments, partially disorganized, are already seen under the microscope to be devoid of the sulfur granules which they always contain when they are normal.

I conclude therefore that the hydrogen sulfide which forms arises from the intracellular sulfur, but it would be very difficult to attribute this conversion to a vital process in the sulfur bacteria. . . .

The progress of the phenomenon is easy to understand. In the absence of air the sulfur combines with hydrogen and hydrogen sulfide is produced. . . .

But what is the mechanism for the combination of hydrogen and sulfur? The preceding experiences have shown that it is not due to the action of the sulfur bacteria, since they weaken when the process begins and die when the process reaches its maximum intensity. It is probable that the process is due to the action of putrefying organisms which appear quite soon in the droplet and multiply profusely around the dead filaments. It is known that they produce hydrogen sulfide during the putrefaction of all organic materials containing sulfur. . . .

My experience therefore leads me to the conclusion . . . that the sulfur bacteria have no part in the reduction of sulfur in the presence of organic matter, and that the formation of hydrogen sulfide is due to a secondary phenomenon, which is fermentation in the absence of air.

It is quite easy to convince oneself that the formation of sulfur granules in the cells of the sulfur bacteria is the result, not of a reduction of sulfate, but of the oxidation of the hydrogen sulfide in the water. Filaments of *Beggiatoa* which are sufficiently pure, immersed in a solution of sulfate, rapidly lose their sulfur granules and do not form any more, so long as they are in this medium. But as soon as a small amount of water containing hydrogen sulfide is added, even in the same sulfate solution if some hydrogen sulfide gas is bubbled in, new sulfur granules appear within the organisms in 2–3 minutes and completely fill the cells within several hours. There is therefore no doubt about the origin of the sulfur granules in the sulfur bacteria.

What is the physiological role of this sulfur which is so avidly and abundantly stored by the living protoplasm of the cells? I have studied this question for a long time, and my conclusions have recently been contested by M. Olivier. My conclusion is that these sulfur granules are transformed by the organism into sulfuric acid. The demonstration of this fact by the procedure I have followed is so simple and certain that it is impossible to see how M. Olivier could have contested the results if he had carried out this procedure.

I began by studying the microchemical reactions which could be used to detect the presence of sulfuric acid and have found that I can use a solution of barium chloride which has been acidified with hydrochloric acid. . . .

Distilled water must never be used with *Beggiatoa*, since they die in this liquid. Fortunately this can be replaced with a liquid which is quite poor in sulfate. That which I have used contains only 0.0014% sulfuric acid. This concentration is less than that necessary to permit the formation, under the influence of a drop of barium chloride solution, of crystals of barium sulfate that are recognizable under the microscope. Crystals are only formed when the concentration of sulfuric acid is above 0.004%. The microchemical reaction is therefore less sensitive than the macroscopic reaction, but I have also found the advantage that under the microscope I only obtain the reaction for sulfuric acid after a considerable enrichment of the initial liquid with acid, and this makes the test independent of the amount of calcium sulfate which originally was present in the water which I was using.

Therefore several floccules of filaments are selected, as similar as possible, that are rich in sulfur granules. They are washed in a cuvette, with frequent changes in the water, and then each floccule is placed in a drop of equal size, and these drops are placed on microscope slides. These are separated into two groups, with the second group containing those filaments that are the richest in sulfur. These serve as the controls and are killed by heating them briefly or by exposing them to an atmosphere of chloroform. The first group of filaments remain alive and their state is carefully observed under the microscope during the experiment.

When drops of each lot were treated with barium chloride solution after 24, 48 hours, and so forth, the results were quite striking and reproducible. Large amounts of barium sulfate crystals were found in the drops containing living *Beggiatoa*, while there was not the slightest reaction in the controls, even though these contained the largest amounts of sulfur granules. It was possible to obtain an approximate idea of the rapidity of the oxidation phenomenon in the living cells by comparing under the microscope the quantity of crystals formed with those formed by known solutions of sulfate.

I have found in such an experiment the following figures. These figures were derived from a drop which contained 0.2–0.3 cc. of a liquid containing 0.0014% sulfuric acid and containing a small floccule of *Beggiatoa.*

After 24 hours	0.0066%
After 48 hours	0.0093%
After 5 days	0.0443%
After 8 days	0.0486%

The amount of sulfuric acid which is formed in the liquid surrounding the living filaments is more than 34 times what it had been at the beginning, while in the control the amount was just at the limit of sensitivity of the microchemical reaction. The amount of acid formed in the controls, due to a pure chemical oxidation of the sulfur of their cells, had a maximum of 0.0014 to 0.004 per cent, or about three times the original value.

Therefore the physiological role of the sulfur bacteria is a purely oxidative one. Hydrogen sulfide of the medium is oxidized in their protoplasm and deposited as sulfur granules in their cells. These sulfur granules are transformed into sulfuric acid and excreted. The protoplasm of the living organism enters actively into this process of combustion and makes it particularly intense. The energy that becomes available in this oxidation is their principle source of energy, as I have shown in special experiments, making them unique beings. . . .

Comment

This paper is one of the early works of Winogradsky on the properties of the chemosynthetic bacteria. He had previously worked on the iron bacteria, and now his work on the sulfur bacteria extended further a whole new field of microbiology, concerning organisms that could derive their energy from the oxidation of inorganic compounds. At the time of the present work he had not yet formulated completely the idea that these organisms used the energy from inorganic oxidations to fix carbon dioxide into organic compounds of their protoplasm. But he showed quite clearly that the sulfur bacteria could convert, under aerobic conditions, hydrogen sulfide into free sulfur, and this into sulfuric acid.

He was working with an organism that is quite large and easy to see under the microscope. Its presence in sulfur waters was well known, and because it occurred in clumps it was easily available for study. Without the pure culture methods which he later worked out, he found it difficult to study this organism in mass culture but succeeded in working out its properties in microscopic culture. Because the sulfur granules were so easy to see, it was quite simple to show that these were only formed when hydrogen sulfide was present under aerobic conditions. Under anaerobic conditions he showed that the reduction of sulfur compounds to hydrogen sulfide took place instead. He inferred without proving it that this reduction process was due to other organisms, which we now know to be those of the genus *Desulfovibrio.* After showing the initial oxidation to sulfur granules, he proceeded to demonstrate that these granules could then be oxidized further to sulfuric acid, and that this process only occurred when the organism was alive. His clever technique for demonstrating the production of sulfuric acid in microcultures is quite notable. The crystals of barium sulfate are distinctive enough to be readily identified, and he was also able to perform crude quantitative experiments on the amount of acid formed.

There were many things left to be done with the sulfur bacteria after these experiments, and it was a long time before the picture was completely clarified. However, Winogradsky left this field to work on the nitrifying and nitrogen fixing bacteria (see below). He profited in this from his earlier work on the sulfur bacteria, since he was able to view the nitrifying bacteria as chemosynthetic organisms quite analogous to the sulfur bacteria.

On the nitrifying organisms

The original French article appears in the Appendix, pages 274–275.

The original French article appears in the Appendix, pages 274–275.

1890 • S. Winogradsky

Winogradsky, S. 1890. Sur les organismes de la nitrification. *Comptes rendus de l'Académie des Sciences*, Vol. 110, pages 1013–1016.

BEFORE SUMMARIZING THE WORK ON nitrification which has occupied me for the past year, I would like to recall several of my previous works which were the point of departure for the present report.

Besides the organisms which are the subject of the present note, two groups of organisms have been studied which have the ability to oxidize inorganic substances. I have designated them by the names sulfur bacteria and iron bacteria.

The first group live in natural waters which contain hydrogen sulfide and do not grow in media lacking this substance. This gas is absorbed extensively and oxidized by their cells and is converted into sulfur granules. These latter are in turn degraded and sulfuric acid is excreted. The second group are able to oxidize iron salts, and their life is also closely connected with the presence of these compounds in their nutrient medium.

My efforts to elucidate the physiological significance of these phenomena have led me to the concept that these inorganic compounds are the fermentable materials (in the largest sense of the word) in the life of these beings, instead of the organic materials which are the fermentable substances for the large majority of microbes. This concept leads to the logical conclusion, confirmed by experience, that these beings comprise a group with certain physiological properties which can be summarized as follows. All of the energy necessary for their vital activity would be furnished by the oxidation of mineral substances, and their dependence on organic compounds for growth would be quite slight. In addition, inorganic compounds of carbon which are not utilizable by other organisms that lack chlorophyll would be used by them as a source of carbon.

The remarkable work of MM. Schloesing and Müntz has thrown light on the role of lower organisms in the process of nitrification. However, although their work makes it highly likely that a special agent exists for nitrification, they have not succeeded in demonstrating the process away from the soil, which is a natural medium with a wide variety of microorganisms. The principal requirement for all microbiological experiments today is the isolation in pure culture of the agent responsible for the process. Because of the difficulties involved, a number of workers have failed to isolate the nitrification ferment, so that the conclusion of MM. Schloesing and Müntz concerning the existence of this ferment has not been confirmed by bacteriologists and botanists.

This question must be clarified first. I have found that the failures of my predecessors are due to the fact that they used media which had been solidified with gelatin, such as are used so often today for the isolation and culture of microbes. The nitrifying organisms will not grow on such media, so that if a mixture of microbes taken from a soil that is in the process of nitrifying are placed in such a medium, all of the organisms that are active die, and one only isolates those which are ineffective. It is possible, with some difficulty, to eliminate one by one all of the foreign species and to obtain pure and in large numbers the nitrifying species, by using a medium that is favorable to it but unfavorable to the other organisms. These cultures are able under the usual microbiological experimental conditions to carry out the nitrification process just as intensely as M. Schloesing has recently shown it to occur in the soil.

This organism has been more difficult to experiment with than any of the other very delicate organisms which I had previously worked with. However, its physiological properties not only confirm my conclusions, but have revealed a new fact which I would like to report to the Academy.

I applied to this study the ideas which I had already acquired concerning the nutrition of organisms which oxidize mineral substances. I cultivated the nitrifying microbe from the beginning in a liquid which did not contain organic matter, but only a natural water that was very pure. Since the addition of organic compounds did not seem to promote its growth, I have used for its culture a mineral solution that is completely devoid of organic carbon. Although this medium does not have any other carbon compounds in it but carbonic acid and carbonates, the action of the nitrifying organism has not diminished in its intensity over several months.

We must conclude that this organism is able to assimilate carbon from carbonic acid, and this conclusion is confirmed by the amounts of organic carbon in the cultures. This demonstrates that there has been an accumulation of organic carbon by the action of this organism.

The nitrifying organism, which is colorless, is able to synthesize completely its cell substance from carbonic acid and ammonia. It carries out these syntheses independently of the light, and without other sources of energy than the oxidation of ammonia. This new fact is contradictory to that fundamental doctrine of physiology which states that a complete synthesis of organic matter cannot take place in nature except through chlorophyll-containing plants by the action of light.

It is hardly likely that the nitrifying organism exhibits a chlorophyllous action, since a release of oxygen has never been observed. Another hypothesis, that it is an amide, perhaps urea, that is the first stage in the synthesis occurring in this organism, seems to me to be the only plausible one.

Further studies on the physiology and morphology of the nitrifying organism are in progress.

Comment

Winogradsky was able to show in a clear way for the nitrifying organisms that they obtained their energy from the oxidation of ammonia and use this energy for the assimilation of carbon dioxide. His earlier studies on the sulfur and iron bacteria had pointed this way, but these organisms had proven harder to work with. This discovery is really one of the most important in physiology, since it shows, as Winogradsky realized, that carbon dioxide is convertible into organic carbon without the intervention of light energy through chlorophyll. With the addition here of a third group of bacteria that could obtain energy from the oxidation of inorganic compounds, the chemosynthetic bacteria appeared to be fairly common.

The process of nitrification turned out to be more complicated than it appears here, and Winogradsky was instrumental in clarifying this picture. He described two genera of bacteria, one which oxidized ammonia to nitrite, and the other which oxidized nitrite to nitrate. This process is important agriculturally, since ammonia is easily lost from the soil, while nitrate is more stable and serves as a good nitrogen source for plants. As he mentioned, the isolation of these organisms in pure culture was quite difficult, mainly because the soil is so rich in bacteria that other forms, which grow much faster than the nitrifying bacteria, will take over on agar plates containing organic media. Further, the nitrifying bacteria seem to be inhibited by organic matter, so that it is necessary to find a substitute for agar or gelatin. Winogradsky later did this, using silica gel, and succeeded in this way in isolating pure cultures of each of the nitrifying bacteria. He was then able to demonstrate this process in pure culture and show that a different organism was responsible for each stage.

The biochemical aspects of chemosynthetic organisms are just beginning to be worked out. We know that the process of carbon dioxide fixation in the sulfur bacteria is quite similar to the process in green plants, using the same enzyme systems. The difference is in the source of energy. The sulfur and nitrifying bacteria derive their energy from the oxidation of these inorganic compounds, and these oxidations are coupled to phosphorylation, giving ATP. The energy from ATP is used in the process of carbon dioxide fixation. Only a small amount of modern work has been done on these interesting organisms, and many new things remain to be discovered.

Enrichment culture studies with urea bacteria

1901 • Martinus W. Beijerinck

Beijerinck, M. W. 1901. Anhäufungsversuche mit Ureumbakterien. *Centralblatt f. Bakteriologie,* Part II, Vol. 7, pages 33–61.

THE PRESENT PHASE IN THE DEVELOPMENT of microbiology can be called the "systematic" phase, because in microbiology, as in every young science, it is first necessary to describe the material to be studied, and to place it into an orderly classification. For this phase of microbiology, the enrichment culture experiment has a special importance. The enrichment culture experiment makes it possible to isolate a large variety of microorganisms which are adapted to particular environmental conditions, and to bring them into development side by side in liquid culture media. In this way it is possible to set up enrichment cultures under conditions which are similar to those as they exist in nature itself, such as in natural fermentations, or in bacterial diseases. But all the uncontrollable variables of nature can be avoided in enrichment cultures and can be replaced by precisely controlled factors. (Footnote: Enrichment culture experiments are important not only scientifically, but pedagogically. I intend to publish at a later time a summary of these ideas as well as a description of the bacteriological methods used in my laboratory.)

A noteworthy property of both the artificial and natural enrichments is that the microbes that exist in the enrichment do not occur in a single variety, such as is always the case when a single colony is isolated and all further cultures made from it. In the enrichment all of the varieties are present that existed in the material used for inoculum and that can grow under the selected conditions. We are therefore able to become acquainted with a species not through a selected isolated variety, but through a number of possible variations of this species. This latter seems to me to be especially important for taxonomic studies, and it can be stated that every description of a new species must begin with an enrichment culture.

Because of our very imperfect understanding of the environmental requirements of the majority of microbes, it is impossible in most enrichment culture experiments to go further than to bring about a relative increase in the numbers of a desired form, without leading to a complete disappearance of the other species present. Often this partial enrichment only occurs at a particular stage of the experiment, whereas earlier and later other forms predominate. Because of this, enrichment culture experiments can be called "perfect" or "imperfect." In a perfect experiment, a single species is isolated in all its varieties. Only a small number of such perfect experiments have been performed, but in the future more of these will be performed, as our understanding improves. It is to be hoped that this idea will soon become widely understood, since the field yet to be studied is very extensive.

The purpose of the following report is to describe such a perfect enrichment experiment using urea. . . .

In a systematic study it was found first that when urea is added by itself to a solution containing the necessary phosphates and other salts, but lacking any other carbon source, it is not decomposed by any microorganisms. If one inoculates such a solution of urea in tap water with any natural material, there is no urea breakdown and no development of microorganisms.

The addition of another carbon source . . . to this mixture produces a nutrient solution for urea decomposers. Even the poorest carbon source for the growth of microbes, such as oxalic acid, is no exception from this. I found therefore that in a solution of the following composition: 100 tap water, 5 urea, 1 ammonium oxalate, 0.025 KH_2PO_4, which was inoculated with garden soil and kept 10 days at 30°C., more than two parts of the urea were decomposed.

The same result was obtained when the ammonium oxalate was replaced with one per cent sodium acetate, so that it is clear that urea can be used as a nitrogen source by certain urea microbes.

If the ammonium oxalate is replaced with one per cent potassium sodium tartrate, two parts of the urea are converted into ammonium carbonate. If ammonium citrate is used, three parts of the urea disappear, and if ammonium malate is used, four parts of the original urea are decomposed. At the same time in every case one or two characteristic urea bacteria appear, their type varying with the carbon source, and these forms can be carried further by inoculating into the same medium again.

As can be seen, in none of the mentioned solutions is all five parts of the urea decomposed, and there clearly exists some relationship between the nutrient value of the carbon source and the quantity of urea decomposed.

If the amount of urea added is reduced to lower levels, it is possible to show that it is all decomposed.

By bacteriological studies of these liquids it can be shown that in no case does there appear any bacterial species such as usually appear when richer media are used, in which organisms such as *Urobacillus pasteurii* develop that are capable of decomposing very large amounts of urea. If a medium is used which contains very favorable carbon and nitrogen sources, such as meat extract, and urea is added to this, then organisms develop after inoculation with garden soil which are able to decompose completely 10–12 per cent urea in a few days.

This last result induced me to carry out a detailed study which enabled me ultimately to isolate a pure culture of the strongest of all urea bacteria, *Urobacillus pasteurii* Miquel. I further found that earlier in the development of this pure culture a series of weaker urea splitters appeared for a while and then disappeared. This led me to the conclusion that all of these bacteria are widespread in nature, but they are only rarely or never observed. . . .

It has been noted above that in our enrichment culture experiments we find not only a dominant species, *Urobaccillus pasteurii*, which can be developed into a pure culture by continued transfer. We also find in the early stages of this culture a mixed flora of other urea decomposing bacteria, so it is possible in the experiment to become acquainted with a variety of urea bacteria. . . .

Although *U. pasteurii* grows on solidified medium only in the presence of ammonium carbonate, the other bacteria are able to grow easily on meat extract-gelatin without further additions, although naturally the organisms that do not split urea can also grow.

During the enrichment experiment, at particular times that are determined by the amount of urea remaining, portions of the liquid are streaked onto meat extract-gelatin plates. The colonies that develop are then tested to see whether or not they can split urea.

Two facts are clear from these observations: First, that urea in 10 per cent solution is able to retard or inhibit the growth of the majority of the non-splitters, even before it has begun to be broken down. Second, that as the concentration of ammonium carbonate increases, the rest of the non-splitters that were not hindered by the urea are now inhibited, while the urea splitters are not. . . .

With the disappearance of the various saprophytic forms, there is a parallel increase in the urea splitters, and the ultimate species that is obtained will depend on the concentration of urea used, and whether or not the soil has been previously pasteurized to kill non-spore-formers.

Comment

Both Beijerinck and Winogradsky made wide use of the enrichment culture technique to isolate species capable of carrying out particular physiological functions. The essentials of an enrichment culture experiment are described in this paper. Unfortunately, Beijerinck never got around to writing a detailed description of the techniques and uses of enrichment cultures for isolating new species and for learning something about the ecological relationships of microorganisms in their environments. The present paper was chosen because it was the most typical example of an enrichment culture experiment in Beijerinck's hands.

The philosophy of such an experiment is revealed by the statement: "Everything is everywhere, the environment only selects." This concept is at the basis of evolution as well as ecology. It assumes that any species which exists can eventually make its way into new environments, and when it does, if the new environment is favorable, it may prosper. Otherwise it will not appear in this environment, because competition from better adapted organisms will be too fierce. If this idea is correct, then the microorganisms that we find in nature are highly adapted to the environments in which they live and grow, and we should look for the reasons for this high adaptation in their genetic make-up and in their physiological functions. The enrichment culture then provides a way for viewing some of these processes of competition in the laboratory, under more carefully controlled conditions.

Practical aspects of this phenomenon are widespread. For instance, many commercial microbiological processes are really enrichment cultures. The making of wine, sauerkraut, pickles, and vinegar are good examples, since usually there is no conscious inoculation of microorganisms into the material. The environment, such as the grape-juice, cabbage, cucumbers or apple cider, then selects the microorganism that will develop and lead to the characteristic product.

Enrichment culture methods have been used widely by biochemists to isolate microorganisms capable of degrading particular chemical compounds in which they were interested. If a microorganism is desired that will degrade gasoline, it is only necessary to set up an enrichment culture using gasoline as the sole carbon and energy source. Usually an organism can be readily isolated, and quite often obtained in pure culture. Inoculation of bacteria into animals is a form of enrichment culture, since only an organism which is pathogenic for the particular species used will grow, and when the animal dies, cultures from its organs will often yield pure cultures of a desired pathogen. The implications of Beijerinck's technique are widespread indeed.

On oligonitrophilic microorganisms

1901 · Martinus W. Beijerinck

Beijerinck, M. W. 1901. Ueber oligonitrophile Mikroben. *Centralblatt f. Bakteriologie*, Part II, Vol. 7, pages 561–582.

BY OLIGONITROPHILES, I MEAN THOSE microbes which are able to develop on nutrient media to which no nitrogen compounds have been added, but in which there has been no effort to remove the last traces of nitrogenous compounds. These organisms have the ability to fix free atmospheric nitrogen and use it for their nutrition. . . .

The elective or enrichment culture of oligonitrophiles in nutrient liquids containing sugar as the organic carbon source were first carried out by Winogradsky, using conditions of anaerobiosis. Under these conditions he always isolated a particular form of butyric acid producing organism, which he called *Clostridium pasteurianum.* He used a medium of 2–4 per cent glucose with the necessary mineral salts and 2–4 per cent calcium carbonate, but devoid of added nitrogen compounds. . . . His cultures were made in closed flasks, so that the air within the flask could not be replenished. Garden soil was used as a source of inoculum. There first developed a rich aerobic flora, which used up the oxygen and made possible the development of the butyric acid anaerobe. He also worked with pure cultures of this species, in which air was excluded, while nitrogen gas was passed through the flask. . . .

My experiments have been performed under conditions quite different from those of Winogradsky, so that by providing an adequate supply of air at all times, the butyric acid bacterium was either suppressed or sharply inhibited. I also used other carbon sources than those he used. With these new methods I have discovered a new genus of oligonitrophilic bacteria which is aerobic. This genus is very easy to distinguish because of the large size of the individual cells. I have named this genus *Azotobacter*, with two species, *A. chroococcum*, which is very common in garden soil, and the other, *A. agile*, which is widespread in the canal water in Delft.

Highly aerobic conditions were obtained in my experiments by using thin layers of medium in large Erlenmeyer flasks, with frequent renewal of air. Although the butyric acid bacterium cannot develop in the complete absence of air, it behaves as a microaerophile, and by providing highly aerobic conditions, its development can be completely suppressed. In addition, I have chosen for my experiments carbon sources that are easily assimilable by *Azotobacter*, but are unassimilable by the butyric acid organism. For this I have found especially suitable: 2–10 per cent mannitol and 0.5 per cent calcium, potassium, or sodium propionate. . . . Sucrose and glucose are not as useful, since these sugars are easily fermented in the absence of nitrogen compounds to butyric acid. However, a weak butyric acid production, at least in the presence of calcium carbonate, is not harmful to my experiments, since the butyrate is readily assimilated by *Azotobacter* as a carbon source.

The ordinary saprophytic bacteria, whose germs are present in massive numbers in the inoculum, cannot reproduce in the crude culture, probably because of the absence of nitrogen compounds, so that these forms can be called "polynitrophiles." . . .

Another way in which my organism can be differentiated from the butyric acid organism is its failure to produce spores. If pasteurized soil is used as inoculum, *Azotobacter* does not develop, while the butyric acid organism can survive 90–95°C., since it produces spores. . . .

Very pure cultures were obtained simply in the following way. A nutrient solution of the following composition was used: 100 g. tap water, 2 g. mannitol, 0.02 g. K_2HPO_4. It was placed in a thin layer in an Erlenmeyer flask, with about 0.1 to 0.2 g. of fresh garden soil, and incubated at 27–30°C. The nutrient solution is weakly alkaline (Footnote: The alkaline reaction is very favorable for the experiment. If KH_2PO_4 is used, the progress of the experiment is much slower), and a precipitate of calcium phosphate occurs. The only nitrogen compounds are those that are present in the tap water, but these are necessary for the success of the experiment, since without them only a slight microbial development occurs, which soon ceases. This holds both for *Azotobacter* and for the butyric acid organism. Larger amounts of nitrogen compounds are harmful, so that there is no development of *Azotobacter* when the nutrient solution contains 10 mg. or more of KNO_3 per liter. Other nitrogen compounds in even smaller quantities are too great to allow *Azotobacter* to compete against the nitrophiles. . . .

In this nitrogen-poor medium, after 2 or 3 days at 30°C. there develops on the surface of the liquid an active film containing the large cells of the very characteristic bacterium *Azotobacter chroococcum*. This film grows without sinking and develops into a membrane. . . . Many other carbon sources can be used instead of mannitol and propionate, with varying degrees of success. . . .

The *Azotobacter* membrane consists of very thick short rods, around 4 microns thick by 5–7 microns long, with rounded ends. They are often connected together and resemble giant diplococci. . . . The majority of the cells are nonmotile, but occasional motile cells are seen. A cell wall is clearly visible as a slime layer of variable thickness. . . .

It is very easy to isolate a pure culture from the active membrane by streaking some of the material on a medium of the following composition: 100 g. distilled water, 2 g. mannitol, 0.02 g. K_2HPO_4, 2 g. agar. The 2 per cent agar contains the necessary mineral substances in sufficient quantities. After incubation at 30°C., the *Azotobacter* is already visible after 24 hours as sticky colonies, which can easily be distinguished from the watery transparent colonies of the nitrophiles. Even although these latter had been suppressed in the enrichment culture, they can develop now because of the presence of nitrogen compounds in the agar. All of the other species stop growing in a few days, while the *Azotobacter* colonies continue to reproduce, enlarging into large white slimy lumps, so that they are easily identified in the mixed culture.

The pure cultures of *Azotobacter* can be cultivated for long periods of time on various other media, such as pea leaf extract gelatin with 2 per cent sugar, or nutrient gelatin. . . .

In liquid media the growth of the pure cultures was markedly stimulated by small amounts of various nitrogen compounds, such as nitrate, ammonium salts, and asparagine, although peptone was very poor. . . .

Comment

Although Winogradsky was the first to isolate a nitrogen-fixing bacterium, the work of Beijerinck reported here has turned out to be more significant. Since most soils are aerobic, the organism *Azotobacter* occurs widely and can be readily isolated by the methods Beijerinck describes. Although the free-living nitrogen fixing bacteria, such as *Azotobacter*, are widespread, their nitrogen fixation is quantitatively not as important as that of the symbiotic root nodule bacteria (see page 220). But *Azotobacter* has been very useful in studies on the mechanism of nitrogen fixation, since it can be cultured so readily.

Although Beijerinck believed that the organism required small amounts of fixed nitrogen for growth, we know now that it can grow in the complete absence of nitrogen compounds, although it grows faster when supplied with fixed nitrogen.

In this first paper Beijerinck did not show a chemical fixation of nitrogen by doing analyses of the culture medium before and after growth to see if the nitrogen content had increased. He did this in a later paper, to make the proof complete.

The present paper also shows another application of the enrichment culture technique.

A new substance necessary for the growth of yeast

1901 • E. Wildiers

Wildiers, E. 1901. Nouvelle substance indispensable au développement de la levure. *La Cellule,* Vol. 18, pages 313–332.

INTRODUCTION

The well-known experiments of Pasteur on the alcoholic fermentation revealed the true nature of yeast and established its functions so well that work since Pasteur's time has revealed no new concepts, but only added details.

If the first complete report of Pasteur on the alcoholic fermentation is read, which appeared in 1860, one is astounded by its monumental scientific achievement, since in only 100 pages Pasteur established with certainty the true theory of fermentation and provided a firm base for the science which was later named biochemistry by Duclaux.

In this monograph Pasteur showed that the cells of the yeast were alive and multiplied, facts that had been entirely unknown before him.* In addition, he proved that the formation of alcohol and carbon dioxide is due to their vital activity.

This illustrious author also studied and clarified the relationship between sugar and its products of decomposition.

* [This is not exactly true. See Schwann and Cagniard-Latour's work (pages 16 and 20).]

Finally, an important part of this first monograph dealt with the nature of the nutrients necessary for the life and reproduction of yeast.

And since that time there has not been a living organism whose chemical processes have been elucidated as clearly as those of yeast, so that Pasteur's data form one of the most beautiful chapters in physiological chemistry. It must be stated that for a whole generation since Pasteur, there have been no data discovered that modify the work of Pasteur.

In the present work we will reveal certain facts which require a correction of the work of Pasteur, but a correction whose complete significance is not yet known.

Pasteur established with a certainty which up until now has not been altered that yeast requires for growth and fermentation only the following materials:

Yeast ash
Ammonium salt
Fermentable sugar

In other words, yeast can convert into living matter a salt of ammonium. Furthermore, yeast can make all of its organic compounds from sugar and minerals.

This fact is of the greatest importance. Because of its synthetic powers yeast must be placed in the group of non-green plants. This contrasts radically the manner of life of yeast cells with our own cells. . . .

STATE OF THE QUESTION

Since Pasteur's time (1859), numerous works have been published on the elements that are necessary, utilizable, and stimulatory for the alcoholic fermentation with different varieties of *Saccharomyces cerevisiae*.

1. The mineral nutrition was one of the first areas that was clarified with certainty by Millon, Mayer, Duclaux, Schutzenberger, Laurent, and Stein. The results of these studies have led today to the classical formulas for culture media of Heyduck, Naegeli, and Raulin. All of these formulas contain a phosphate, a sulfate, a potassium salt, and a magnesium salt. Calcium and sodium salts appear to be superfluous.

2. With regard to nitrogen, Pasteur showed first that an ammonium salt in the form of a tartrate was sufficient.

Today, the study of fermentable sugars has led to some very interesting work, but our study does not lead us into this area.

The question which we consider is therefore clear. In a mineral medium with a fermentable sugar, can the yeast multiply or ferment? Does it require another element? If so, what is this element?

The answer to these questions, revealed by our experiments, is that there is indeed another chemical element which is needed besides the mineral medium and sugar. The nature of this chemical element is unknown, but certain properties of it have been discovered which should serve us in further research.

PERSONAL EXPERIMENTS

Preliminary. Using the classical media for the culture of yeast, we had the intention of studying the synthesis of substances containing phosphorus (phosphates, nucleins, lecithins) and their successive appearance in the culture medium.

This special purpose required us to remove from our culture medium the maximum amount of phosphorus impurities possible. In addition we wished to remove above all that unknown phosphorus which would be added in the yeast used as inoculum.

In order to accomplish this, we inoculated with the minimum number of viable cells in the most appropriate culture media.

To our great astonishment, our cultures did not grow at all or grew only very slowly. We consulted various specialists and because of their counsel we employed various additional methods—aeration of the culture medium, addition of asparagine, addition of invert sugar—but without success.

Furthermore, our yeast was quite viable, since when it was inoculated in the same manner and in the same proportions into sterile beer wort, we obtained abundant fermentation and multiplication. . . .

During our preliminary experiments, it was easy to show that a rapid fermentation could be induced in any ordinary culture medium if a little larger amount of yeast was used as inoculum. Several series of experiments performed with the intention of clarifying this point showed us that the amount of inoculum was indeed at the crux of the question.

We had here the guiding light which could lead us in our research into the unknown.

We will reveal here in a stepwise manner the experiments and the results which were obtained.

EXPERIMENTAL TECHNIQUE

1. Culture medium. We have employed all of the widely used culture media. Since we have shown that the particular medium used is of no importance, we have finally arrived at the following medium:

Water	200 ml.
Sucrose	20 g.
Magnesium sulfate	500 mg.
Potassium chloride	500 mg.
Ammonium chloride	500 mg.
Disodium monohydrogen phosphate	500 mg.
Calcium carbonate	100 mg.

2. Except for certain control experiments, the yeast employed in all this work was a top yeast of the type *Saccharomyces cerevisiae* I (Hansen) cultivated in pure culture in sterile beer wort in a culture flask with a ground glass top. This flask always remained sterile and was transferred to fresh medium every 15 days.

3. The culture media were sterilized, allowed to stand for 24 hours to become oxygenated, and then inoculated with varying amounts of living yeast from an inoculum flask. The culture was inoculated into a round bottom flask having connected to it the apparatus of Crispo. The whole unit had been weighed quite accurately on a balance to the nearest decigram. The apparatus of Crispo is a tube with two enlargements which contain sulfuric acid, which absorbs the water vapor which leaves the culture. The loss in weight of the whole unit is therefore due only to carbon dioxide. The culture was then placed in an incubator at the optimum temperature for growth, 28°C. Weighings were made every day in order to determine the loss in carbon dioxide and, indirectly from this, the amount of sugar destroyed during the fermentation. The loss in carbon dioxide is almost equal to 50% of the sugar destroyed. A loss of 10 g. of carbon dioxide represents about 20 g. of sugar destroyed. This value in grams or decigrams is given in the columns of the various tables of our experiments.

We used for each culture about 250 ml. of medium. Later we used only about one-half of this amount.

CHAPTER I

Existence of a special substance necessary for the growth of yeast.

First experiment.

The mineral medium of Pasteur or analogous mineral media do not permit the growth of the yeast if it is inoculated in very small amounts. . . .

The minimum quantity of yeast suspension that brought about the growth in a mineral medium was determined for each suspension used. The following experiments show how definite the limiting concentration is. . . .

Two flasks were inoculated with two drops of yeast suspension, and two others with five drops, using the same pipette. The values in the table give the loss in weight in grams for the respective cultures.

Days	*2 drops*	*2 drops*	*5 drops*	*5 drops*
2	0	0	0.5	0.4
3	0	0	2.2	1.9
5	0	0	5.0	5.5
15	0	0	—	—

Days	*1 cc. boiled yeast +2 drops living yeast*	*2 cc. boiled yeast +2 drops living yeast*	*3 cc. boiled yeast +2 drops living yeast*	*4 cc. boiled yeast +2 drops living yeast*	*5 cc. boiled yeast +2 drops living yeast*
2	0	0	0.5	1.2	2.5
3	0	0	1.0	2.1	4.7
4	0	0	1.2	3.0	5.6

These results are quite clear-cut. Similar experiments have been performed a large number of times and have given the same results. . . .

Second experiment.

The following question arises immediately. Is this minimum quantity of yeast cells due to a requirement for the living cells themselves or due to a chemical substance present in them? . . .

To prove this, we prepared a boiled suspension of yeast which was added in increasing quantities to the mineral medium. Then these flasks were inoculated with two drops of a pure culture of living yeast [see above].

It is therefore clear that it is not the living yeast cells which are necessary to bring about growth. . . .

Third experiment.

Is it the cellular bodies in this suspension of boiled yeast which contain the active substance, or is it the liquid in which they are suspended?

To determine this, a filtrate of boiled yeast completely lacking cellular bodies was prepared with a Chamberlain filter. In addition, a suspension of yeast was prepared by washing the cells a number of times in hot water, and these washed yeast cells were used. The results of this experiment follow: [see below]

This result is clear. The liquid in which the yeast were boiled contains the indispensable principle. . . .

The fermentable properties of yeast water have been known for a long time (1828), and since the work of Pasteur, it has been reported many times that yeast water is an excellent culture medium. . . .

In our work we will be now concerned with a soluble substance in yeast water, or perhaps, a whole series of substances. We know where these substances are to be found. Our concern will now be with the physical and chemical properties of these substances and methods for their isolation.

It will be convenient to give this unknown substance a name, while awaiting the exact determination of its

Days	*Filtrate of 5 cc. of boiled yeast +2 drops of live yeast*	*Filtrate of 20 cc. of boiled yeast +2 drops of live yeast*	*Precipitate of 5 cc. of same boiled yeast + 2 drops live yeast*	*Precipitate of 20 cc. boiled yeast + 2 drops live yeast*
2	0.5	0.6	0	0
3	2.0	4.1	0	0
4	3.2	5.2	0	0
5	5.1	—	0	0
15	—	—	0	0

chemical nature. We shall call this substance "bios." Let us hope that a chemical name can soon be given to this substance! . . . [In the next series of experiments Wildiers determines some of the chemical properties of "bios." He shows that the substance is insoluble in absolute alcohol or ether but soluble in 80% alcohol. He shows that when the yeast extract is ashed, the activity disappears, indicating that it is not a mineral substance but something organic. He shows that "bios" is not destroyed by boiling for 30 minutes in 5% sulfuric acid (v/v) but is destroyed by boiling in 10% sulfuric acid (v/v). "Bios" is destroyed by boiling for 30 minutes in 1% NaOH. It is not precipitable by lead acetate. Several other inorganic precipitants also do not precipitate it. A number of known organic substances were tested to see whether they had "bios" activity, and were all inactive: urea, asparagine, aniline, tyrosine, adenine, guanine, thymus nucleic acid (DNA), creatine, products of peptic or tryptic digestion of the chemically pure proteins edestin and ovalbumin. "Bios" dialysed easily through a membrane, indicating that it was a small molecule. It was present in meat extract, commercial peptone, and beer wort (barley extract) as well as in the cells of yeast. Further, "bios" was not found in the fermentation liquid after a yeast had grown in a mineral medium, but only within the cells of the yeast, indicating that the substance was not excreted by the cells but was actually retained within them. Other species of yeast were tested to see if they also required "bios," and a number of other species did.]

GENERAL CONSIDERATION

In the first place, our work is interesting because it modifies the absolute law established by Pasteur that, aside from sugar, the yeast requires no other nutrients except inorganic substances.

However, in discovering an unknown substance that is indispensable for the life of yeast, our work makes uncertain a very important concept of biology.

If this unknown substance is an organic nitrogen compound, as seems very probable, it will no longer be possible to state that yeast can synthesize all types of proteinaceous materials from sugar, ammonium, and sulfates.

But, if yeast is incapable of synthesizing all its proteins, in the same way that it is incapable of synthesizing sugar, its synthetic chemistry diverges considerably from that of green plants and approaches that of the animals.

This would be an important discovery for all those interested in the study of the nitrogen nutrition of animals or of the chemistry of their nitrogenous compounds.

Indeed, when we state today that the animal organism cannot synthesize all of its proteinaceous substances, we mean this in a restricted sense. It is quite certain that the animal, which is able to synthesize urea, uric acid, and amides in general, is probably able to synthesize three-fourths of the nitrogenous compounds that are present in its proteins. A fraction of the unknown nitrogen compounds in proteins cannot be synthesized by our cells. . . .

Therefore, if there exists in the cell of the yeast a certain fraction of materials that cannot be synthesized, possibly a group of substances that are especially difficult to synthesize, it may be that the same substances are not synthesized by both the yeast and the animal.

We have no intention of stating that this hypothesis is a certainty. We are merely stating that the question could be extended in this direction. As long as our unknown substance has not been characterized, there will remain a doubt concerning the accepted concepts regarding the synthesis of proteins by yeast.

It is with the desire of communicating these findings to the most expert and competent chemists that we have published this work.

Comment

This work is important for several reasons. First, it extended considerably the work of Pasteur and others on the nutrition of yeasts. Second, it served to explain a controversy that had raged between Pasteur and Liebig. When Pasteur published his famous paper in 1860, he seemed to have settled for all time the question of whether yeast was a cause or an effect of the fermentation (see page 31). He did this by growing the yeast in a mineral medium in which there were no organic substances except pure sugar. One of the key points of Liebig's argument had been that the proteinaceous materials in beer wort or grape juice were unstable and broke down, and during this breakdown sugar was converted into alcohol, and the yeast that was formed was merely a product of this protein degradation (see page 24). When Pasteur showed that proteins were not necessary for fermentation, he seemed to settle this question. However, Liebig could not repeat Pasteur's experiment and could only get fermentation when proteinaceous materials were present. It is probable that Liebig used smaller amounts of yeast as inoculum than Pasteur (the latter used that amount that would fit on the head of a pin, which is much larger than Liebig or Wildiers used) and did not get fermentation. Pasteur was probably carrying over into his fermentation flasks sufficient amounts of "bios" to bring about fermentation. So Wildiers' work explains this controversy.

The most important aspect of this work is that it shows that there are unknown organic substances that are necessary for growth. These substances, which we now call vitamins, have proved to have far-reaching implications in nutrition and biochemistry. We know that animals require a number of different vitamins and that many microorganisms also require them. The discovery of these vitamins was given a sound beginning with the work of the present paper. The chemical characterization of the vitamins could not occur until the proper biochemical techniques had been developed. Various vitamins were first identified as growth factors for animals, while others were first shown to be required by microorganisms. But the microorganisms have been very important in vitamin research, because they have been useful as assay organisms for vitamins in a way similar to that used by Wildiers. Further, some microorganisms (especially yeast) produce large amounts of vitamins and have been useful as sources of them. In 1933, Williams and co-workers studied "bios" further and identified one of the substances in it as pantothenic acid. Other substances identified in bios include thiamine, nicotinic acid, biotin, pyridoxine, and para-aminobenzoic acid.

Fixation of free atmospheric nitrogen by *Azotobacter* in pure culture

1908 • M. Beijerinck

Beijerinck, M. 1908. Fixation of free atmospheric nitrogen by *Azotobacter* in pure culture. *Kon. Akademie van Wetenschappen*, Amsterdam, Vol. 11, pages 67–74.

WHEN CARBOHYDRATES ARE USED AS sources of carbon in *Azotobacter* cultures, there existed until now some doubt whether the fixation of free nitrogen which occurred was originally effected by *Azotobacter* itself or by other bacteria found in symbiosis with it, because *Azotobacter* in pure culture with carbohydrates and free nitrogen gas only does not show any considerable development.

For this reason I was formerly of the opinion that in such cultures *Bacillus radiobacter*, a species closely allied to the root-nodule bacteria, and which is never absent in enrichment cultures of *Azotobacter*, would be the real cause of the nitrogen fixation.

Continued research, however, rendered this supposition more and more improbable, and the facts which are now to be stated have proved beyond any doubt that the said faculty belongs indeed to *Azotobacter* itself.

These facts concern the very peculiar relation between *Azotobacter* and the salts of the organic acids, more particular to calcium malate.

1. *Calcium malate as source of carbon.*

When into a wide Erlenmeyer jar a nutrient liquid is introduced of the composition: 100 tap water, 2 calcium malate, 0.05 K_2HPO_4, with addition of some 10–20 cc. canal water, or a small amount of soil for inoculation, care being taken that the layer of liquid in the jar be not thicker than 2–5 cm., on cultivation in an incubator at 30°C., usually after 2–3 days, a floating *Azotobacter* film appears, consisting of strongly motile individuals, relatively soon developing a considerable thickness. Hereby so much calcium carbonate is produced that it forms a closed, floating layer, so to say a cover, on the surface of the liquid.

If some of this film is inoculated into another jar containing the same medium, corresponding phenomena are seen when the culture conditions are alike. At first sight already, there can be no doubt that under these circumstances fixation of considerable quantities of nitrogen must take place, and chemical analysis proves that this is really the case. . . .

2. *Quantity of the fixed nitrogen.*

. . . the analysis of the cultures is performed as follows.

The whole quantity of the liquid, in which are present the calcium carbonate formed by oxidation from the malate or the other organic salt, together with the undecomposed malate, the salt of the volatile acid, and the bacteria, is treated with a known quantity of normal hydrochloric acid and the carbon dioxide is expelled by heating. Following this, a titration with normal alkali and phenolphthaleine as indicator, shows how much calcium carbonate is produced and consequently how much of the organic salt is oxidized.

After addition of a little sulfuric acid the liquid is evaporated to dryness and after Kjeldahl's method examined for nitrogen, while in each of the materials used the rate of nitrogen is stated separately. . . .

The data show that the amount of nitrogen which can be fixed in the crude culture is at most 4.9 and 2.8 mg. per gram of oxidized calcium salt, obtained respectively with calcium propionate and calcium acetate, while 2.6 mg. was fixed per gram of calcium malate, and 1.8 mg. per gram for calcium lactate. It seems that the fixation goes on more rapidly at the beginning than later in the course of the experiment, whence it follows that when little of the organic salt is used, proportionately more nitrogen is fixed than by larger amounts. . . .

Comment

This paper carries further the work that Beijerinck began on *Azotobacter*, showing chemically that there was actually a fixation of nitrogen during the growth of the bacteria. See page 237 for his earlier work on this subject.

Unity and diversity in the metabolism of micro-organisms

1924 • Albert J. Kluyver

Kluyver, A. J. 1924. Eenheid en verscheidenheid in de stofwisseling der microben'. *Chemisch Weekblad*, Vol. 21, page 266. Translation from *Albert Jan Kluyver, His Life and Work*, 1959, North-Holland Publishing Company, Amsterdam.

*My daughter Alice is a student in High School. One of the prescribed courses is General Science. The section on Bacteria left her with a vague impression of a world teeming with deadly germs awaiting an opportunity to infect mankind. It seems probable that this malignant conception of bacteria is very generally held. In reality civilisation owes much to the microbe.**

* (From the Preface of A. I. Kendall, *Civilisation and the Microbe* (Boston, 1923).)

WHEN, SOME TIME AGO, I RECEIVED the esteemed invitation to deliver a lecture on some biochemical subject before the Netherlands' Chemical Society, I gladly accepted because a true microbiologist may never pass up the opportunity to contribute to the vindication of the smallest living organisms. For there is an altogether too prevalent notion that the microbes, as irreconcilable enemies of man, plant, and animal, deserve attention only in order to make it possible the better to combat them. However understandable such a concept may be in view of the beneficial effects resulting from the brilliant discoveries of the role of microbes in numerous diseases, it nevertheless does not detract from the fact that it represents an extremely warped picture of reality.

But I shall not use the time at my disposal to show you how, in a microbe-less world, the conditions for human life on earth would soon no longer be realized, so that man possesses at least as many friends as enemies in the domain of the microbes. Taking advantage of the fact that I am addressing a chemically trained audience, I shall rather limit myself to an attempt at striking a sensitive chord by discussing the chemical potentialities of micro-organisms. I trust that a sober review of these capacities may suffice to make you regard these smallest living beings with a little more sympathy in the future.

The most specific characteristic of living matter resides in its metabolic properties. It is an empirical fact that the maintenance of life requires a continuous supply of special chemical substances which, in the living cell, undergo transformations that lead to their partial excretion in altered form.

You are sufficiently familiar with the notion that this applies to the higher organisms; the idea that it is equally true for the microbes follows immediately from the consideration that in so many cases the presence of microbes is forced upon us by the very fact that we observe chemical changes. When milk turns sour, when sugar- or protein-containing solutions start to ferment, it is first of all metabolic activities that draw our attention, and, after half a century of microbiological research, will lead to the inference that microbes are present.

As the title of today's lecture indicates, it is my intention to discuss the unity and diversity in microbial metabolism. But I shall take the liberty of changing the sequence and first to dwell upon the diversity. Later on I shall then try to indicate some aspects of the unity by which this diversity is tied together. In doing so I shall also follow in the main the historical development of microbiology.

When, owing to Pasteur's pioneer investigations, the idea had become established that fermentation, putrefaction, and mineralization were none other than metabolic processes of microscopically small organisms, microbiologists obviously considered it their first duty to make a survey of the multitude of different causative agents.

Chemically the diversity manifested itself in two ways. It was observed on the one hand that one and the same medium might yield a variety of metabolic products; on the other that different microbes differ greatly in their requirements for particular chemical substances. Let me first illustrate the former aspect with some examples.

If I inoculate a series of flasks, each containing a sterile solution of 5 per cent glucose in yeast extract, with pure cultures of a film-forming yeast *(Mycoderma cerevisiae)*, and with some common molds, such as *Aspergillus niger* and *Citromyces glaber*, respectively, a chemical analysis of the cultures at the end of their development will yield different results. Confining the investigation to a determination of the fate of the consumed sugar, it will be found that the film yeast has oxidized it to carbon dioxide and water; hence the over-all metabolism resembles that of animals. In contrast we shall find that *Asp. niger* and *C. glaber* have produced sizable amounts of oxalic and citric acid, respectively, in addition to carbon dioxide. Consequently this simple experiment reveals metabolic differences although these are not as yet very pronounced. Much more spectacular is the result of a larger experiment in which flasks with the same medium are inoculated with pure cultures of *Saccharomyces cerevisiae* (baker's yeast), *Lactobacillus delbrückii* (the so-called "domesticated" lactic acid bacterium), *Lactobacillus fermentum* (a "wild" lactic acid bacterium), *Bacterium coli*, *Bacterium aerogenes*, *Bacterium typhosum*, *Granulobacter saccharobutyricum*, and *Granulobacter butylicum*, and incubated under exclusion of air. After some time it will be evident that the yeast has converted the sugar largely into ethanol and carbon dioxide; *L. delbrückii* into lactic acid; *L. fermentum* into lactic and acetic acids, ethanol and carbon dioxide; *B. coli* into lactic, acetic, and succinic acids, carbon dioxide, and hydrogen; *B. aerogenes* into the same products with, in addition, 2,3-butylene glycol; *B. typhosum* into formic, acetic, and lactic acids, and ethanol; *G. saccharobutyricum* into butyric and acetic acids, carbon dioxide, and hydrogen; *G. butylicum* into butanol, acetone, carbon dioxide, and hydrogen. Thus a remarkable diversity emerges.

Nevertheless, this diversity in products obtainable from a single substrate is almost negligible in comparison with the differences that various microbes exhibit with respect to their nutrient requirements.

For their studies on micro-organisms the early microbiologists depended more or less on the fortuitous appearance of special types. But gradually they became aware of correlations between the initial composition of the medium and the microbes therein encountered. To mention just one example: it soon became clear that true yeasts are found only in sugar media. After the introduction of pure culture methods had permitted a closer investigation of metabolic activities, it became increasingly evident that substances which represent excellent foodstuffs for one type of microbe may be less so, if not entirely unsuitable, for others. Once this had been recognized, the idea obviously occurred that the investigator possessed a powerful tool for encouraging the development of certain microbes at the expense of others present in the inoculum. Although ultimately this method of elective cultures harks back to Pasteur and Raulin, it is in the hands of Beijerinck and Winogradsky that it has indubitably produced the richest harvest. Well-nigh overwhelming is the metabolic diversity revealed by the consistent application of their "enrichment cultures."

Based on Berthelot's studies, showing that the increase in bound nitrogen of fallow soils was unquestionably the result of a biological process, Winogradsky (1902) succeeded in isolating one of the causative agents, *Clostridium pasteurianum*, by means of the elective culture method in 1893. It turned out to be an organism that can grow only in the absence of air. The following remarkable experiment throws a clear light on its extraordinary metabolism. A solution containing glucose and all other elements necessary for growth with the exception of nitrogen is inoculated with spores of this bacterium. The culture is placed in a container which is thereupon evacuated in order to remove the toxic oxygen. Even after prolonged incubation one does not observe any changes. At this point gaseous nitrogen, freed from the last traces of oxygen and of nitrogenous compounds, is admitted. After 24 hours a vigorous fermentation ensues, and in a few days the sugar has completely disappeared, the bacteria have grown profusely, and nitrogen has been fixed. This demonstrates the remarkable phenomenon of a living organism that is roused to life from a latent state by the inert gaseous nitrogen!

Closely related to this curious organism is the bacterium, *Azotobacter chroococcum*, that was discovered a few years later by Beijerinck and Van Delden. It, too, fixes nitrogen, and to an even greater extent. But in contrast to *C. pasteurianum* it can do so only if an ample supply of oxygen ensures a rapid oxidation of organic substrates. . . .*

Yet other specialists among bacteria demand our attention. How curious a group of organisms are not the urea bacteria that more or less intensively convert urea into ammonium carbonate! Special mention deserves *Urobacillus pasteurii* which under suitable conditions completely hydrolyses a 10 per cent urea solution and thus can tolerate with impunity the alkaline reaction of a 16 per cent ammonium carbonate solution.**

I may not omit an organism such as *Bacillus oligocarbophilus*, for which Beijerinck and Van Delden showed that it can derive its nutrients from the traces of organic matter that never seem to be lacking in laboratory air. Also the group of bacteria that feed on hydrocarbon vapors, studied by Söhngen (1903), may not be neglected.

But even this diversity is still limited in one respect. The carbon, essential for the growth of the organisms so far mentioned, must be supplied in the form of an organic substance, though the nature of the latter may vary enormously. This implies that in the end these organisms derive their food from other organisms, hence their designation as heterotrophs.

Nevertheless, it is almost forty years ago since Winogradsky expressed the idea that even in this regard the diversity of microbial metabolism is not limited.† To be sure, it had long ago been established that chlorophyll-containing microbes could build up their cellular constituents from carbon dioxide and other minerals, by virtue of the absorbed solar energy. But that colorless microbes, without the aid of radiant energy, could do so too is a concept so daring that even today it is surprising that a human mind has ventured to propose it.

A genius like Winogradsky (1887) did not hesitate, however, to conclude, as early as 1887, from his simple and ingenious experiments that the colourless *Beggiatoa alba* frequently encountered in sulphur spring waters can grow in darkness in a strictly mineral medium. The organic cellular constituents would be synthesized from carbon dioxide, a process made possible because the necessary energy would be obtained from the oxidation of hydrogen sulphide, first to sulphur, and next to sulphate.‡ In addition to *Beggiatoa* this would also apply to a large group of bacteria, partly colour-

* [See Beijerinck, page 237.]
** [See Beijerinck, page 234.]
† [See Winogradsky, page 231.]
‡ [See Winogradsky, page 227.]

less, partly purple, that exhibit the common characteristic of depositing free sulphur in their cells.

Subsequent experiments have amply confirmed the correctness of Winogradsky's concept. Consciously executed elective cultures have shown that, in addition to the relatively large organisms that deposit colloidal sulphur in the form of droplets and as a reserve material in their cells, and which had long been known to naturalists, there exists another large group of small bacteria that, though producing sulphur extracellularly when grown in sulphide media, yet must be reckoned as belonging to the physiological group of sulphur bacteria on account of their metabolic behaviour which is similar to that of *Beggiatoa.* These bacteria, discovered by Nathansohn, but studied more particularly by Beijerinck, Lieske (1912), and Jacobsen (1912, 1914), have recently been much discussed in America; permit me to give you another striking example of the chemical proficiency of the microbes. In a series of papers the American microbiologist, Waksman (1922), has reported on the astounding properties of an organism, designated as *Thiobacillus thiooxidans*, that can also oxidize sulphur powder to sulphuric acid, and get along with carbon dioxide as the sole carbon source. It differs from *Thiobacillus thioparus*, earlier described by Beijerinck and by Jacobsen, especially by its insensitivity to acid which is so pronounced that in certain media the *p*H reaches 0.6. Nay, J. G. Lipman, the well-known editor-in-chief of "Soil Science," has communicated in a recent paper that *T. thiooxidans* can still effect a perceptible sulphur oxidation in solutions containing 5 per cent sulphuric acid.

It should not surprise you that this powerful ability to produce sulphuric acid has been exploited in different directions in the U.S.A. I shall mention but one example because, in a sense, it implies an attack on one of the branches of chemical industry. Lipman and Waksman have already carried out experiments on a rather extensive scale to test the possibility of circumventing the commercial manufacture of superphosphate by fertilizing the soil with a mixture of natural phosphate and free sulphur, and thus causing the localized formation of sulphuric acid, hence also of superphosphate, through the activity of *T. thiooxidans.* In certain soils this treatment appears to have given fully satisfactory results.

As far as an impressive chemical performance is concerned the last-mentioned bacterium is still exceeded, however, by Beijerinck's *T. denitrificans* which can grow in the complete absence of oxygen and gaseous carbon dioxide in a medium containing sulphur, potassium nitrate, chalk, and a small amount of phosphate; thus it should be able to produce organic matter from these compounds in deeper soil layers.

I could display before you several other specialists which resemble the sulphur bacteria in being able to grow in purely mineral media. Time limitations prevent me from dwelling on them; I shall merely mention some other examples, such as the bacteria that oxidize nitrite to nitrate, or ammonia to nitrite, and for which glucose is almost as toxic as is sublimate for other organisms. Let me finally also refer to the various types of organisms for which mixtures of hydrogen and oxygen, methane and oxygen, and hydrogen and nitrous oxide constitute nutrients.

If we consider all these data it is well-nigh impossible not to be impressed by the enormous diversity in microbial metabolism. If we further survey the trends in microbiology during the past few decennia it is hard to avoid the conclusion that the attempts to demonstrate this diversity have largely dominated the investigations. It was not so much the microbes themselves that were the starting point for the studies; rather was the mere suspicion that certain chemical transformations might occur in nature a sufficient impetus to formulate the hypothesis that there would be microbes to accelerate them. And the correctness of the hypothesis was substantiated in nearly every case by those who had learned to utilize the elective culture method. The very consideration that the cycle of matter on earth is closed, and that consequently the naturally occurring hydrocarbons must eventually be converted into carbon dioxide and water, led Söhngen to the discovery and isolation of an important group of bacteria, members of the genus *Mycobacterium*, that can use these substances as carbon source.

I can cite an even more striking instance. The studies of the above-mentioned autotrophic bacteria, of which the sulphur-, nitrite-, and ammonia-oxidizing bacteria are the best-known examples, unconsciously led Nathansohn to the bold hypothesis that still other reactions, proceeding very slowly at ordinary temperatures, might serve as a basis for microbial metabolism even if the oxidizable compound is not known to occur in nature. Thus he could ascertain the fact that the exclusively man-made thiosulphate can serve as the major nutrient for particular types of sulphur bacteria.

These examples may suffice to show how gradually the investigations became subservient to what might be called a "microbiological imperialism." It became a contest to open up new, seemingly inaccessible areas for the microbes. Barely had one specialist among microorganisms been discovered when another even more spectacular one was announced; the microbiological theatre resembled one grand naturalistic vaudeville show.

However great may be our admiration for those who, through their intuition and ingenious experimentation, have guided this flight of general microbiological discoveries, and however indispensable has been this thorough exploration of the microbial world for the continued development of microbiology, in the long run this approach could not persistently be satisfying. Even while the main current of microbiological research was concentrated on the diversity, a field of study gradually developed in which a search for unity in the diversity became apparent. It would be unjust to depict this trend as an entirely new and recent development. Far from it; many classical figures in microbiological science, and its founder, Pasteur, in the very first place, must receive honourable mention in this connection. Nevertheless, one may safely say that the attempt to view microbial metabolism in the light of the outcome of general physiological research has been deliberately initiated by the German physiologist and hygienist Rubner, only at the beginning of the twentieth century. And we may immediately add that thus far this approach has found very little response. This regrettable fact can be readily explained by the circumstance that microbiology has hitherto developed primarily as an applied science. By far the majority of medical, technological, and agricultural microbiologists have studied the role of microbes in the important and eminently practical problems they had to face. And of the relatively very small number of investigators who did have the opportunity and desire to study microbes for their own sake, the majority was under the spell of the diversity; quite understandably so in view of the fascination of this type of work.

If now I am going to embark on the attempt to give you a glimpse of the unity that can be discovered in microbial metabolism, I do so with the full knowledge that it is but a meagre account I have to offer. That nevertheless I have ventured to solicit your attention for this topic finds its explanation in the fact that I shall thus have the opportunity to make you realize that this still presents an immense field of endeavour. It seems to me that the solution of the problems in this area is indispensable to elevate microbiology from its status as a largely descriptive branch of science to a higher plane, and equally so to the application of microbiology to many situations. And I am firmly convinced that only a close cooperation between microbiologists and accomplished chemists can lead to advances in this respect.

Let me then first of all show you how the unity in the divergent metabolic processes of the microbes finds expression in the fact that we recognise in them the same general trends that have come to light as a result of investigations of the metabolism of higher organisms. Without discussing this aspect in detail, the following remarks may serve to illustrate this point. Studies of the metabolism of higher organisms have unequivocally shown that one can always distinguish two types of processes. Part of the foodstuffs is converted into cell materials, the latter being used either for growth or for replacement of degenerated cellular constituents. Another fraction of the food appears to owe its significance largely to the therein accumulated chemical energy which is unleashed by the living cell, thus enabling it to carry out energy-requiring functions; this part is ultimately degraded, generally producing heat. These two processes are differentiated as assimilation and dissimilation.

Doubtless the dissimilatory process is the more essential characteristic of life. Whereas other typical vital phenomena, such as growth, reproduction, internal or external motion, may frequently be lacking, dissimilation is absent only in some stages of latent life, resulting, for example, from dessication.

Consequently it had been established that in the case of higher organisms the perpetuation of life is tied to a continuous conversion of chemical into other forms of energy, and Rubner demonstrated that the first law of thermodynamics applies to the energy transformations in the animal body, while Atwater showed this also for man.

As far as the nature of the energy-providing reactions is concerned, Lavoisier had already pointed to the significance of the respiratory process, *i.e.*, the slow combustion taking place in living organisms, which reveals itself through their requirement for oxygen. The validity of this concept was emphasized when Rubner experimentally established his principle of "isodynamic substitution." He showed that, up to a point, proteins, fats, and carbohydrates can replace one another, in a weight ratio that is inversely proportional to the heat of combustion of these substances, in the adult animal for the maintenance of the same vital functions; in other words that, within certain limits, equality in chemical energy corresponds to equality in food value.

As early as 1885 Rubner pointed out that in all probability it would be possible that also in the metabolism of microorganisms a conversion could be designated which derives its significance for the organism entirely from the resultant energy liberation. Now

Pasteur, who, in 1860, had discovered the first instance of organisms that can multiply in the complete absence of oxygen, had intuitively recognized the connexion between the absence of respiration in these organisms and the fermentation process that characterizes their mode of life.*

Meanwhile the mutual substitution of respiration and fermentation was initially considered largely from a material angle, so that the term fermentation was used to indicate a respiration with bound oxygen. The need to consider this substitution primarily on the basis of energetics was first formulated with sufficient clarity by Rubner as a logical consequence of his attempt to interpret the metabolism of any and all living creatures from a single point of view. But not until 1902 did he begin his series of micro-calorimetric measurements of the metabolism of various microbes which corroborated the validity of these hypotheses. Since that time it has been satisfactorily established that the metabolic activities of every microbe comprise the processes whereby new cell material is synthesized, as well as dissimilatory processes characterized by the fact that chemical energy is utilized by the living cell for its energy-requiring functions, and finally appears as heat.

A number of the most important dissimilatory processes encountered among various groups of micro-organisms is summarized in Table I. The first ten represent oxidative dissimilation reactions of which the upper three

* [See Pasteur, page 39.]

TABLE I

MICROBIAL DISSIMILATORY PROCESSES †

	Examples of organisms:
A. *Oxidative processes*	
1. $C_6H_{12}O_6 + 6O_2 = 6CO_2 + 6H_2O + 676$ cal.	various fungi
2. $C_2H_5OH + 3O_2 = 2CO_2 + 3H_2O + 325\frac{1}{2}$ cal.	mycodermic yeasts ('Kahmhefen')
3. $CH_2NH_2COOH + 1\frac{1}{2}O_2 = 2CO_2 + H_2O + NH_3 +$ 142 cal.	many aerobic bacteria
4. $C_2H_5OH + O_2 = CH_3COOH + H_2O + 116\frac{1}{2}$ cal.	acetic acid bacteria
5. $CH_4 + 2O_2 = CO_2 + 2H_2O + 210$ cal.	methane oxidizing bacteria
6. $H_2 + \frac{1}{2}O_2 = H_2O + 68\frac{1}{2}$ cal.	hydrogen oxidizing bacteria
7. $NH_3 + 1\frac{1}{2}O_2 = HNO_2 + H_2O + 79$ cal.	nitrite forming bacteria
8. $KNO_2 + \frac{1}{2}O_2 = KNO_3 + 22$ cal.	nitrate forming bacteria
9. $H_2S + \frac{1}{2}O_2 = H_2O + S + 67$ cal.	sulphur bacteria
10. $S + 1\frac{1}{2}O_2 + H_2O = H_2SO_4 + 141$ cal.	sulphur bacteria
B. *Fermentative processes*	
11. $C_6H_{12}O_6 = 2C_2H_5OH + 2CO_2 + 25$ cal.	yeasts
12. $C_6H_{12}O_6 = 2C_3H_6O_3 + 28$ cal.	lactic acid bacteria
13. $C_6H_{12}O_6 = C_4H_8O_2 + 2CO_2 + 2H_2 + 15$ cal.	butyric acid bacteria
14. $C_6H_{12}O_6 + \frac{1}{2}H_2O = CH_3CHOHCOOH + CO_2 + H_2 + \frac{1}{2}CH_3COOH + \frac{1}{2}C_2H_5OH + 8$ cal.	coliform bacteria
15. $(CH_3COO)_2Ca + H_2O = CaCO_3 + CO_2 + 2CH_4 +$ 19 cal.	methane bacteria
16. $CO(NH_2)_2 + 2H_2O = (NH_4)_2CO_3 + 6$ cal.	urea splitting bacteria
17. $C_2H_5OH + 2.4KNO_3 = 1.2N_2 + 1.2K_2CO_3 + 0.8CO_2 + 3H_2O + 293$ cal.	denitrifying bacteria
18. $C_2H_5OH + 1\frac{1}{2}CaSO_4 = 1\frac{1}{2}H_2S + 1\frac{1}{2}CaCO_3 + \frac{1}{2}CO_2 + 1\frac{1}{2}H_2O + 14$ cal.	sulphate reducing bacteria

† This Table makes no attempt whatever at completeness. In particular, the list of oxidative dissimilation processes with organic substrates can be expanded indefinitely. As for the fermentative dissimilation processes, the most important types known today are represented, with the exception of the anaerobic decomposition of protein breakdown products. These have been omitted because it is very difficult to denote them by means of simple equations. See. however, *Arch. f. Hyg.* 66: 209 (1908).

occur also in higher animals, the others being typical dissimilatory processes of the earlier mentioned microbes. The dissimilations listed under the heading "fermentative" are examples of transformations that satisfy the energetic requirements of organisms that live temporarily or permanently in the absence of oxygen.

These fermentative processes understandably are characterized by a considerably smaller caloric effect, and the correctness of the energetic approach to these transformations is reflected in the fact that per unit weight of the causative organism comparatively much more food is used, and much greater amounts of metabolic products are formed during fermentative existence. Particularly clearly is this phenomenon revealed by organisms which, depending on circumstances, derive their energy either from an oxidative or from a fermentative dissimilation. This is true, for example, for yeast; and Pasteur had already established that the sugar consumption per unit weight is considerably greater during anaerobic than during aerobic cultivation; in the latter case part of the sugar is also oxidized.*

Now it cannot be doubted that the energetic interpretation of metabolism can also be used for the clarification and systematization in other directions. In the first place it forces the investigator to develop a clearer picture than is customary of the metabolism of the organism studied. To mention an example: it is usually stated without further specification that the extensively investigated *B. coli* can grow in meat extract broth both with and without sugar, and that it is a facultative anaerobe, implying that it can develop both in the presence and in the absence of oxygen. Many a microbiologist does not realize, however, that the dissimilation of this bacterium consists in an oxidative degradation of proteinaceous split products on the one hand, but in a fermentative decomposition of carbohydrates or sugar alcohols on the other, so that the presence of the last-mentioned substances is a prerequisite for anaerobic growth. This is in contrast to bacteria of the *Proteus*-group which are usually similarly characterized in the literature, although they appear to possess the property of obtaining energy from a fermentation of protein degradation products so that they can lead an anaerobic existence even in the absence of carbohydrates. While this example indicates that the vague designation of an organism as a facultative anaerobe should be replaced by a rational consideration of potential dissimilation processes in order to permit a proper insight into the conditions needed for development, the following example may further illustrate this.

The organisms that can be assembled in the group of genuine lactic acid bacteria are often designated as facultatively anaerobic. In a sense this is permissible because the majority can indeed develop in the presence as well as in the absence of air. It would, however, be a glaring error if one were to infer from this fact that these organisms, like the above-mentioned facultative anaerobes, can display both an oxidative and a fermentative metabolism. On the contrary, even in the presence of air the lactic acid bacteria do not utilize oxygen in their metabolism. They are therefore purely fermentative organisms; carbohydrates or sugar alcohols are essential for their growth; and they differ from other strictly fermentative microbes only in the fact that free oxygen does not completely inhibit their development.

In this connection it is not without significance to point out that the ab-

* [See Pasteur, page 41.]

sence of an oxidative metabolism among the lactic acid bacteria is undoubtedly related to a property that Beijerinck in 1893 established for all lactic acid bacteria, *viz.*, the absence of the enzyme catalase which decomposes hydrogen peroxide into water and oxygen, and which had been found in all cells of higher plants and animals, and until then also in all microbes that had been tested. Later Orla-Jensen pointed out that the exclusively anaerobic butyric acid bacteria also lack catalase, and we have established the same situation in the case of anaerobic protein-decomposing bacteria. The widely accepted, though still hotly disputed (Dakin, 1922) hypothesis that in all aerobic organisms, the higher as well as the lower, catalase plays a role in the transfer of oxygen to the oxidizable substrate is undoubtedly supported by these facts. Conversely it seems likely that a study of the occurrence of catalase in a newly isolated microbe can be fruitfully employed to determine the presence of an oxidative metabolism.

The great significance of the establishment of the nature of a microbe's dissimilation processes, these primarily characterizing its metabolism, also resides in the fact that the natural relationships of micro-organisms are unquestionably expressed in their metabolism. (Footnote: But—as Orla-Jensen too points out correctly—it does not follow that identical metabolic processes may not also be encountered among phylogenetically independent evolutionary lines.) This idea has been used in an eminent manner by Orla-Jensen (1909) in his sketch of a general system of classification of bacteria. By assigning every microbe to one of the natural groups, whose delineations become progressively clearer, it will frequently be possible to predict its properties on the basis of its relationships, and it will also become feasible to indicate more rational culture conditions for it. For example, one frequently hears complaints concerning the difficulty of cultivating numerous streptococci; this must in the first place be ascribed to the fact that the investigators fail to realize that they are working with organisms belonging to the group of genuine lactic acid bacteria.

Table II indicates the distribution of dissimilatory processes among some important groups of microbes. It shows that, as a rule, oxidative processes do not exhibit any great specificity as far as the nature of the oxidizable substrate is concerned. Not infrequently the oxidizing capacity extends to nitrogen-containing compounds, to carbohydrates and sugar alcohols, and to organic acids, although in the series, acetic acid bacteria, moulds, aerobic sporeformers, and the *Pseudomonas* group, (Footnote: Although the organisms assembled in this group exhibit physiological similarities, they do not form a "natural group"; see the previous footnote.) a decreasing tendency towards carbohydrate oxidation and an increasing one towards the oxidation of nitrogen-containing compounds may be detected. In contrast, the fermentative processes are generally more dependent on a specific substrate, even though in some groups several potentialities may co-exist. . . .

It is not only in energetic respect that we may discern a unity in the metabolism of micro-organisms. Also in material respect there exists a much greater unity than was assumed not so long ago. This has been shown by recent studies; again, the circumstances do not permit me to document this statement extensively. But it may be pointed out that the investigations of Neuberg and collaborators have

TABLE II. DISSIMILATORY PROCESSES OF SOME IMPORTANT GROUPS OF MICRO-ORGANISMS

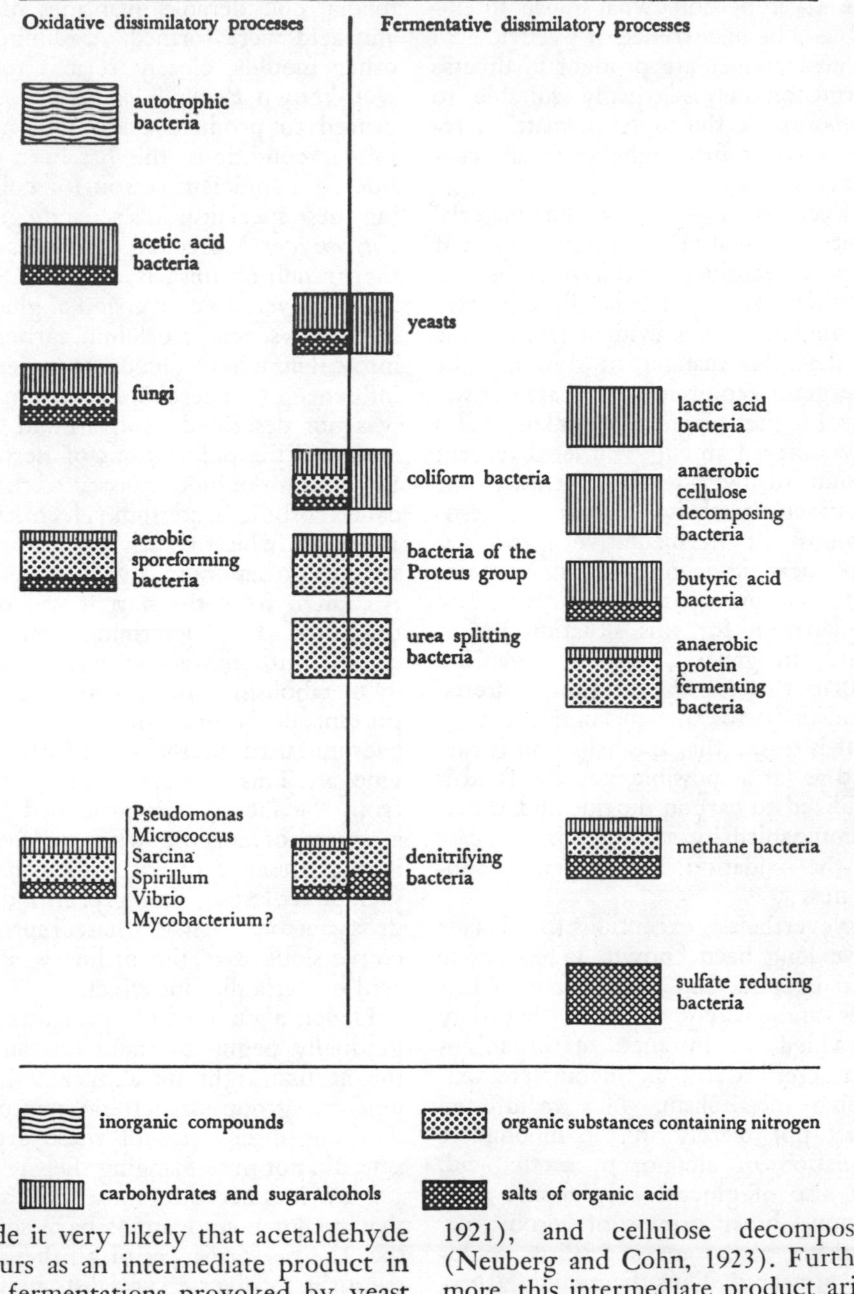

inorganic compounds — organic substances containing nitrogen

carbohydrates and sugaralcohols — salts of organic acid

made it very likely that acetaldehyde occurs as an intermediate product in the fermentations provoked by yeast (cf. Fuchs, 1922), coli bacteria (Neuberg and Nord, 1919), butyric acid bacteria (Neuberg and Arinstein, 1921), and cellulose decomposers (Neuberg and Cohn, 1923). Furthermore, this intermediate product arises in butyric acid fermentations both of substrates with 6 and 3 carbon atoms per molecule, which at least renders the formation of butyric acid with its 4 carbon atoms from compounds of the C_3 type somewhat more intelligible. The occurrence of pyruvic acid as an intermediate product in diverse fermentations is equally suitable to demonstrate the unity in material respect of at first sight very different processes.

Even stronger does the material unity in metabolic reactions appear if now we consider the microbes that are characterized by a typically oxidative dissimilation. It is evident from Table II that this manner of fulfilling the energetic requirements, characteristic of all higher plants and animals, is also encountered amongst rather divergent groups of microbes. In contrast to the relatively extensive chemical investigations of fermentative processes, these aerobic decompositions have not yet been much studied, however. The explanation for this situation lies at hand; in general, aerobic organisms utilize the energy of the proffered foodstuffs to the maximum extent, that is to say that the oxidation is carried as far as possible, and the food is oxidized to carbon dioxide and water, accompanied by ammonia in the case of the oxidation of nitrogenous substrates.

Nevertheless, exceptions to this rule have long been known. When acetic acid bacteria had been discovered, it was immediately apparent that they provided an instance of organisms characterized by an incomplete oxidative metabolism. This manifested itself not merely in the incomplete oxidation of alcohol to acetic acid, but also of glucose to gluconic acid. Among other groups of aerobic organisms, too, some specialists gradually appeared. Thus the mould, *Aspergillus niger*, acquired some reputation as the causative agent of what has unfortunately been called an oxalic acid fermentation; it was found that if the organism is grown on sugar-rich media, considerable quantities of oxalic acid were formed. A number of other moulds, closely related to the well-known *Penicillium glaucum*, appeared to produce citric acid under similar conditions; this has been considered a sufficient reason for collecting these specialists in a separate genus, *Citromyces*. Moreover, as early as 1886 the French chemist, Boutroux (1886), had discovered a conversion of glucose, in the presence of calcium carbonate, into calcium keto-gluconate under the influence of an aerobic bacterium that was not described. And around 1900 appeared the publications of Bertrand (1904) on the biochemistry of the so-called sorbose bacterium *(Acetobacter xylinum)* which rightly caused a sensation also amongst organic chemists. Although from the start it was obvious that this bacterium should be classed with the acetic acid bacteria, its metabolism showed nonetheless important deviations from that of the bacteria used in the manufacture of vinegar. This appeared, for instance, from the fact that it produced large amounts of sorbose when cultivated in the presence of sorbitol; mannitol yielded fructose; and glycerol dihydroxyacetone. All of these represent conversions that the ordinary acetic acid bacteria did not effect.

Hence a number of specialists had gradually begun to stand out amidst the at first sight metabolically quite uniform group of aerobic microbes. And the discoverers of these organisms did not tire of singing their praise.

Now it is certainly most interesting that recent investigations have shown that the performances of all these apparently so diverse specialists may be correlated. Firstly, the experiments of Currie (1917), and particularly those of Molliard (1920, 1922), showed that by changing the culture conditions it is possible at will to cause *A. niger* to convert an important fraction of the sugar into gluconic, or citric, or oxalic acid. Consequently one could make this organism, hitherto known only as an oxalic acid specialist, produce substances that until that time had been known only as specific products of acetic acid bacteria, and of *Citromyces* species, respectively. This very fact made it quite likely that the abovementioned compounds are none other than intermediate products of the oxidative degradation of glucose to carbon dioxide and water. The difference between the specialists would then merely consist in the fact that one organism would stop at an earlier intermediate stage.

Perhaps it may appear to you that the interpretation has always lain at hand that metabolic products such as citric, oxalic, and obviously gluconic acids are normal metabolic products in the oxidation of glucose. This notion had not been generally accepted, however. This appears from the fact that Butkewitsch (1922, 1923) who, during the past several years, has carried out extensive experimental investigations on citric and oxalic acid formation by moulds, has independently reached the same conclusion only quite recently. Experiments carried out in the Delft laboratory during the last year on the metabolism of various acetic acid bacteria have also led to the same concept of a stepwise oxidative degradation. To begin with, De Leeuw probably recovered the bacterium used by Boutroux, and long since lost; it turned out to be an acetic acid bacterium which we have called *Acetobacter suboxydans*. This organism also shows a close relationship to Bertrand's sorbose bacterium, although it clearly differs from the latter in some respects. The investigations have shown that *A. suboxydans* can carry out the same mild oxidations of different sugar alcohols as have already been mentioned in connection with the sorbose bacterium. But it also appeared that *A. suboxydans* possesses a still weaker oxidative capacity than *A. xylinum*, as evidenced, for example, by the fact that the former oxidizes calcium gluconate only to ketogluconate, while the latter can also oxidize the gluconate to carbonate. Besides, *A. suboxydans*, in contrast to *A. xylinum*, cannot be induced to oxidize substances like dihydroxyacetone and potassium ketogluconate. Thus the latter bacterium appears capable of further oxidizing its characteristic metabolic products under suitable conditions. And this shows up the intermediate nature of these incomplete oxidations.

Still more clearly does the correctness of these concepts follow from the fact that we (Kluyver and De Leeuw, 1924) have succeeded in inducing acetic acid bacteria that under ordinary culture conditions completely oxidize substrates like glycerol and mannitol to bring about incomplete oxidations by the use of methods similar to those employed by Molliard in his studies on *Asp. niger*. A few words may here be devoted to a discussion of these methods through which oxidations may be halted at various intermediate stages. Molliard paid attention primarily to the quantitative regulation of the amounts of N, P, and K in the culture medium. In connection with earlier observations of Beijerinck and Hoyer, we made use of the fact that the nature of the nitrogenous foodstuffs often determines whether a particular substrate, hence also a particular intermediate product, will or will not be further oxidized. This also reveals a very striking relation between dissimilation and assimilation, and perhaps may open a way for penetrating more deeply into these problems. After all, it is quite remarkable that *A. suboxydans* can grow perfectly well in a mineral medium containing a few per cent mannitol if nitrogen is supplied in the form of an ammonium salt. It is thus certain that this bacterium can use ammonia nitrogen for the synthesis of its proteins. But it appears also that this organism does not grow in the same medium if glucose is substituted for mannitol, whereas, in the presence of peptone, glucose is very rapidly oxidized, and the bacteria then multiply profusely. Glucose is therefore fully adequate as an energy-providing substrate. Thus we see how two substances, each one utilizable in its own function, can furnish an inadequate combination as food. Might it perhaps be possible to ascribe this unexpected phenomenon to a difference in the decrease in free energy of the first stages of the oxidation of mannitol and glucose, respectively, a difference that might cause ammonia nitrogen to be used for the synthesis of bacterial protein in the one, but not in the other case? We don't know; but we can learn from this example that the occurrence of a biological oxidation does not depend only on the nature and condition of the living cells and the presence of an oxidizable substrate, but that it is conditioned by very subtle modifications in the composition of the medium. . . .

It seems to be a practicable task to determine the successive intermediate oxidation stages of physiologically important compounds such as glucose, glycerol, etc., with numerous microorganisms that exhibit a low oxidative capacity. At the same time one may attempt to ascertain in a quantitative manner the ease with which the various steps are accomplished. It is hardly doubtful that such investigations will increase our comprehension of the nature of microbial metabolism. But by virtue of the above-mentioned unity it will also be eminently useful for our understanding of the metabolic activities of higher organisms which are much less amenable to experimentation. However this may be, I hope that my discussion may have convinced you that a study of microbial metabolism offers many an intriguing problem to the chemist, particularly in view of both the existing diversity and the manifest unity.

Comment

Kluyver was Beijerinck's successor to the Chair of Microbiology at the Technical University at Delft, Holland. He was just the right man to carry on the tradition of general microbiology that Beijerinck had established, and through his own work and that of his students, has carried this tradition down to the present.

The men of the Delft school of general microbiology were pioneers in an era when most workers were too fascinated by applied problems in medicine, agriculture, and industry to worry about the chemosynthetic and photosynthetic microorganisms or those exhibiting unusual fermentations. Yet it was an understanding of the nature of the fermentative and oxidative processes of these specialists, as Kluyver calls them, which has led to much of our current understanding of intermediary metabolism. In the paper presented here, Kluyver surveyed these diverse organisms with the view that there was an underlying unity, and when this unity was understood, it would help in understanding problems of general metabolism, even in higher organisms.

This view has been shown to be correct, in the light of the last 30 years' work, and much of the credit for this is due to the influence of Kluyver, both through the present paper and through a later one with H. J. L. Donker, "Unity in Biochemistry."

The basic metabolic pathways of most organisms are similar. But most of the intermediates that are formed during the oxidation (either anaerobically or aerobically) of glucose do not accumulate and could not be easily detected. However, some microorganisms seem to carry on certain steps in the pathway at faster rates than other steps, so that intermediates do accumulate. It is thus possible to isolate these intermediates and characterize them chemically. Such experiments have often revealed for the first time a possible intermediate step, and later work, using more refined techniques, has shown that this particular intermediate step occurs even in organisms that do not accumulate the product. In this way, through the years, biochemistry has often advanced.

There are other aspects of the unity of microorganisms, and Kluyver covers these briefly. Considerations of energy metabolism show that different organisms can use different energy sources, but there is some underlying thermodynamic foundation. We know today this is due to the fact that most energy-yielding reactions are accompanied by the synthesis of high energy phosphate compounds, predominantly adenosine triphosphate (ATP).

For the microbial physiologist, Kluyver's paper is important since it outlines a vast area of unsolved problems in understanding the behavior of different organisms. The importance of enrichment culture techniques for revealing the existence of diverse organisms in natural situations was emphasized, showing that the ecological relationships of microorganisms have a biochemical basis.

Appendix

Mémoire sur la fermentation appelée lactique (Extrait par l'auteur) *

1857 · Louis Pasteur

J'ai été conduit à m'occuper de la fermentation à la suite de mes recherches sur les propriétés des alcools amyliques et sur les particularités cristallographiques fort remarquables de leurs dérivés. J'aurai l'honneur de présenter ultérieurement à l'Académie des observations qui offriront une liaison inattendue entre les phénomènes de la fermentation et le caractère de dissymétrie moléculaire propre aux substances organiques naturelles . . .

Les conditions matérielles de la préparation et de la production de l'acide lactique sont bien connues des chimistes. On sait qu'il suffit d'ajouter à de l'eau sucrée de la craie, qui maintient le milieu neutre, plus une matière azotée, telle que le caséum, le gluten, les membranes animales, etc., pour que le sucre se transforme en acide lactique. Mais l'explication des phénomènes est très-obscure; on ignore tout à fait le mode d'action de la matière plastique azotée. Son poids ne change pas d'une manière sensible. Elle ne devient pas putride. Elle se modifie cependant et elle est continuellement dans un état d'altération évidente, bien qu'il serait difficile de dire en quoi il consiste.

Des recherches minutieuses n'ont pu jusqu'à présent faire découvrir dans ces opérations le développement d'êtres organisés. Les observateurs qui en ont reconnu ont établi en même temps qu'ils étaient accidentels et nuisaient au phénomène.

Les faits paraissent donc très-favorables aux idées de M. Liebig.[1] A ses yeux, le ferment est une substance excessivement altérable qui se décompose et qui excite la fermentation par suite de l'altération qu'elle éprouve elle-même, en ébranlant par communication et désassemblant le groupe moléculaire de la matière fermentescible. Là, selon M. Liebig, est la cause première de toutes les fermentations et l'origine de la plupart des maladies contagieuses. Cette opinion obtient chaque jour un nouveau crédit. On peut à cet égard consulter le Mémoire de MM. Fremy et Boutron sur la fermentation lactique, les pages qui traitent de la fermentation et des ferments dans le bel ouvrage que M. Gerhardt a laissé en mourant, enfin le Mémoire tout récent de M. Berthelot sur la fermentation alcoolique. Ces travaux s'accordent à rejeter l'idée d'une influence quelconque de l'organisation et de la vie dans la cause des phénomènes qui nous occupent. Je suis conduit à une manière de voir entièrement différente.

Je me propose d'établir dans la première partie de ce travail que, de même qu'il existe un ferment alcoolique, la levûre de bière, que l'on trouve partout où il y a du sucre qui se dédouble en alcool et en acide carbonique, de même il y a un ferment particulier, une levûre lactique toujours présente quand du sucre devient acide lactique, et que, si toute matière plastique azotée peut transformer le sucre en cet acide, c'est

* Mémoire sur la fermentation appelée lactique. *Comptes rendus des séances de l'Académie des Sciences*, Tome 45, p. 913–916.

[1] Il résulte des recherches historiques récentes de M. Chevreul, insérées au *Journal des Savants*, que Stahl avait déjà émis des idées analogues à celles de M. Liebig sur les causes de la fermentation alcoolique.

qu'elle est pour le développement de ce ferment un aliment convenable.

Il y a des cas ou l'on peut reconnaître dans les fermentations lactiques ordinaires, au-dessus du dépôt de la craie et de la matière azotée, des portions d'une substance grise formant quelquefois zone à la surface du dépôt. Son examen au microscope ne permet guere de la distinguer du caséum, du gluten désagrégés, etc., de telle sorte que rien n'indique que ce soit une matière spéciale, ni qu'elle ait pris naissance pendant la fermentation. C'est elle néanmoins qui joue le principal rôle. Je vais tout d'abord indiquer le moyen de l'isoler, de la préparer à l'état de pureté.

J'extrais de la levûre de bière sa partie soluble en la maintenant quelque temps à la température de l'eau bouillante avec quinze à vingt fois son poids d'eau. La liqueur est filtrée avec soin. On y fait dissoudre environ 50 grammes de sucre par litre, on ajoute de la craie et l'on sème sans le milieu une trace de la matière grise dont j'ai parlé tout à l'heure, en la retirant d'une bonne fermentation lactique ordinaire. Dès le lendemain, il se manifeste une fermentation vive et régulière. Le liquide, parfaitement limpide à l'origine, se trouble, la craie disparaît peu à peu, en même temps qu'un dépôt s'effectue et augmente continûment et progressivement au fur et à mesure de la dissolution de la craie. En outre, on observe tous les caractères et tous les accidents bien connus de la fermentation lactique. On peut remplacer dans cette expérience l'eau de levûre par la décoction de toute matière plastique azotée, fraîche ou altérée selon les cas. Voyons maintenant les caractères de cette substance dont la production est corrélative des phénomènes compris sous la dénomination de *fermentation lactique*. Son aspect rappelle celui de la levûre de bière quand on l'étudie en masse et égouttée ou pressée. Au microscope, elle est formée de petits globules ou de petits articles très-courts, isolés ou en amas constituant des flocons irréguliers. Ses globules, beaucoup plus petits que ceux de la levûre de bière, sont agités vivement du mouvement brownien. Lavée à grande eau par décantation, puis délayée dans de l'eau sucrée pure, elle l'acidifie immédiatement, progressivement mais avec une grande lenteur, parce que l'acidité gêne beaucoup son action sur le sucre. Si l'on fait intervenir la craie qui maintient la neutralité du milieu, la transformation du sucre est fort accélérée; et lors même que l'on opère sur très-peu de matière, en moins d'une heure le dégagement du gaz est manifeste et la liqueur se charge de lactate et de butyrate de chaux. Il faut très-peu de cette levûre pour transformer beaucoup de sucre. Ces fermentations doivent s'effectuer de préférence à l'abri de l'air, sans quoi elles sont gênées par des végétations ou des infusoires parasites. . . .

La fermentation lactique est donc aussi bien que la fermentation alcoolique ordinaire un acte corrélatif de la production d'une matière azotée qui a toutes les allures d'un corps organisé mycodermique probablement très-voisin de la levûre de biére. Mais les difficultés du sujet ne sont qu'a moitié résolúes. Sa complication est extrême. L'acide lactique est bien le produit principal de la fermentation à laquelle il a donné son nom. Il est loin d'être le seul. On le trouve constamment accompagné d'acide butyrique, d'alcool, de mannite, de matière visqueuse. La proportion de ces matières est soumise aux plus capricieuses variations. Il y a une circonstance mystérieuse relative à la mannite. Non-seulement la proportion qui s'en forme est sujette aux plus grandes variations; M. Berthelot vient d'établir, en outre, que si l'on remplace le sucre par la mannite dans la fermentation lactique, toutes les autres conditions demeurant sensiblement les mêmes, la mannite fermente en donnant de l'alcool, de l'acide lactique et de l'acide butyrique. Comment dès lors concevoir qu'il puisse y avoir formation de mannite dans des cas de fermentation lactique, puisque, peut-on croire, elle devrait se détruire au fur et à mesure de sa production?

Étudions avec plus de soins que nous ne l'avons fait les proprietés chimiques de la nouvelle levûre. J'ai dit que lavée à grande eau et placée dans de l'eau sucrée pure, elle acidifiait progressivement le liqueur. La transformation du sucre devient, dans ces conditions, de plus en plus pénible, à mesure que le liquide prend lui-même une plus grande acidité. Or, si l'on analyse la liqueur, ce qui ne peut être accompli avec succès qu'après la saturation des acides par la craie et la destruction ultérieure du sucre en excès par le levûre de bière, on trouve dans le liquide évaporé, et en proportion variable, la mannite d'une part, de l'autre la matière visqueuse. Ainsi donc la levûre lactique lavée mise en présence du sucre le transforme en divers produits parmi lesquels il y a toujours de la mannite, mais c'est à la condition que le liquide puisse devenir promptement acide; car si l'on répète exactement la même expérience avec la précaution d'ajouter un peu de craie afin que le milieu reste constamment neutre, ni gomme, ni mannite ne prennent naissance, ou mieux ne peuvent persister, parce que, on va le voir, les conditions de leur propre transformation se trouvent réunies.

J'ai rappellé tout à l'heure que M. Berthelot avait prouvé qu'en substituant la mannite au sucre dans la fermentation lactique, cette matiere fermentait. Or il est facile de se convaincre que dans les cas nombreux de fermentation de la mannite, c'est la levûre lactique qui prent naissance et produit le phénomène. Si l'on mêle à une solution de mannite pure de la craie en poudre et de la levûre lactique fraîche et lavée, au bout d'une heure déjà le dégagement gazeux et la transformation chimique de la mannite commenceront. Il se forme de l'acide carbonique, de l'hydrogène, et la liqueur renferme de l'alcool, de l'acide lactique, de l'acide butyrique, tous les produits de la fermentation de la mannite.

Quant a l'acide butyrique, l'expérience prouve que la levûre lactique agit directement sur le lactate de chaux en donnant du carbonate de chaux et du butyrate de chaux. Mais l'action s'exerce d'abord sur le sucre, et tant qu'il y en a dans la liqueur, la levûre le fait fermenter de préférence à l'acide lactique.

Dans des communications très-prochaines, j'aurai l'honneur de présenter à l'Académie l'application des idées générales et des nouvelles méthodes d'expérimentation de ce travail à d'autres fermentations.

Animalcules infusoires vivant sans gaz oxygène libre et déterminant des fermentations *

1861 · Louis Pasteur

On sait combien sont variés les produits qui se forment dans la fermentation appelée *lactique*. L'acide lactique, une gomme, la mannite, l'acide butyrique, l'alcool, l'acide carbonique et l'hydrogène, apparaissent simultanément ou successivement en proportions extrêmement variables et tout à fait capricieuses. J'ai été conduit peu à peu à reconnaître que le végétal-ferment qui transforme le sucre en acide lactique est différent de celui ou de ceux (car il en existe deux) qui déterminent la production de la matière gommeuse, et que ces derniers à leur tour n'engendrent pas d'acide lactique. D'autre part j'ai également reconnu que ces divers végétaux-ferments ne pouvaient dans aucune circonstance, s'ils étaient bien *purs*, donner naissance à l'acide butyrique.

Il devait donc y avoir un ferment butyrique propre. C'est sur ce point que j'ai arrêté depuis longtemps toute mon attention. La communication que j'ai l'honneur d'adresser aujourd'hui à l'Académie se rapporte précisément à l'origine de l'acide butyrique dans la fermentation appelée lactique.

Je n'entrerai pas ici dans tous les détails de cette recherche. Je me bornerai d'abord à enoncer l'une des conclusions de mon travail: c'est que *le ferment butyrique est un infusoire*.

J'étais bien éloigné de m'attendre à ce résultat, à tel point que pendant longtemps j'ai cru devoir appliquer mes efforts à écarter l'apparition de ces petits animaux, par la crainte où j'étais qu'ils ne se nourrissent du ferment végétal que je supposais être le ferment butyrique, et que je cherchais à découvrir dans les milieux liquides que j'amployais. Mais n'arrivant pas à saisir la cause de l'origine de l'acide butyrique, je finis par être frappé de la coincidence que mes analyses me montraient inévitable, entre cet acide et les infusoires, et inversement entre les infusoires et la production de cet acide, circonstance que j'avais attribuée jusque-là à l'utilité ou à la convenance que l'acide butyrique offrait à la vie de ces animalcules.

* Animalcules infusoires vivant sans gaz oxygène libre et déterminant des fermentations. *Comptes rendus des séances de l'Académie des Sciences,* Tome 52, p. 344–347.

Depuis lors, les essais les plus multipliés m'ont convaincu que la transformation du sucre, de la mannite et de l'acide lactique en acide butyrique, est due exclusivement à ces infusoires, et qu'il faut les considérer comme le véritable ferment butyrique.

Voici leur description: Ce sont de petites baguettes cylindriques, arrondies à leurs extrémités, ordinairement droites, isolées ou réunies par chaînes de deux, de trois, de quatre articles et quelquefois même davantage. Leur largeur est de $0^{mm},002$ en moyenne. La longueur des articles isolés varie de $0^{mm},002$ jusqu'à $0^{mm},015$ ou $0^{mm},02$. Ces infusoires s'avancent en glissant. Pendant ce mouvement, leur corps reste rigide ou éprouve de légères ondulations. Ils pirouettent, se balancent ou font trembler vivement la partie antérieure et postérieure de leur corps. Les ondulations de leurs mouvements deviennent très-évidentes dès que leur longueur atteint $0^{mm},015$. Souvent ils sont recourbés à une de leurs extrémités, quelquefois à toutes deux. Cette particularité est rare au commencement de leur vie.

Ils se reproduisent par fissiparité. C'est évidemment à ce mode de génération qu'est due la disposition en chaînes d'articles qu'affecte le corps de quelques-uns. L'article qui en traîne d'autres après lui s'agite quelquefois vivement comme pour s'en détacher.

Bien que les corps de ces Vibrions aient une apparence cylindrique, on les dirait souvent formés d'une suite de grains ou d'articles très-courts à peine ébauchés. Ce sont sans nul doute les premiers rudiments de ces petits animaux.

On peut semer ces infusoires comme on sèmerait de la levûre de bière. Ils se multiplient si le milieu est approprié à leur nourriture. Mais ce qui est bien essential à remarquer, on peut les semer dans un liquide ne renfermant que du sucre, de l'ammoniaque et des phosphates, c'est-à-dire des substances cristallisables et pour ainsi dire toutes minérales, et ils se reproduisent corrélativement à la fermentation butyrique qui apparaît très-manifeste. Le poids qui s'en forme est notable, bien que toujours minime, comparé à la quantité totale d'acide butyrique produit, comme cela se passe pour tous le ferments.

L'existence d'infusoires possédant le caractère des ferments est déjà un fait qui semble bien digne d'attention; mais une particularité singulière qui l'accompagne, c'est que ces animalcules infusoires vivent et se multiplient à l'infini sans qu'il soit nécessaire de leur fournir la plus petite quantité d'air ou d'oxygène libre.

Il serait trop long de dire ici comment je me suis arrangé pour que les milieux liquides où ces infusoires vivent at pullulent par myriades ne renferment absolument pas d'oxygène libre dans leur intérieur ou à leur surface, ce que j'ai d'ailleurs soigneusement constaté. J'ajouterai seulement que je n'ai pas voulu présenter mes résultats à l'Académie sans en avoir rendu témoins plusieurs de ses Membres, qui m'ont paru reconnaître la rigueur des preuves expérimentales que j'ai mises sous leurs yeux.

Non-seulement ces infusoires vivent sans air, mais l'air les tué. Que l'on fasse passer dans la liqueur où ils se multiplient un courant d'acide carbonique pur pendant un temps quelconque, leur vie et leur reproduction n'en sont aucunement affectées. Si, au contraire, dans des conditions exactement pareilles, on substitue au courant d'acide carbonique un courant d'air atmosphérique, pendant une ou deux heures seulement, tous périssent, et la fermentation butyrique liée à leur existence est aussitôt arrêtée.

Nous arrivons donc à cette double proposition:

1° *Le ferment butyrique est un infusoire.*

2° *Cet infusoire vit sans gaz oxygène libre.*

C'est, je crois, le premier exemple connu de ferments animaux, et aussi d'animaux vivant sans gaz oxygène libre.

Le rapprochement du mode de vie et des propriétés de ces animalcules avec le mode de vie et les propriétés des ferments végétaux qui vivent également sans le concours du gaz oxygène libre, se présente de lui-même, aussi bien que les conséquences qu'il est permis d'en déduire, relativement à la cause des fermentations. Cependant je veux réserver les idées que ces faits nouveaux suggèrent jusqu'à ce que j'aie pu les soumettre à la lumière de l'expérience.

De l'atténuation du virus du choléra des poules *

1880 · Louis Pasteur

Des divers résultats que j'ai eu l'honneur de communiquer à l'Académie sur l'affection vulgairement appelée *choléra des poules*, je prends la liberté de rappeler les suivants:

1° Le choléra des poules est une maladie virulente au premier chef.

2° Le virus est constitué par un parasite microscopique au'on multiplie aisément par la culture, en dehors du corps des animaux que le mal peut frapper. De là la possibilité d'obtenier le virus à l'état de pureté parfaite et la démonstration irréfutable qu'il est seul agent de maladie et de mort.

3° Le virus offre des virulences variables. Tantôt la maladie est suivie de la mort; tantôt, après avoir provoqué des symptômes morbides d'une intensité variable, elle est suivie de guérison.

4° Les différences que l'on constate dans la puissance du virus ne sont pas seulement le résultat d'observations empruntées à des faits naturels: l'expérimentateur peut les provoquer à son gré.

5° Comme cela arrive, en général, pour toutes les maladies virulentes, le choléra des poules ne récidive pas, ou plutôt la récidive se montre à des degrés qui sont en sens inverse de l'intensité plus ou moins grande des premières atteintes de l'affection, et il est toujours possible de pousser la préservation assez loin pour que l'inoculation du virus le plus virulent ne produise plus du tout d'effet.

6° Sans vouloir rien affirmer présentement sur les rapports des virus varioleux et vaccinal humains, il est sensible par les faits précédents que, dans le choléra des poules, il existe des états du virus qui, relativement au virus le plus virulent, font l'office du vaccin humain relativement au virus varioleux. Le virus vaccin proprement dit donne une maladie bénigne, la vaccine, qui préserve d'une maladie plus grave, la variole. Pareillement, le virus du choléra des poules présente des états de virulence atténuée qui donnent la maladie et non la mort, et dans de telles conditions que, après guérison, l'animal peut braver l'inoculation d'un virus très virulent. La différence est grande cependant, à certains égards, entre les deux ordres de faits, et il n'est pas inutile de remarquer que, sous le rapport des connaissances et des principes, l'avantage est du côté des études sur le choléra des poules: tandis qu'on discute encore sur les relations de la variole et de la vaccine, nous avons la certitude que le virus atténué du choléra dérive du virus très virulent propre à cette maladie, qu'on passe directement du premier de ces virus au second, en un mot, que leur nature fondamentale est la même.

Le moment est venu de m'expliquer sur l'assertion capitale qui fait le fond de la plupart des propositions précédentes, à savoir qu'il existe des états variables de virulence dans le choléra des poules: étrange résultat assurément, quand on songe que le virus de cette affection est un organisme microscopique qu'on peut manier à l'état de pureté parfaite, comme on manie la levûre de bière ou le mycoderme du vinaigre. Et pourtant, si l'on considère de sang-froid cette donnée mystérieuse de la virulence variable, on ne tarde pas à reconnaitre qu'elle est probablement commune aux diverses espèces de ce groupe des maladies virulentes. Ou donc est l'unicité dans l'un ou l'autre des fléaux qui composent ce groupe? Pour ne citer qu'un exemple, ne voit-on pas des épidémies de variole très graves à côté d'autres presque bénignes, sans que les différences puissent être attribuées à des conditions extérieures, de climat ou de constitution des individus atteints? Ne voit-on pas également les grandes contagions s'éteindre peu à peu pour reparaître plus tard et s'éteindre de nouveau?

La notion de l'existence d'intensités variables d'un même virus n'est donc pas faite, à la rigueur, pour surprendre le médecin ou l'homme du monde, quoiqu'il y ait un im-

* De l'atténuation du virus du choléra des poules. *Comptes rendus des séances de l'Académie des Sciences,* Tome 91, p. 673–680.

mense intérêt à ce qu'elle soit scientifiquement établie. Dans le cas particulier qui nous occupe, le mystère apparaît surtout dans cette circonstance que, le virus étant un parasite microscopique, les variations dans sa virulence sont à la merci de l'observateur. C'est ce que je dois établir avec rigueur.

Prenons pour point de départ le virus du choléra dans un état très virulent, le plus virulent possible, si l'on peut ainsi dire. Antérieurement, j'ai fait connaître un curieux moyen de l'obtenir avec cette propriété. Il consiste à aller recueillir le virus dans une poule qui vient de mourir, non de la maladie aiguë, mais de la maladie chronique. J'ai fait observer que la choléra se présente quelquefois sous cette dernière forme. Les cas en sont rares, quoiqu'il ne soit pas très difficile d'en recontrer des exemples. Dans ces conditions, la poule, après avoir été très malade, maigrit de plus en plus et résiste à la mort pendant des semaines et des mois. Lorsqu'elle périt, ce qui a lieu peu de temps après que le parasite, localisé jusque-là dans certains organes, a passé dans le sang et s'y cultive, on observe que, quelle qu'ait été la virulence originelle du virus au moment de l'inoculation, celui qu'on extrait du sang de l'animal qui a mis un si long temps à mourir est d'une virulence considérable, qui tue ordinairement dix fois sur dix, vingt fois sur vingt.

Cela posé, faisons des cultures successives de ce virus, à l'état de pureté, dans du bouillon de muscles de poule, en prenant chaque fois la semence d'une culture dans la culture précédente, et essayons la virulence de ces cultures diverses. L'observation démontre que cette virulence ne change pas d'une manière sensible. En d'autres termes, si nous convenons que deux virulences sont identiques lorsque, en opérant dans les mêmes conditions sur un mêmes nombre d'animaux de même espèce, la proportion de la mortalité est la même dans le même temps, nous constaterons que pour nos cultures successives la virulence est la même.[1]

Dans ce que je viens de dire, j'ai passé sous silence la durée de l'intervalle d'une culture à la culture voisine, ou, si l'on veut, la durée de l'intervalle d'un ensemencement à l'ensemencement suivant, et son influence possible sur les virulences successive. Portons notre attention sur ce point, quelque minime que paraisse son importance. Pour un intervalle d'un à huit jours, les virulences successives n'ont pas changé. Pour un intervalle de quinze jours, même résultat. Pour un intervalle d'un mois, de six semaines, de deux mois, on n'observe pas davantage de changement dans les virulences. Toutefois, à mesure que l'intervalle grandit, on croit saisir parfois, à certains signes de peu de valeur apparente, comme un affaiblissement du virus inoculé. Par exemple, la rapidité de la mort, sinon la proportion dans la mortalité, subit des retards. Dans les diverses séries inoculées, on voit des poules qui languissent, très malades, souvent très boîteuses, parce que le parasite, dans sa propagation à travers les muscles, a atteint ceux de la cuisse; les péricardites traînent en longeur; des abcès apparaissent autour des yeux; enfin le virus a perdu, pour ainsi dire, de son caractère foudroyant. Allons donc encore au delà des intervalles précités, avant la reprise et le renouvellement des cultures. Portons leurs durées à trois, à quatre, à cinq, à huit mois et plus, avant d'étudier la virulence des développements du nouvel être microscopique. Cette fois, la scène change du tout au tout. Les différences dans les virulences successives, qui jusque-là ne s'accusaient pas ou qui s'accusaient d'une manière douteuse, vont se traduire maintenant par des effets considérables.

Avec de tels intervalles dans les ensemencements, ii arrive que, à la reprise des cultures, au lieu de virulences identiques, c'est-à-dire de mortalité de dix poules sur dix poules inoculées, on tombe sur des mortalités descendantes de neuf, huit, sept, six,

---

[1] L'égalité dans la virulence, étant ainsi définie, ne doit pas être considérée comme une donnée absolue, parce qu'elle se trouve fonction du nombre des animaux inoculés. Que la mortalité soit la même dans deux séries de dix animaux, notre convention nous invite à dire que la virulence est la même pour les deux virus inoculés; une différence aurait pu s'accuser si l'on eût opéré, non sur deux séries de dix animaux, mais sur deux séries de cent. Que deux virus, inoculés chacun séparément à cent poules, fournissent des mortalités de soixante sujects dans un cas et de cent dans l'autre: l'épreuve, reprise sur dix, et dix poules seulement, pourra conduire, même dans plusieurs expériences successives, à l'égalité des virulences, si l'on s'en tient à notre convention sur la manière d'évaluer cette égalité. Or, nous voyons qu'en réalité elles differeraient dans les rapports de 60 à 100.

Toutefois, il faut adopter une convention, parce que, dans ce genre d'études, on est forcément limité par la convenance de ne pas pousser trop loin le nombre des victimes et de ne pas exagérer outre mesure la dépense toujours très grande de ces expériences.

cinq, quatre, trois, deux, une sur dix, et quelquefois même la mortalité est absente, c'est-à-dire que la maladie se manifeste sur tous les sujets inoculés et que tous guérissent. En d'autres termes, dans un simple changement du mode de culture du parasite, dans le seul fait d'éloigner les époques des ensemencements, nous avons une méthode pour obtenir des virulences progressivement décroissantes, et finalement un vrai virus vaccinal, qui ne tue pas, donne la maladie bénigne et préserve de la maladie mortelle.

Il ne faudrait pas croire que pour toutes ces atténuations les choses se passent avec une fixité et une régularité mathématiques. Telle culture que attend depuis cinq ou six mois son renouvellement peut montrer une virulence toujours considérable, tandis que d'autres de même origine seront déjà très atténuées après trois ou quatre mois d'attente. Nous aurons bientôt l'explication des ces anomalies, qui ne sont qu'apparentes. Souvent même il y a comme un saut brusque d'une virulence encore fort grande à la mort du parasite microscopique et pour un intervalle de peu de durée : en passant d'une culture à la suivante, on est surpris par l'impossibilité de tout développement; le parasite est mort. La mort du parasite est d'ailleurs une circonstance habituelle et constante toutes les fois qu'avant la reprise des cultures on laisse s'écouler un temps suffisant.

Et maintenant, l'Académie connaît le véritable motif du silence dans lequel je me suis renfermé et pourquoi j'ai réclamé la liberté d'un délai avant de l'informer de ma méthode d'atténuation. Le temps était un élément de ma recherche.

Au cours des phénomènes, que devient donc l'organisme microscopique? Change-t-il de forme, d'aspect, en changeant de virulence d'une manière aussi profounde? Je n'oserais pas affirmer qu'il n'existe pas certaines correspondances morphologiques entre le parasite et les virulences diverses qu'il accuse, mais je dois avouer qu'il m'a été jusqu'ici impossible de les saisir et que, si elles se montrent réellement, elles disparaissent, pour l'oeil armé du microscope, devant la petitesse si grande du virus. Les cultures sont pareilles pour toutes les virulences. Si l'on croit parfois apercevoir de faibles changements, ils semblent bientôt n'être qu'accidentels, car ils s'effacent ou se produisent en sens inverse dans des cultures nouvelles.

Ce qui est digne de remarque, c'est que, si l'on prend chaque variété de virulence comme point de départ de nouvelles cultures successives faites à intervalles rapprochés, la variété de virulence se conserve avec son intensité propre. S'agit-il, par exemple, d'un virus atténué qui ne tue plus qu'une fois sur dix, il garde cette virulence dans ses cultures si les intervalles des ensemencements ne sont pas exagérés. Chose également intéressante, quoiqu'elle soit dans le sens général des observations précédentes, un intervalle d'ensemencement qui suffit pour faire périr un virus atténué respecte un virus plus virulent qui peut bien en être atténué de nouveau, mais qui n'en meurt pas nécessairement.

Au point où nous sommes arrivés, une importante question se présente, celle de la cause de la diminution de la virulence.

Les cultures du parasite se font nécessairement au contact de l'air, parce que notre virus est un être aerobie et qu'à l'abri de l'air son développement n'est pas possible. Il est donc naturel de se demander tout d'abord si ce ne serait pas dans le contact de l'oxygène de l'air que réside l'influence affaiblissante de la propriété de virulence. Ne se pourrait-il pas que le petit organisme qui constitue le virus, restant abandonné en présence de l'oxygène de l'air pur, dans le milieu de culture où il vient de se multiplier, subisse quelques modifications qui se montreraient permanentes quand on soustrairait l'organisms à l'influence modificatrice? On peut, il est vrai, se demander en outre si quelque principe de l'air atmosphérique, autre que l'oxygène, principe chimique ou fluide, n'interviendrait pas dans l'accomplissement du phénomène, dont l'incomparable étrangeté autorise toutes les suppositions.

Il est aisé de comprendre que la solution de ce problème, au cas où elle relèverait de notre première hypothèse, celle d'une influence de l'oxygène de l'air, est assez facilement accessible à l'expérience : si l'oxygène de l'air, en effet, est l'agent modificateur de la virulence, nous pourrons vraisemblablement en avoir le preuve par les effets de la suppression de sa présence.

A cette fin, pratiquons nos cultures de la manière suivante. Une quantitè convenable de bouillon de poule étant ensemencée par notre virus très virulent, remplissons-en des tubes de verre aux deux tiers, aux trois quarts, etc., de leur volume; puis fermons ces tubes à la lampe d'émailleur. A la faveur de la petite quantité d'air restée dans le tube, le développement du virus va commencer, circonstance qui se traduit pour l'oeil par un trouble croissant du liquide; le progrès de la culture fait peu à peu disparaître tout l'oxygène contenu dans le tube. Alors le trouble tombe, le virus se dépose sur les parois et le liquide de culture s'éclaircit. Il faut deux ou trois jours pour que cet effet se produise. Le petit organisms est désormais à l'abri du contact de l'oxygène et il restera dans cet état aussi longtemps que le tube ne sera pas ouvert.[1] Que va-t-il advenir cette fois de sa virulence? Pour plus de sûreté dans notre étude, nous aurons préparé un grand nombre de tubes pareils, et simultanément un nombre égal de flacons de la même culture, mais librement exposés au contact de l'air pur. Nous avons dit ce qu'il advient de ces cultures exposées au contact de l'air; nous savons qu'elles éprouvent une atténuation progressive de leur virulence : nous n'y reviendrons pas. Parlons seulement des cultures en tubes fermés, à l'abri de l'air. Ouvrons-les, l'un, après un intervalle d'un mois, et après avoir fait une culture par ensemencement d'une portion de son contenu essayons-en la virulence, l'autre après un intervalle de deux mois, et ainsi de suite pour un troisième, un quatrième, etc., tube, après des intervalles de trois, de quatre, de cinq, de six, de sept, de huit, de neuf, de dix mois. C'est là que je me suis arrêté pour le moment. Il est remarquable, l'expérience le prouve, que les virulences sont toujours semblables à celle du début, à celle du virus qui a servi à préparer les tubes fermés. Quant aux cultures exposées à l'air, on les trouve mortes ou en possession des plus faibles virulences.

La question qui nous occupe est donc résolue : c'est l'oxygène de l'air qui affaiblit et éteint la virulence.[2]

Vraisemblablement, il y a ici plus qu'un fait isolé: nous devons être en possession d'un principe. On doit espérer qu'une action inhérente à l'oxygène atmospherique, force naturelle partout présente, se montrera efficace sur les autres virus. C'est, dans tous les cas, une circonstance digne d'intérêt que le grande généralité possible de cette méthode d'atténuation de la virulence, qui emprunte sa vertu à une influence d'ordre cosmique, en quelque sorte.[3] Ne peut-on pas présumer dès aujourd'hui que c'est à cette influence qu'il faut attribuer, dans le présent comme dans le passé, la limitation des grandes épidémies?

Les faits que je viens d'avoir l'honneur de communiquer à l'Académie suggèrent des inductions nombreuses, prochaines ou éloignées. Sur les unes et les autres, je suis tenu à une grande réserve. Je ne me croirai autorisé à les présenter au public que si je parviens à les faire passer à l'état de vérités démontrées.

---

[1] Avec le temps l'aspect des tubes fermés change beaucoup, en ce sens qu'après leur agitation ils deviennent presque limpides. Les granulations dans lesquelles se résolvent les premiers articles du développement initial prennent une réfringence pareille à celle de l'eau et ne troublent le liquide que d'une manière insensible. Sont-ce de véritables germes qu'on puisse comparer, par exemple, aux corpuscles germes de la bactéridie charbonneuse? Je ne le crois pas. Il n'est pas probable que notre parasite donne lieu à de véritables germes. S'il était suivi de germes, on comprendrait difficilement que, soit au contact de l'air, soit en tubes fermés, il perdît à la longue toute vitalité, toute faculté de reproduction. En outre, lorsqu'il y a germes véritables, ceux-ci supportent une température plus élevée que l'organisme en voie de développement, sous sa forme d'articles. Rien de pareil n'a lieu pour le microbe du choléra des poules. Les vieilles cultures conservées au contact de l'air (je n'ai pas éprouvé encore les autres) périssent même à des températures inférieures à celles qui atteignent les cultures récentes. C'est un caractère habituel du groupe des microcoques.

[2] Puisque, à l'abri de l'air, l'atténuation n'a pas lieu, on conçoit que, si dans une culture au libre contact de l'air (pur) il se fait un dépôt du parasite en quelque épaisseur, les couches profondes soient à l'abri de l'air, tandis que les superficielles se trouvent dans de tout autres conditions. Cette seule circonstance, jointe à l'intensité de la virulence, quelle que soit, pour ainsi dire, la quantité du virus employé, permet de comprendre que l'atténuation d'un vase ne doit pas nécessairement varier proportionnellement au temps d'exposition à l'air.

[3] J'ai passé sous silence, dans cette Note, une question ardue dont l'étude m'a pris un temps considérable. Je m'étais persuadé (à vrai dire, je ne sais pourquoi) que tous les faits d'atténuation que j'observais s'expliqueraient, d'une manière plus conforme aux lois naturelles, dans l'hypothèse de mélanges en proportions variables et déterminées de deux virus, l'un très virulent, l'autre très atténué, que par l'existence d'un virus à virulence progressivement variable. Après m'être pour ainsi dire acharné à la recherche d'une démonstration expérimentale de cette hypothèse de deux seuls virus, j'ai fini par acquérir la conviction que telle n'était pas la vérité.

Sur un microbe invisible antagoniste des bacilles dysentériques* (Note [1])

1917 · *F. d'Herelle*

Des selles de divers sujets convalescents de dysenterie bacillaire, et dans un cas de l'urine, j'ai isolé un microbe invisible doué de propriétés antagonistes vis-à-vis du bacille de Shiga. Sa recherche est particulièrement aisée dans les cas d'entérite banale consécutive à une dysenterie; chez les convalescents ne présentant pas cette complication la disparition du microbe anti suit de très près celle du bacille pathogène. Malgré de nombreux examens, je n'ai jamais trouvé de microbes antagonistes, ni dans les selles de dysentériques à la période d'état, ni dans les selles de sujets normaux.

L'isolement du microbe anti-Shiga est simple : on ensemence un tube de bouillon avec quatre à cinq gouttes de selles, on place à l'étuve à 37° pendant 18 heures puis on filtre à la bougie Chamberland L_2. Une petite quantité d'un filtrat actif ajoutée, soit à une culture en bouillon de bacilles de Shiga, soit à une émulsion de ces bacilles dans du bouillon ou même dans de l'eau physiologique, provoque l'arrêt de la culture, la mort des bacilles puis leur lyse qui est complète après un laps de temps variant de quelques heures à quelques jours suivant l'abondance plus ou moins grande de la culture et la quantité de filtrat ajoutée.

Le microbe invisible cultive dans la culture lysée de Shiga car une trace de ce liquide, reportée dans une nouvelle culture de Shiga, reproduit le même phénomène avec la même intensité : j'ai effectué jusqu'à ce jour, avec la première souche isolée, plus de 50 réensemencements successifs. L'expérience suivante donne d'ailleurs la preuve visible que l'action antagoniste est produite par un germe vivant : si l'on ajoute à une culture de Shiga une dilution d'une culture précédente lysée, de façon que la culture de Shiga n'en contienne qu'un millionième environ, et si, immédiatement après, on étale sur gélose inclinée une gouttelette de cette culture on obtient, après incubation, une couche de bacilles dysentériques présentant un certain nombre de cercles d'environ I^{mm} de diamètre, où la culture est nulle; ces points ne peuvent représenter que des colonies du microbe antagoniste : une substance chimique ne pourrait se concentrer sur des points définis. En opérant sur des quantités mesurées, j'ai pu voir qu'une culture lysée de Shiga contient de cinq à six milliards de germes filtrants par centimètre cube. Un trois-milliardième de centimètre cube d'une culture précédente en Shiga, c'est-à-dire un seul germe, introduite dans un tube de bouillon, empêche la culture du Shiga même ensemencé largement; la même quantité ajoutée à 10^{cm3} d'une culture de Shiga la stérilise et la lyse en cinq ou six jours.

Les diverses souches du microbe anti que j'ai isolées n'étaient primitivement actives que contre le bacille de Shiga; par culture en symbiose avec les bacilles dysentériques type Hiss ou Flexner j'ai pu, après quelques passages, les rendre antagonistes pour ces bacilles. Je n'ai obtenu aucun résultat en opérant sur d'autres microbes : bacilles typhiques et paratyphiques, staphylocoques, etc. L'apparition d'une action antagoniste contre le bacille de Flexner ou celui de Hiss s'accompagne d'une diminution puis d'une perte du pouvoir contre le Shiga, ce pouvoir reparaît d'ailleurs avec son intensité primitive àpres quelques cultures en symbiose; la spécificité de l'action antagoniste n'est donc pas inhérente à la nature même du microbe invisible, mais acquise dans l'organisme du malade par la culture en symbiose avec le bacille pathogène.

En l'absence de bacilles dysentériques le microbe anti ne cultive dans aucun milieu, il n'attaque pas les bacilles dysentériques tués par la chaleur, par contre il cultive

---

* Sur un microbe invisible antagoniste des bacilles dysentériques. *Comptes rendus des séances de l'Académie des Sciences*, Tome 165, p. 373-375.

[1] Séance du 3 septembre 1917.

parfaitement dans une émulsion en eau physiologique de bacilles lavés : il résulte de ces faits que le microbe antidysentérique est un bactériophage obligatoire.

Le microbe anti-Shiga n'exerce aucune action pathogène sur les animaux d'expérience. Les cultures lysées de Shiga sous l'action du microbe invisible, qui sont en réalité des cultures du microbe anti, jouissent de la propriété d'immuniser le lapin contre une dose de bacilles de Shiga tuant les témoins en cinq jours.

J'ai recherché si l'on pouvait mettre en évidence un microbe anti chez les convalescents de fièvre typhoïde : dans deux cas, une fois dans l'urine, l'autre fois dans les selles, j'ai réussi à isoler un microbe filtrant doué de propriétés lytiques nettes vis-à-vis du bacille paratyphique A, mais toutefois moins marquées que chez le microbe anti-Shiga. Ces propriétés se sont atténuées dans les cultures suivantes.

En résumé, chez certains convalescents de dysenterie, j'ai constaté que la disparition du bacille dysentérique coïncidait avec l'apparition d'un microbe invisible doué de propriétés antagonistes vis-à-vis du bacille pathogène. Ce microbe, véritable microbe d'immunité, est un bactériophage obligatoire; son parasitisme est strictement spécifique, mais s'il est limité à une espèce à un moment donné, il peut s'exercer tour à tour sur divers germes par accoutumance. Il semble donc que dans la dysenterie bacillaire, à côté d'une immunité antitonique homologue, émanant directement de l'organisme du sujet atteint, il existe une immunité antimicrobienne hétérologue produite par un microorganisme antagoniste. Il est probable que ce phénomène n'est pas spécial à la dysenterie, mais qu'il est d'un ordre plus général car j'ai pu constater des faits semblables, quoique moins accentués, dans deux cas de fièvre paratyphoïde.

Sur les organismes de la nitrification *

1890 · S. Winogradsky

Je voudrais, avant de résumer les recherches que je poursuis depuis un an sur la nitrification, rappeler ceux de mes travaux antérieurs qui en ont été le point de départ.

Ils ont eu pour objet l'étude de deux groupes d'organismes qui, comme ceux dont il est question dans cette Note, ont pour fonction d'oxyder des substances inorganiques, et que j'ai désignés par les noms de *sulfobactéries* et de *ferrobactéries*.[1]

Les premières habitent les eaux naturelles contenant de l'hydrogène sulfuré et refusent de vivre dans des milieux qui en sont exempts. Ce gaz est avidement absorbé et oxydé par leurs cellules, qui se remplissent de soufre, lequel est brûlé à son tour et excrété à l'état d'acide sulfurique. Les secondes ont pour fonction d'oxyder les sels ferreux, et leur vie est aussi étroitement liée a la présence de ces composés dans leur milieu nutritif.

Mes efforts pour élucider la signification physiologique de ces phénomènes m'ont conduit à la conception que ces corps inorganiques tiennent lieu, dans la vie de ces êtres, de la matière fermentescible (au sens large du mot) qui, pour la très grande majorité des microbes, est de la matière organique. De là resulte une conséquence logique, confirmée par l'expérience, c'est que ces êtres ont un ensemble de qualités physiologiques qui se résume ainsi : toute l'énergie nécessaire à leur travail vital leur étant fournie par la combustion de corps minéraux, la dépense en matière organique

---

* Sur les organismes de la nitrification. *Comptes rendus des séances de l'Académie des Sciences*, Tome 110, p. 1013–1016.

[1] *Sur les sulfobactéries (Botan. Zeitung*, 1887).—*Sur les ferrobactéries (Ibid.*, 1888). —*Contribution à l'étude physiologique et morphologique des bactéries*, Ier fasc.; Leipzig, 1888.—*Recherches physiol. sur les sulfobactéries (Ann. de l'Institut Pasteur*, t. III, n° 2).

pendant leur végétation est extrêmement faible, et des composés carbonés, incapables de nourrir les autres organismes dépourvus de chlorophylle, leur suffisent comme source de carbone.

Les travaux remarquables de MM. Schloesing et Müntz ont les premiers mis en lumière le rôle des organismes inférieurs dans la nitrification. Mais, tout en ayant rendu très probable l'existence d'un agent spécial de la nitrification, ils n'ont pas enseigné à l'isoler du sol, ce milieu naturel si riche en microbes divers. C'est que l'isolement et la culture à l'état pur, qui sont aujourd'hui l'exigence principale de toute expérience microbiologique, présentent, quand il s'agit du ferment nitrificateur, des difficultés assez grandes pour que nombre de savants y aient échoué, et la conclusion de MM. Schoesing et Müntz relativement à l'existence d'un ferment nitrique spécial n'a pas été, en somme, confirmée jusqu'ici par les bactériologistes et botanistes.[1]

Il fallait d'abord trancher cette question. Je me suis assuré que les échecs de mes prédécesseurs tenaient à l'emploi des milieux de culture gélatinises, si utilisés aujourd'hui pour l'isolement et la culture des microbes. L'organisme nitrificateur refuse d'y croître, si bien qu'en faisant passer par ce milieu un mélange de microbes empruntés à un sol en pleine nitrification, on tue tous les êtres qui y sont actifs et l'on ne récolte que de ceux qui sont inefficaces. Une étude suivie des conditions de culture favorables aux premiers et défavorables aux seconds m'a permis, mais non sans difficultés, d'éliminer une à une toutes les espèces étrangères et d'avoir des végétations abondantes et pures de l'espèce nitrifiante. Celle-ci montrait et conservait, dans les conditions usuelles de l'expérience microbiologique, une action aussi intense qu'on pouvait de désirer, en prenant pour comparaison les expériences récentes de M. Schloesing sur la nitrification dans la terre.

L'étude des propriétés physiologiques de cet être, qui se prêtait mieux à l'expérimentation que les organismes très délicats de mes travaux antérieurs, a non seulement justifié mes prévisions, mais m'a conduit à la découverte d'un fait nouveau que je voudrais principalement signaler à l'Académie.

Appliquent à cette étude les notions déjà acquises sur la nutrition des organismes comburants de substances minérales, je cultivais le microbe de la nitrification, dès le début, dans un liquide ne renfermant, en fait de matière organique, que ce qu'une eau naturelle très pure peut en contenir. L'addition de composés hydrocarbonés ne paraissant pas favoriser sa végétation, j'ai été conduit à essayer de préférence, pour sa culture, une solution minérale exempte de toute trace de carbone organique. Dans ce liquide, qui ne donnait à l'organisme d'autres composés carbonés que l'acide carbonique et les carbonates, ni l'abondance de sa multiplication ni l'intensité de son action ne parurent diminuer pendant plusieurs mois.

La conclusion s'imposait que cet organisme est en état d'assimiler le carbone de l'acide carbonique, et elle fut rendue irréfutable par des dosages de carbone organique dans ses cultures; ils démontrèrent qu'une accumulation de carbone organique, par son action, est un fait constant.

Le microbe de la nitrification, qui est un organisme incolore, est ainsi capable d'une synthèse complète de sa substance aux dépens de l'acide carbonique et de l'ammoniaque. Il accomplit cette synthèse indépendamment de la lumière, et sans autre source de force que la chaleur dégagée par l'oxydation de l'ammoniaque. Ce fait nouveau est en contradiction avec cette doctrine fondamentale de la Physiologie, qu'une synthèse complète de la matière organique n'a lieu dans la nature que dans les plantes à chlorophylle par l'action des rayons lumineux.

Il est peu probable que l'action du ferment nitrifiant soit une action chlorophyllienne, car on n'observe jamais avec lui de dégagement d'oxygène; une autre supposition, celle qu'un amide, de l'urée peut-être, est la première étape de la synthèse opérée par lui, serait ici la seule qui me parût plausible.[2]

Cette question ainsi que d'autres concernant la physiologie et la morphologie du ferment nitrique sont encore à l'étude.

---

[1] Voir l'historique contenu dans mon Mémoire: *Recherches sur les organismes de la nitrification* (*Ann. de l'Institut Pasteur*, t. IV, n° 4).

[2] Un Mémoire, traitant ces questions avec plus de détail, paraîtra dans le prochain numéro des *Annales de l'Institut Pasteur*.